Advances in Geographical and Environmental Sciences

Series Editor

R. B. Singh, University of Delhi, Delhi, India

Advances in Geographical and Environmental Sciences synthesizes series diagnostigation and prognostication of earth environment, incorporating challenging interactive areas within ecological envelope of geosphere, biosphere, hydrosphere, atmosphere and cryosphere. It deals with land use land cover change (LUCC), urbanization, energy flux, land-ocean fluxes, climate, food security, ecohydrology, biodiversity, natural hazards and disasters, human health and their mutual interaction and feedback mechanism in order to contribute towards sustainable future. The geosciences methods range from traditional field techniques and conventional data collection, use of remote sensing and geographical information system, computer aided technique to advance geostatistical and dynamic modeling.

The series integrate past, present and future of geospheric attributes incorporating biophysical and human dimensions in spatio-temporal perspectives. The geosciences, encompassing land-ocean-atmosphere interaction is considered as a vital component in the context of environmental issues, especially in observation and prediction of air and water pollution, global warming and urban heat islands. It is important to communicate the advances in geosciences to increase resilience of society through capacity building for mitigating the impact of natural hazards and disasters. Sustainability of human society depends strongly on the earth environment, and thus the development of geosciences is critical for a better understanding of our living environment, and its sustainable development.

Geoscience also has the responsibility to not confine itself to addressing current problems but it is also developing a framework to address future issues. In order to build a 'Future Earth Model' for understanding and predicting the functioning of the whole climatic system, collaboration of experts in the traditional earth disciplines as well as in ecology, information technology, instrumentation and complex system is essential, through initiatives from human geoscientists. Thus human geosceince is emerging as key policy science for contributing towards sustainability/survivality science together with future earth initiative.

Advances in Geographical and Environmental Sciences series publishes books that contain novel approaches in tackling issues of human geoscience in its broadest sense – books in the series should focus on true progress in a particular area or region. The series includes monographs and edited volumes without any limitations in the page numbers.

More information about this series at http://www.springer.com/series/13113

Seema Sahdev · R. B. Singh ·
Manish Kumar
Editors

Geoecology of Landscape Dynamics

Springer

Editors
Seema Sahdev
Department of Geography
University of Delhi
New Delhi, Delhi, India

R. B. Singh
Department of Geography
University of Delhi
New Delhi, Delhi, India

Manish Kumar
Department of Geography
University of Delhi
New Delhi, Delhi, India

ISSN 2198-3542 ISSN 2198-3550 (electronic)
Advances in Geographical and Environmental Sciences
ISBN 978-981-15-2096-9 ISBN 978-981-15-2097-6 (eBook)
https://doi.org/10.1007/978-981-15-2097-6

This Springer imprint is published by the registered company Springer Nature Singapore Pte Ltd.
The registered company address is: 152 Beach Road, #21-01/04 Gateway East, Singapore 189721, Singapore

Preface

The term "Geoecology" was first used by Carl Sauer in 1919. The analysis of landscape dynamics is one of the focus areas in Geoecology. Geosystems along with bioecosystem and anthrop ecosystem play a vital role in the formation of landscape Ecosystems. The structure of landscape has an important influence on various ecological processes occurring in the landscape. Various natural geological or biological processes have changed the landscape structure which have significant impacts on the ecology of landscapes. The analysis of landscape changes has been occurring at a large scale in the last century. Landscape dynamics are major challenges as their complex interactions of environmental factors and driving forces alter the composition of spices and these dynamics are reflected on different landscape patterns. Principle agents of the changes are urbanization and natural calamities. This dynamic phenomenon necessitates landscape monitoring and assessment of changes in spatial pattern over time. Hence, there is a need to identify driving forces for landscape changes for ensuring the sustainability of natural resources. Understanding the pattern of landscape ecology in the context of increasing human dominance in the biosphere enables the land managers to make decisions about sustainable land management. Thus ecological, economic and social patterns and processes in landscape change with scale are central to project sustainable landscapes.

The Geoecology has been effectively applied to deal with various types of problems of geo-ecosystem such as natural resource management, ecosystem and landscape characterization, landscape degradation, soil, environment, climate change and vegetation modelling, providing new insights to identify the influence of the changes on biodiversity, ecological stability and land use and land cover, generally on the course of landscape processes and characteristics.

The book aims to identify the ecological indicators of the quality of functioning and the structure of the formation mechanism for geographical landscape integration and put forth the application related to natural resources management in various environments such as mountain, plains, river basin, island, coastal, etc. The book highlights the indicators which can be used for quantifying sustainability in a geospatially explicit manner.

The book discusses land inventory mapping concept for green building to curb the global warming in cities, morphometric analysis of soil resources and inventory for meeting the challenges of land degradation. The book highlights the recent application areas of spatial information techniques in the field of land use, land cover dynamics, monitoring and modelling of urban sprawl, hydrological modelling, sustainable land management options, sustainable energy development and sustainable economic growth.

The book includes the research work by professors, planners' scientists and research scholars from various universities and institutions of India as well as from other countries of world, such as Lomonosov Moscow State University, University of Delhi, Indian Institute of Remote Sensing, Friendship University of Russia, University of Jammu, Indian Institute of Technology, Indian Meteorological Department, National Institute of Hydrology, Kumaun University, Uttarakhand Space Application Centre, Kamla Nehru Institute of Physical and Social Science, National Institute of Technology, University of Calcutta, Galgotias University, Banaras Hindu University, Gorakhpur University, Indian Institute of Science, Jawaharlal Nehru University, G.B Pant National Institute of Himalayan Environmental and Sustainable Development, Jamia Millia Islamia, and Punjab University.

New Delhi, India

Seema Sahdev
R. B. Singh
Manish Kumar

Contents

Chapter 1
Role of Data Indices for UN Sustainable Development Goals Implementation in Russia

Elena Milanova, Nikolay Dronin and Aleksandra Nikanorova

Abstract In 2015, the UN adopted 17 Sustainable Development Goals (SDGs) and 169 corresponding targets. To reach these goals, it is needed to use different quantitative and qualitative indicators based on global, national, and country data. One of the useful indicators is the Environmental Performance Index (EPI) that grades countries' performance in relation to two principal environmental policy objectives: the Protection of Ecosystem Vitality and the Environmental Health. With regard to these principal objectives, the EPI covers nine problems' areas: Climate and Power production sector, Agriculture, Biodiversity Loss, Fisheries, Forestry, Water quality, Health Impacts, Air pollution, and Water sanitation. According to the 2016 EPI, Russia occupies the 32nd position in this rating. 40% of population in Russia is living at the territory occupying 15% of total country area with unfavorable environmental conditions because of air and water pollution, deforestation, biodiversity worsening and forests' cover loss, poor waste management. The World Wildlife Fund estimates annual loses of the country reached about 1 billion dollars from the illegal export of wood. The measures to improve energy efficiency policy, to decrease air pollution were studied for Baikal model region. It is expected that the SDG indicators will become a common reference point for national and subnational monitoring in the Russian Federation.

Keywords Sustainable development goals · Global indices · Indicators · Russia · Baikal lake

---

E. Milanova (✉) · N. Dronin · A. Nikanorova
Faculty of Geography, Lomonosov Moscow State University, Moscow, Russia
e-mail: elena.v.milanova@gmail.com

N. Dronin
e-mail: ndronin@gmail.com

A. Nikanorova
e-mail: Alexandra.nikanorova@gmail.com

S. Sahdev et al. (eds.), *Geoecology of Landscape Dynamics*,
Advances in Geographical and Environmental Sciences,
https://doi.org/10.1007/978-981-15-2097-6_1

1.1 Section

In 2015, the UN adopted historical Document "Transforming our world: the 2030 Agenda for Sustainable Development" with 17 Sustainable Development Goals (SDGs) and 169 corresponding targets. The SDGs comprise an integrated balance of all three domains of sustainable development: economy, society, and natural environment. The use of different quantitative and qualitative indicators based on global, national, and country data indices is presumed to be needed to reach the SDGs.

Many global (the World Bank, OECD, SCOPE, etc.) and some regional (the European Commission) organizations are actively working and promoting their own sets of indicators for sustainable development. The organizations have proposed the indicators for systems at different scales: global, regional, national, local, industry as well as at the level of individual settlements and enterprises.

The multiscale approach (country and municipal) with the use of global indices is implemented to analyze the ranking of Russia as well as to assess the ecological indicators influencing the measuring and planning of ecological activities at municipal level for a model region—Baikal region.

1.2 Study Area

The research was held for Russia to evaluate the rating in accordance with scores of global indices at country level and at municipal level for the Baikal region as a model study area located in the east part of Russia around the famous Baikal Lake and characterized by enormous potential for ecological wealth and resources with the huge world natural significance (UNESCO World Heritage site).

1.3 Data and Methodology

1.3.1 The Role of Global Indices and Indicators

The indicators displayed from the primary data allow the evaluation of efforts of governments, ministries, and other bodies taken to improve the situation and to change the economic, social, or environmental variables to reach sustainable development. During the last years in addition to the thematic indicators, the system of integral indices has been developed and applied in practice. Index is the aggregate or weighted indicators' system, based on several other thematic simple indicators or data. The use of indices is acceptable if its causal connection is well understood.

Indicators and indices are used as the basis to justify the decision through the quantification and simplification; to help to interpret the changes; to allow identifying weaknesses in ecological management; to facilitate access to information for different categories of users; to facilitate the exchange of scientific and technical information. The indicators of sustainable development serve as the tool to support decision-making and planning in nature resources usage, to perform important functions to support the development plans, and to make easier communications. Therefore, the status of environmental indicators shall inform and attract public attention to certain environmental threats. This often mobilizes people for self-adoption of the necessary measures or treatment to authorities or private companies representatives.

One of the world-known ecological complex indexes is the Environmental Performance Index 2016 (EPI). It was elaborated with international efforts and presented in two manuscripts published by Yale University (Hsu et al. 2013, 2016). The 2016 EPI suggests a new paradigm to imply to nations' environmental policy. The state of a nation's environment is no longer an autonomous issue as it inevitably impacts its neighbors. Pollution does not recognize national borders—moreover, it could be even higher beyond state borders. This is also true for global dimension. Local activities could be superposed leading to stronger adverse effects at the global level. Implementation of the 2016 EPI serves for the purpose to portrait global environment through the prism of nations' environmental health. The index reflects these realities revealing global synergies of regional and local environmental impacts. It also shows areas where minor progress or even some deterioration is taking place.

The Environmental Performance Index (EPI) is based on more than 20 indicators referring to national and global environmental statistics (Hsu et al. 2016). The index presents aggregation of these parameters. The EPI ranks countries' performance in two areas of high-priority environmental policy objectives: the Protection of Ecosystem Vitality and the Environmental Health. The list of indicators used for the EPI is represented in Table 1.1. Within the two objectives, the EPI scores national performance in nine problematic areas: Climate and Power production, Biodiversity loss, Fisheries, Forestry, Agriculture, Water quality, Health Impacts, Air pollution, and

Table 1.1 The list of indicators of Environmental Performance Index (EPI) (*data source* Hsu et al. 2016)

Ecosystem vitality (divided into 9 issues)						Environmental health (divided into 3 issues)		
Issues								
Climate and energy	Biodiversity and habitat	Fisheries	Forests	Agriculture	Water resources	Health impact	Air quality	Water sanitation
Indicators (see the list below)								
1, 2	3, 4, 5, 6, 7	8	9	10, 11	12	13	14, 15, 16, 17	18, 19

List of indicators: 1. Carbon Intensity, per kWh; 2. Trend in Carbon Intensity; 3. Species Protection (National); 4. Species Protection (Global); 5. Terrestrial Biome Protection (National); 6. Terrestrial Biome Protection (Global); 7. Marine Protected Areas; 8. Fish Stocks; 9. Tree Cover Loss; 10. Nitrogen Balance; 11. Nitrogen Use Efficiency; 12. Wastewater Treatment; 13. Environmental Risk Exposure; 14. Household Air Quality; 15. Air Pollution Avg. Exposure to Fine Particulate Matter; 16. Air Pollution Fine Particulate Matter Exceedance; 17. Air Pollution, Avg. Exposure to NO_2; 18. Unsafe Drinking Water; 19. Unsafe Sanitation

Water sanitation. For each of the problems, the indicator is calculated on the base of global and national data to measure countries' progress toward the SDGs.

In 2016, 180 countries were evaluated and the new scoring altered the Index's previous application. Finland now took the top position, followed by Iceland, Sweden, Denmark, and Slovenia.

The other important index is the Global Innovation Index (GII), which has been elaborated by the business School for the World (INSEAD) since 2007, then with the support of Cornell University and World Intellectual Property Organization (WIPO) (Dutta et al. 2016).

In modern economic environment, a major driving force of growth is innovation, especially in the technological sphere. In 2016, 128 countries were evaluated by GII's 82 indicators. Assessment included several blocks important for innovation development: Human capital and research; Ecological sustainability; Market sophistication; Business sophistication, Innovation linkages, Knowledge and Technology output, Creative output. Each of the blocks was ranked by different indicators, which were calculated in percentages (Dutta et al. 2016).

The indicators of GII can be used for environmental evaluation of countries regardless of their specific circumstances. They present important statistics on the state of environment so far. A spatial and temporal coverage by datasets is a crucial characteristic for their applicability for the purpose of GII. The main ecological indicators assess country energy efficiency and energy consumption, greenhouse emissions and effect, waste management, access to clean drink water and other issues.

1.3.2 Country and Municipal Levels of Ecological Problems' Evaluation

As it was evaluated by the indices for country level, one of the main ecological problems is the increasing energy efficient use as well as preventing the green gase emissions. Russia is still characterized by very high energy intensity of its economy (exceeded the average world index by 2–3 times) and huge direct losses of energy due to outdate energy infrastructure (estimated at 15% for electric power and up to 50% for heat). Russia has to accept more serious obligations to deal with global climate change due to its high industrial emissions and huge potential to convert environmental issues into political, economic, and social problems, especially, at the local level.

The multiscale approach (at country and municipal levels) is implemented to rank Russian position in accordance with global indices as well as to assess the influence of ecological indicators on measuring and planning of ecological activities at municipal level for a model region.

According to the country's statistics, municipal sector consumes over 30% of the total energy produced in the Russian Federation. Therefore, this sector is responsible for the large portion of GHG emissions in the country. On the other hand, this provides a great opportunity for a radical reduction of the emissions by modernizing of energy distribution and consumption systems and introducing relatively cheap and simple means for energy saving at individual consumer's level.

Energy efficiency (EE) and green energy policies should be introduced at municipal level countrywide. There is a long list of proved measures for modernization of energy distribution system in municipal sector such as the utilization of wood waste in municipal heating systems, deployment of solar batteries, applying new standards of buildings insulation, introduction of heat regulating devices and others.

The EE policies should be elaborated for each Russian region depending on its natural (primary climate) conditions, social–economic development, and available energy resources. There could be indeed very different energy efficiency programs as well as prospects for reduction of energy consumption in Russian regions.

1.4 Results and Discussion

1.4.1 Ecological Indices and Indicators for Russia Score Evaluation

To evaluate the ecological situation, the multilevel approach was used for country and municipal scales. According to the two discussed indices (EPI and GII), Russia has the potential to build up the level of energy efficiency and decrease carbon emissions, which will have positive effect on its economy. The reduction of greenhouse gas emissions, primarily CO_2, is the most important task. Different measures can be used to reduce the emissions, especially in industry and transport sector as well as in fuel and energy consumption. Countries adopted measures to reduce greenhouse gases perform further efficiency improvements.

In the EPI rating 2016, Russia occupied the 32nd place with score 83, 52, it had improved its rating for 24% for 2 years from the previous investigation and the EPI publication. The main Russian ecological problems assessed in the framework of the investigation of ecological efficiency indices and collection data were the air pollution with CO_2 and NO_2, carbon intensity, wastewater treatment, forest cover losses, problems of protected areas, biodiversity and terrestrial habitat conservation and water species protection, etc.

By general GII assessment, Russia occupied the 43rd place with score 38.5. During calculation, GII was divided into Innovation Input Sub-Index (Russia was on 44th place) and Innovation Output Sub-Index (Russia—47th place). In the framework of the general ecological sustainability, Russia has 94th rank for GDP/unit of energy use, PPP$/kg oil eq—114th place, Environmental performance—32nd place, ISO 14001 environmental certificates/bn PPP$ GDP—91st place.

1.4.2 Multiscale Evaluation of Ecological Problems for Sustainable Development Planning

At country scale, one of the ways to improve Russian rating of EPI and GII is the adoption of the Government Act of the best available technology (BAT) implementation to decrease negative impact from environmentally hazardous enterprises. BAT is the new approach to sustainable development, which facilitates the implementation of green economy (green growth) approach based on the comprehension of significant role of environmental factors for future human well-being.

In Russia, the best available technology has been emerging in recent years, thus its realization in practice was complicated and slowed due to gaps in legislation and some contradictions in legal system. Notwithstanding, the new tendency of active adoption and implementation of BAT is widely observed in Russia today. First of all, the Federal Law "On the protection of the environment" (from 07.21.2014, no. 219-FZ) was modified: the legal definition of BAT and its principles, categories of environmentally dangerous enterprises, requirements for obligatory execution of programs for environmental efficiency improvement, rates of negative impact on the environment were legally defined by the adjustments of the law. At the same time, Rosstandard signed the Order No. 1920 on the 3 December 2014 to establish the BAT Bureau with the status of governmental regulation body. Its objectives are to remove administrative barriers and reduce excessive regulation as the important recommendations for BAT Reference Books creation. These documents mention the importance of the enterprises division in the categories, and the application to them adequate different measures.

Particularly important is the BAT incorporation in the sphere of energy efficiency policy development. Improvements of the energy saving and the air pollution reduction systems are the essential elements for resource preservation.

It is expected that from 2015 to 2017, the BAT implementation in Russia facilitates the carrying out of the public record of all the enterprises, their division in categories, publishing the Reference Guides on BAT, and adoption of all BAT regulations. In the coming time, around 300 Russian enterprises will be determined as the key pollutants and from 2019, the use of BAT will be obligatory for them. The same year, the BAT must be employed for designing of new businesses. Until 2025, all enterprises of the first category will get comprehensive environmental permits. As a result, during the 1st period (2015–2021), it is expected to decrease negative impact in the country for not less than 15%, during the 2nd (2021–2026)—for 45–50%, for the 3rd (2026–2031)—for 75–80%, in the 2033–2040s, around 15 000 enterprises shall correspond the BAT requests.

The system of environmental law will be based on motivation principles: the benefits for the payments; set-off cost for the enterprise if it takes measures to decrease the negative impact; the provision of tax privileges for businesses obeyed the BAT and others. Listed approaches for the BAT development will facilitate the growth of enterprises' efficiency as well as have a positive effect on ecological conditions in Russia.

The other way to improve energy efficiency policy and to decrease air pollution was studied at municipal scale (Milanova 2012) for the Baikal region, including the Irkutsk region and the Republic of Buryatia. This region was considered as the model study area due to the fact that it is treated as the region of the world natural heritage and the necessity to improve the municipal energy usage avoiding the risks to nature and people's well-being.

Baikal region is located in the east part of Russia near the famous Baikal Lake and characterized by uniqueness of biodiversity, possesses an enormous potential of ecological wealth and resources for tourism and recreation development, medical healthy sources and sites. 20% of the planet's freshwater is accumulated in the Baikal Lake. Under the conditions of ecological crisis, the value of virgin nature will be continuously growing; therefore, the conservation of the Baikal region biodiversity is the most important factor of the world community sustainable ecological development. Nowadays, the Baikal region has a well-developed network of nature conservation areas (more than 20 sites, which occupy 3 mln ha). They include state nature reserves (part of them are biosphere reserves with status of UNESCO World Heritage sites), national parks, specially protected sites, nature monuments, healthy sources and sites.

Several ecologically orientated energy efficiency (EE) projects were implemented in the region, with support of the government and different ecological funds. The aims were to set up the system of rational energy use and to improve the quality of people's life. In the framework of the projects, the priority was given to the wider use of biomass (wood waste), gas, and electricity for heating instead of using coal.

The projects helped for the better ecological situation in the region because the level of local high-ash coal consumption and the volume of GHG emissions were decreased at the same as the problem of wood waste was resolved.

One another approach which was used during the projects is the installations of solar batteries. The Republic of Buryatia and the Irkutsk region have the same number of sunshine hours as the South Europe regions, it accounts for 2200–2500 h annually and the solar energy can be widely utilized. Additionally, some projects were devoted to the minimization of irrational resource consumption what was reached by the means of heat and energy metering facilities installation. These measures enabled to reduce the volume of energy and hot water use up to 20%, which gave the expenditure savings for local people and set up optimal thermal regime in their houses.

Reached energy efficiency savings can re-invest into social infrastructure and facilitate the better profitable municipal budget, which all together would lead to a further building-up of capacity for sustainable development at local level (Milanova and Zaitsev 2013). The additional benefits are the reduction of GHG emissions and positive effects on the environment and socio-economic conditions.

Therefore the general strategy of EE at municipal level should include approaches to attract investors, to conduct political reforms at different levels, to form conditions that legally allow re-investments of energy efficiency savings into municipal development projects as the innovation of technical facilities and municipal management systems.

The local sustainable community development indicators were used to evaluate the EE project results. In compliance with ecologically oriented approach, the sustainable communities' success formulas are defined by two parameters:

(1) community improvements (better environment, ecologically responsible business, optimized energy and resources consumption, improved life quality), and
(2) enhancing positive community processes (widening of ecological resources management, expanding multi-stakeholders cooperation, growing of people awareness on nature status, increasing transparency of local ecological policy and public involvement in its implementation and decision-making).

The system of monitoring and indicators for evaluation of projects' influence on community sustainable development and people's well-being is elaborated. It includes different aspects of healthy community:

- environment improvement: pollution/waste reduction, creation of recreation areas/nature parks, development of ecotourism and ecologically responsible business;
- energy and other nature resources efficient usage: usage of ecologically safe fuel (wood waste, alternative sources), reinvestment of energy savings into decision of community socio-ecological problems and introduction of new technologies;
- ecological education: altered ecologically responsible behavior models, wide public involvement into ecological activities;
- people health improvement: better environment conditions, better drinking;
- water quality, public recreation facilities;
- people well-being improvement: reduction of energy rates and people expenses, new jobs creation and lower unemployment level, youth involvement into community activity and reduction of youth outflow, common growth of economic well-being in the municipal communities.

1.5 Conclusion

The global indices and indicators are the tools to evaluate sustainable development goals implementation at all levels (from local community to the country and the whole world). The rating of Russia was assessed on the base of Environmental Performance Index and Global Innovation Index to assist the indication of appropriate way to solve such important ecological problems as energy efficiency and air pollution by GHG reducing. At municipal level, the influence of energy efficiency projects was evaluated for the model Baikal region (UNESCO heritage site) on the base of elaborated indicators, which show the projects' influence on sustainable communities' development and improvement of people's life quality.

The BAD is considered as a new approach to sustainable development in Russia. Based on the principles of minimum environmental impact, it enables the formation of ecological sustainability. The only condition to support the strategy of ecologically sustainable development according to the SDGs at country and municipal level is the efficient assistance for country and regional development from authorities (governments, business and social organizations partnerships and cooperation) and the world community.

References

Dutta S, Lanvin B, Wunsch-Vincent S (eds) (2016) The global innovation index 2016: winning with global innovation. 2016. Johnson Cornell University, INSEAD The Business School for the World, World Intellectual Property Organization, Ithaca, Fontainebleau, and Geneva

Hsu A, Esty D, Levy M, de Sherbinin A et al (2016) The 2016 environmental performance index report. Yale Center for Environmental Law and Policy, New Haven, CT. http://dx.doi.org/10.13140/RG.2.2.19868.90249

Milanova E (2012) Sustainable development: global challenges–local solution. Probl Reg Ecol 2012(3):166–169 (in Russian)

Milanova E, Zaitsev A (2013) Energy efficiency municipal policy in Russia. In: Abstracts of Internationaler Kongress Fachmesse "Okologische, Technologische und Rechtliche Aspekte der Lebensversorgung", Hannover, November 2013, pp 26–29

Chapter 2
Seismic Vulnerability Assessment of NCT of Delhi Using GIS-Based Multiple Criteria Decision Analysis

Seema Sahdev, Manish Kumar and R. B. Singh

Abstract Seismic vulnerability mapping is one of the crucial issues of planning. The study demonstrates the use of geographic information system (GIS)-based multiple criteria decision analysis (MCDA) for the identification of areas of seismic vulnerability. For this purpose, various thematic maps have been prepared and with the help of the weighting criteria final seismic vulnerability assessment map was prepared using ArcGIS software. Six parameters: Seismic Micro Zonation, Lineament, Bedrock Depth, Depth of Water Level, built up density and Population Density have been taken for assessment. According to their importance weights have been generated for each criterion by comparing them with each other. Final seismic vulnerability assessment map was prepared by using these weights and criteria. To estimate the collective effects of causal factors stepwise multivariate regression analysis was also carried out. The analysis of causal factors of Seismic vulnerability such as Seismic Micro Zonation, Lineament, Bedrock Depth, Depth of Water Level, built up density and Population Density collectively explains 78% variations in it. Seismic vulnerability mapping helps planner to provide proper plan for prevention, mitigation and preparedness (PMP), where funds and efforts are used to solve the fundamental causes of vulnerability.

Keywords Seismic vulnerability · GIS · Multiple criteria decision analysis · Multivariate regression analysis

2.1 Introduction

In simple terms, vulnerability is one of the defining components of disaster risk. The characteristics of vulnerability may determined by physical, social, economic and

---

S. Sahdev (✉) · M. Kumar
Department of Geography, Kalindi College, University of Delhi, Delhi 110008, India
e-mail: seemasahdev@yahoo.com

R. B. Singh
Department of Geography, Delhi School of Economics, University of Delhi, Delhi 110007, India
e-mail: rbsgeo@hotmail.com

S. Sahdev et al. (eds.), *Geoecology of Landscape Dynamics*,
Advances in Geographical and Environmental Sciences,
https://doi.org/10.1007/978-981-15-2097-6_2

environmental factors or processes which increase the susceptibility of an individual, a community, assets or systems to the impacts of hazards. Different natural hazards such as drought, storm, landslide, food scarcity and earthquake cause severe damage of life and property. In order to cope up these hazards effectively, some positive factors are needed to enhance the ability of the people and the society to increase their resilience and reduce their susceptibility. It generally originates some miles beneath the earth surface and forms the origin of seismic focus from where vibration spread in all directions. Epicentre is a point on the earth's surface where the vibration reaches immediately. This point is just above the origin or seismic focus. In fact, epicentre is a point where the shock of the earthquake is experienced first and thereafter it spreads in all directions in the form of waves as seen in still water where a stone is thrown into it and water surface forms circular waves round the point of impact. With the invention of geospatial technologies the vulnerability and hazard assessment has become more precise and reliable. It has paved the wave for mapping, zoning and risk scoring.

In the present study, geospatial techniques have been used to access spatio-temporal nature of geophysical risk. However, it is important to note that the use of geospatial technologies alone is not sufficient to sort out the complexities associated with the earthquake risk management and final decision cannot be taken based on that.

Multiple criteria decision analysis is the process of evaluating a finite set of conflicting and incommensurable criteria and alternatives on the basis of quantitative, qualitative or both. As the problem of earthquake is multidimensional involving multiple criteria and conflicting objectives, its assessment is taken as multiple criteria decision analysis problem and specialized tools and techniques are used to evaluate the complexities involved.

Overall preference values are measured by the preferences values of the alternatives on permissible scale (Malczewski 1999; Jankowski et al. 2001). MCDA techniques have been used in many researches on risk assessment, hazard and vulnerability. Flood vulnerability areas in Turkey were analyzed with the help of MCDA (Yalcin 2002). MCDA was also applied to access the effectiveness of alternative retrofit options in seismic risk mitigation (Giovinazzi et al. 2006), vulnerability of volcanic risk (Aceves-Quesada et al. 2007), evaluation of fire risk (Vadrevu et al. 2009), seismic hazard for a city in India, Bangalore (Anbazhagan et al. 2011), landslide vulnerability in Romania (Armas 2011). Earthquake parameters were also studied and analyzed by Erden and Karaman (2011). They prepared hazard maps for Kucukcekmece region by incorporating GIS and MCE. To access social vulnerability two seismic risks for the municipality of Vila Franca do Campo in Portugal, Martins et al. (2012) present a GIS-based MCDA. It shows that there has been a substantial use of GIS-based MCDA techniques for disaster risk assessment at various places. Articles are being published focusing this topic with source trend in risk assessment studies based on MCDA. Saaty's (1980) Analytical hierarchy process (AHP) is the MCDA technique integrated into GIS for earthquake risk assessment. Spatial multi-criteria analysis and ranking tool (SMART) based on the concept of AHP is employed for developing seismic hazard and vulnerability maps. Thereafter these maps are combining to develop seismic map as done in the linear multiplicative combination methods. NCT of Delhi is one of the probable earthquake hazard prone areas of India. So, in this study attempt is being made to access the seismic vulnerability mapping of NCT of Delhi.

2.2 Study Area

National Capital Territory (NCT) of Delhi lies in Seismic Zone IV of the Seismic Zoning Map of India (Fig. 2.1). It spreads between Lat 28° 24′01″ and 28° 53′00″ N and Long 76° 50′24″ and 77° 20′37″ E occupying an area of 1482 km^2. In geographical terms, NCT of Delhi region is located on folded crustal ramp represented by basement rocks of Delhi Super Group which is bounded by two regional faults, viz, Mahendragarh–Dehradun Subsurface Fault (MDSSF) in the west and Great Boundary Fault (GBF) in the East Delhi. The ramp across the 'fore deep' trending NNE–SSW is placed side by side to Himalayan thrust belt. The other structural element of the belt is NW–SE trending Delhi–Sargodha Ridge (DSR) which passes through Delhi surrounded by basins on both sides. These are Bikaner Basin in southwest and Sahaspur Basin in the north.

Since Delhi city is in a seismically active region, it is imperative to study and analyze the probable earthquake hazard to the city on a scientific basis. Though records are not available to show the earthquake pattern for the decades back, in history the first earthquake in Delhi city took place on 17 July 1720 as recorded. In his Urdu poem 'bhûcâl-nâma' Nazir Akbarabadi (1740–1830) described about high-intensity earthquake felt in the Agra–Mathura region on 25 August 1803. This earthquake caused some damage in the Qutub Minar, which is located in Delhi. On

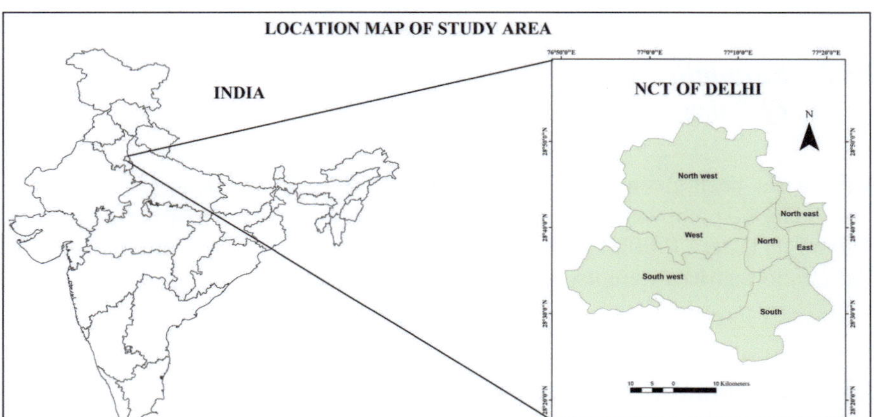

Fig. 2.1 Location map of study area

27 August 1960, an earthquake of magnitude 6 occurred having its epicentral tract in the Delhi–Gurgaon region. An earthquake of magnitude of 4 took place on 28th July 1994 around Delhi causing minor damage to a few old structures. Generally, all these earthquakes have their sources in the Indo-Gangetic plains near Delhi city. It has also been observed that even earthquake occurring at distant places in Himalayan plate boundary Uttarakhand are felt in Delhi though no severe effect is noted.

2.3 Materials and Methods

2.3.1 Data Collection and Integration

In order to develop seismic vulnerability assessment map, various thematic layers were generated using ArcGIS Software. Seismic Micro Zonation, Lineament, Bedrock Depth, Depth of Water Level, Built up density and population density were obtained from Indian Metrological Organisation (IMD), Census of India, Geological Society of India(GSI).

2.3.2 Selection and Preparation of Criteria Maps

In the present study, six criteria were selected. The main criteria which are used for analysis are Seismic Micro Zonation, Lineament, Bedrock Depth, Depth of Water Level, Built up density and Population Density (Fig. 2.2). These criteria were used in the preparation of criteria maps.

2.3.3 Suitability Scoring/Ranking and Development of Pairwise Comparison Matrix

A score has been given to each criteria as per their appropriateness. For this purpose, Saaty's 9 point weighing scale-based pairwise comparison matrix was prepared (Table 2.1) (Saaty 1980). For developing a ration matrix different criteria are used. These ratio matrices were used to create relative weights.

2.3.4 Computation of the Criterion Weights

After preparation of ratio matrix, the computations of the criteria weights were done (Malczewski 1999).

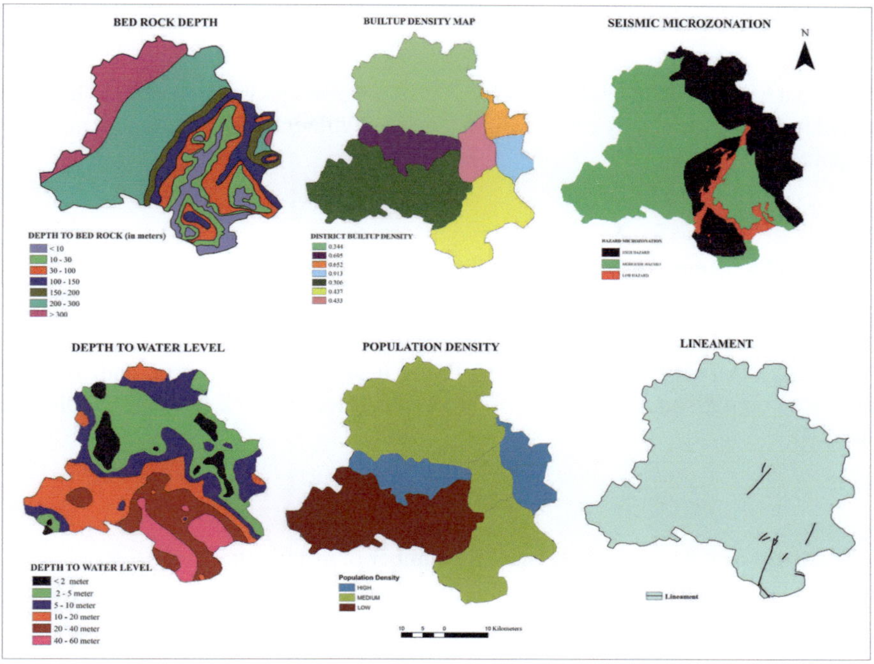

Fig. 2.2 Criteria maps (*Source* IMD, Census of India and GSI)

Table 2.1 Nine point weighting scale for pairwise comparison

Intensity of importance	Description
1	Equal importance
2	Equal to moderate importance
3	Moderate importance
4	Moderate to strong importance
5	Strong importance
6	Strong to very strong importance
7	Very strong importance
8	Very to extremely strong importance
9	Extremely importance

Source Saaty (1980), ESRI (1996)

2.3.5 *Rasterization of Criteria Maps*

Computation is less complex in vector data format as compared to raster data format therefore criteria maps were created in raster data format (Chang 2006).

2.3.6 *Integration of Maps*

After rasterization in ArcGIS software, these raster maps were combined in raster calculator and by multiplying its weight, and final vulnerability assessment map was prepared.

2.3.7 *Preparation of Vulnerability Assessment Map*

A pairwise comparison matrix was developed with the help of the existing criteria (Table 2.2). All criteria were normalized by using the ratio matrix, and by using pairwise comparison method weights were computed for each criterion (Tables 2.3 and 2.4).

$$\lambda = 6.18$$

λ should be equal or greater than the number of criteria under consideration and satisfies this condition.

$$\mathrm{CI} = 0.036$$

Table 2.2 Pairwise comparison matrix

Criteria	Seismic Micro Zonation (a)	Lineament (b)	Bedrock Depth (c)	Depth of Water Level (d)	Built up density (e)	Population density (f)
Seismic Micro Zonation	1	2	4	5	7	8
Lineament	0.5	1	2	4	5	7
Bedrock Depth	0.25	0.5	1	2	4	5
Depth of Water Level	0.2	0.25	0.5	1	2	4
Built up density population density	0.14	0.2	0.25	0.5	1	2
Aspect	0.13	0.14	0.2	0.25	0.5	1
Total	2.22	4.09	7.95	12.75	19.5	27

Source Prepared by the authors

Table 2.3 Normalized pairwise comparison matrix and computation of criterion weights

Criteria	Seismic Micro Zonation (a)	Lineament (b)	Bedrock Depth (c)	Depth of Water Level (d)	Built up density (e)	Population density (f)	Weights (a + b + c + d + e + f)/6
Seismic Micro Zonation	0.45	0.49	0.50	0.39	0.36	0.30	0.42
Lineament	0.23	0.24	0.25	0.31	0.26	0.26	0.26
Bedrock Depth	0.11	0.12	0.13	0.16	0.21	0.18	0.15
Depth of Water Level	0.09	0.06	0.06	0.08	0.10	0.15	0.09
Built up density	0.06	0.05	0.03	0.04	0.05	0.07	0.05
Population density	0.06	0.03	0.03	0.02	0.02	0.04	0.03
Total	1.00	1.00	1.00	1.00	1.00	1.00	1.00

Source Prepared by the authors

Table 2.4 Computation of consistency vector

Criterion	Weighted sum vector	Consistency vector
Seismic Micro Zonation	[(1)(0.42) + (2)(0.26) + (4)(0.15) + (5)(0.09) + (7)(0.05) + (8)(0.03)] = 2.58	2.58/0.42 = 6.14
Lineament	[(0.5)(0.42) + (1)(0.26) + (2)(0.15) + (4)(0.09) + (5)(0.05) + (7)(0.03)] = 1.59	1.59/0.26 = 6.12
Bedrock Depth	[(0.25)(0.42) + (0.5)(0.26) + (1)(0.15) + (2)(0.09) + (4)(0.05) + (5)(0.03)] = 0.92	0.92/0.15 = 6.13
Depth of Water Level	[(0.2)(0.42) + (0.25)(0.26) + (0.5)(0.15) + (1)(0.09) + (2)(0.05) + (4)(0.03)] = 0.54	0.54/0.09 = 6.00
Built up density	[(0.14)(0.42) + (0.2)(0.26) + (0.25)(0.15) + (0.5)(0.09) + (1)(0.05) + (2)(0.03)] = 0.31	0.31/0.05 = 6.20
Population density	[(0.13)(0.42) + (0.14)(0.26) + (0.2)(0.15) + (0.25)(0.09) + (0.5)(0.05) + (1)(0.03)] = 0.195	0.195/0.03 = 6.50

Source Prepared by the authors

$$\text{and}\, CR = 0.03.$$

CR (= 0.03) < 0.10 indicates a reasonable level of consistency in the pairwise comparisons.

The final seismic vulnerability assessment map (Fig. 2.3) was prepared by

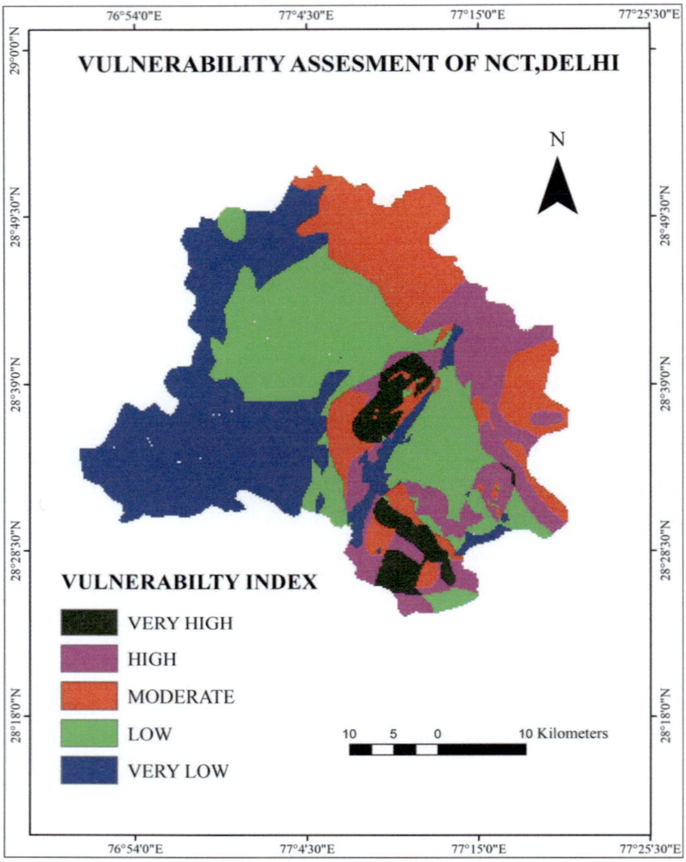

Fig. 2.3 Vulnerability assessment map

applying the following formula:

$$\begin{aligned} \textbf{\textit{Vulnerability index}} = & ([\text{seismic Micro Zones}] * 0.42) + ([\text{Lineament}] * 0.26) \\ & + ([\text{bedrock depth}] * 0.15) + ([\text{depth of water level}] * 0.09) \\ & + ([\text{built up density}] * 0.05) + ([\text{pop density}] * 0.03) \end{aligned}$$

The final seismic vulnerability assessment map reveals that the study area was divided into five different seismic zones. The area under very high, high, moderate, low, very low stands at 101.30 km^2, 285.77 km^2, 369.79 km^2, 520.19 km^2 and 602.26 km^{2,} respectively, (Table 2.5). Approximately large area falls under low and very low suitable category medium, area under moderate suitability and other area under very high and high suitable categories.

Table 2.5 Area under different suitability categories

Vulnerability assessment zones	Area (in km^2)
Very high	101.30
High	285.77
Moderate	369.79
Low	520.19
Very low	602.26

Source Prepared by the authors

Table 2.6 Model summary of stepwise regression

Model	R	R square	Adjusted R square	Std. error of the estimate	Significant
1	0.785[a]	0.617	0.562	7.94687	0.012
2	0.819[b]	0.670	0.560	7.96231	0.362
3	0.831[c]	0.691	0.505	8.44636	0.589
4	0.843[d]	0.711	0.421	9.13158	0.626
5	0.843[e]	0.711	0.537	8.16755	0.997
6	0.877[f]	0.769	0.538	8.15889	0.372

Note Significance at 0.05 Level
[a]Predictors: (Constant), Seismic Micro Zonation
[b]Predictors: (Constant), Seismic Micro Zonation, Lineament
[c]Predictors: (Constant), Seismic Micro Zonation, Lineament, Bedrock Depth
[d]Predictors: (Constant), Seismic Micro Zonation, Lineament, Bedrock Depth, Depth of Water Level
[e]Predictors: (Constant), Seismic Micro Zonation, Lineament, Bedrock Depth, Depth of Water Level, Built up density
[f]Predictors: (Constant), Seismic Micro Zonation, Lineament, Bedrock Depth, Depth of Water Level, Built up density, Population Density

2.4 Modelling of Seismic Vulnerability Assessment

Stepwise multivariate regression analysis was also carried out to estimate the collective effects of causal factors, (Table 2.6). It is assumed that the relationships between variables are linear in the stepwise multivariate regression. The $r2 = 0.893$ shows highest correlation coefficient which collectively explains the 87.7% of seismic vulnerability considering all the causal factors in the stepwise regression. The Seismic Micro Zonation explains the highest proportion (78.5%) and the only significant causal factor.

2.5 Conclusion

In this study, an integrated approach of GIS-based multiple criteria decision analysis was employed in order to identify seismic vulnerability zones. This method not only offers a scientific way to find the vulnerable zones but also provides a methodology for assessing the vulnerability zones as well as cost-benefit analysis for the same. It provides comprehensive and satisfactory database for vulnerability mapping and in turn will help in solving any specific problem. We all know if a country gets hit by any natural calamity it is very difficult for the country to overcome it, so it is necessary to get prepared for any natural calamity so that we can save our people, property and the resources. This vulnerability assessment map can encourage public participation in the mitigation plan and a better decision can be taken against any kind of hazards happening in the area of NCT, Delhi. Primarily the areas which are more susceptible to vulnerability can be cured with a help of this vulnerability assessment map.

References

Aceves-Quesada JF, Díaz-Salgado J, López-Blanco J (2007) Vulnerability assessment in a volcanic risk evaluation in Central Mexico through a multicriteria-GIS approach. Nat Hazards 40:339–356. https://doi.org/10.1007/s11069-006-0018-6

Anbazhagan P, Gajawada P, Parihar A (2011) Seismic hazard map of Coimbatore using subsurface fault rupture. Nat Hazards 60:1325–1345

Armas I (2011) An analytic multicriteria hierarchical approach to assess landslide vulnerability. Case study: Cornu Village, Subcarpathian Prahova Valley/Romania. Z Geomorphol 55:209–229

Chang KT (2006) Raster data analysis. Introduction to geographic information system. Tata McGraw Hill, New Delhi

Environmental Systems Research Institute (ESRI) Inc. (1996) Working with the ArcView spatial analyst. Redlands, California, USA

Erden T, Karaman H (2011) Analysis of earthquake parameters to generate hazard maps by integrating AHP and GIS for Kucukcekmece region. Nat Haz Earth Syst Sci 12:475–483

Giovinazzi S, Lagomarsino S, Pampanin S (2006) Vulnerability methods and damage scenario for seismic risk analysis as support to retrofit strategies: an European perspective. In: Proceedings of the New Zealand society for earthquake engineering conference, Napier. http://ir.canterbury.ac.nz/handle/10092/439

Jankowski P, Andrienko N, Andrienko G (2001) Map-centered exploratory approach to multiple criteria spatial decision making. Int J Geogr Inf Sci 15:101–127

Malczewski J (1999) GIS and multicriteria decision analysis. Wiley, New York, p 392

Martins VN, Silva DS, Cabral P (2012) Social vulnerability assessment to seismic risk using multicriteria analysis: the case study of Vila Franca does Campo (São Miguel Island, Azores, Portugal). Nat Hazard 3:26–45

Saaty TL (1980) The analytic hierarchy process. McGraw-Hill, New York

Vadrevu KP, Eaturu A, Badarinath KVS (2009) Fire risk evaluation using multicriteria analysis a case study. Environ Monit Assess. http://www.colorado.edu/geography/leyk/geog_5113/readings/vadrevu_etal_2009_EnvMonAss.pdf

Yalcin G (2002) Analyzing flood vulnerable areas with multicriteria evaluation [dissertation]. Middle East Technical University, Ankara. http://www.isprs.org/istanbul2004/comm2/papers/154.pdf

Chapter 3
Human Capital Impact for Sustainable Economic Growth

Vladimir M. Matyushok, Nina M. Baranova and Leonid V. Sorokin

Abstract Sustainable economic growth is one of three essential components of the United Nations sustainable development concept together with the society and environment. The up going urbanization force human and economic development. The human capital is one of the important components for growing industrialization, innovations, scientific research, education and formation of the higher skilled workers. The main aim of the study is to provide an analysis of the human capital impact on sustainable economic growth. We used the methods of econometric modelling and cluster analysis on the base of data collections: United Nations Development Programme and the World Bank World Development Indicators (WDI). Cluster analysis was done for the Human Development Index (HDI) and WDI for countries with different levels of social and economic development together with ecological situations. With the help of spatial data analysis of population density and urbanization for the studied countries, we can separate them into different groups. As a result of the analysis, the countries with high HDI and Gross national income (GNI) per capita has a clusterization in different groups due to the role of scientific research and industrialization as in the case with a high volume of trade (exports and imports) per capita and high urbanization. The main difference between them is that the role of innovation in economic growth can generate the same or higher HDI as in the case of a high volume of trade (exports and imports) per capita. But it is possible to separate them with the help of population density and urbanization. It looks like countries with a high volume of trade (exports and imports) per capita have the highest level of urbanization. Based on this approach it is possible to separate the countries with the same HDI value and different income or similar income and different HDI values.

V. M. Matyushok · N. M. Baranova · L. V. Sorokin (✉)
Department of Economic and Mathematical Modelling, Peoples' Friendship University of Russia (RUDN University), Moscow, Russia
e-mail: sorokin_lv@rudn.univesity

V. M. Matyushok
e-mail: matyushok_vm@rudn.univesity

N. M. Baranova
e-mail: baranova_nm@rudn.univesity

S. Sahdev et al. (eds.), *Geoecology of Landscape Dynamics*,
Advances in Geographical and Environmental Sciences,
https://doi.org/10.1007/978-981-15-2097-6_3

Keywords Human development · Economic growth · Urbanization · Inequality · Research and development

3.1 Introduction

The Economic Growth was one of the central research topics of the twentieth century. The economic model of long-run economic growth was developed independently by Robert Solow (1956) and Trevor Swan (1956) in 1956. Robert Solow and Trevor Swan considered capital accumulation, labour or population growth, and increase in productivity, commonly referred to as technological progress and the main factors of the long-run economic growth (Acemoglu 2009; Swan 1956; Solow 1956, 1957). In 1987 Solow was awarded the Nobel Prize in Economics. Economists use the Solow–Swan model to estimate the independent effects on economic growth of technological change, capital, and labour (Haines and Sharif 2006).

The main reason for the economy slowing down in the present time can be a deficit of investment resources and modern technologies, scarcity of professional staff, insufficient development of competition and business environment weaknesses. The President of the Russian Federation charged the government and the leading business associations up to May of the next year to develop a substantive road map for 2025. Its implementation should enable already in the 2019–2020's changed to economic growth.

Sustainable economic growth is one of the third essential components of the United Nations sustainable development concept together with society (Dodds et al. 2017) and the environment. In the present time, more than half of the world's population lives in cities, generating more than 80 percent of global GDP (Human Development Report 2015, p. 5). The up going urbanization force the human and economic development (World Rankings 2016). The human capital (Human Development Report 2015) is one of the important components for growing industrialization (Kiely 1998), innovations (Cornell University 2016), scientific research (Research and development expenditure 2016), education and formation of the highly skilled workers (Knowledge Economy Index 2016).

3.2 Factors that Can Slow Down Sustainable Economic Development

The constantly growing human capital is one of the main conditions for the country's sustainable economic development. According to the World Bank (2016) estimation, only 16% of economic growth is provided by physical capital, about 20%—by natural capital, and 64% is driven by human capital. In the Russian Federation in 2015, we can see just the opposite conditions: 63%—by natural capital; all others in equal parts by physical capital and by human capital. So, for the Russian Federation, the human capital is the main source of innovation economic growing.

In the present time, not only new technologies and investments are required to ensure sustainable development of the country but first of all the high-tech sector, social innovations, a change in the priorities and goals of the development of civilization, and most importantly, a readiness to avoid immediate benefits for the sake of future generations.

To study the problems that can hamper human development, it is necessary to study the different groups of countries and establish its relationship with the main indicators of economic development.

3.3 Methods and Data

We used the methods of econometric modelling and cluster analysis on the base of data collections: United Nations Development Programme (Human Development Report 2015, 2016) and the World Development Indicators (WDI) of World Bank (2016).

Cluster analysis was done for the Human Development Index (HDI) and WDI for countries with different levels of social and economic development together with ecological situations.

For the 3D data visualization and data plot, the OriginPro 8.6 programme was used. With the help of spatial data analysis of population density (Earthdata SEDAC 2016) and urbanization for the studied countries, we can separate them in the different groups.

The World Development Indicators (World Bank WDI XL 2016) was completed for 2016 and was included in the Human Development Report 2016 with data set (Human Development Data 2016). The analysis of the Human Capital Impact for Sustainable Economic Development was done on the basis of 2015–2016 data sets.

In this paper, we make an attempt to provide analyse of the paradoxes mentioned above and consider their role in the development of Human Capital.

3.4 Main Paradoxes

In the Human Development Report 2015 by United Nations Development Programme (Human Development Report 2015) three main paradoxes were underlined:

- Rising inequality between global wealth and the world's population distribution;
- The growing urbanization is one of the forcing factors of global GDP growth but this does not automatically mean the higher income;
- There is no automatic link between income and human development.

In this research, we focus on the relationship of these paradoxes with the Human Development and their joint contribution to Sustainable Economic Development.

3.4.1 Paradox 1. Rising Inequality

Inequality can be observed in many countries. We can refer to Goal 10 of the UN Sustainable Development Goals (UN SDG Goal 10 2016) the well-known relation exists then income inequality between countries may have been reduced at the same time the inequality within countries has risen. This means that "…economic growth is not sufficient to reduce poverty if it is not inclusive and if it does not involve the three dimensions of sustainable development—economic, social and environmental" (UN SDG Goal 10 2016).

UN Human Development Report 2015 shows that the level of income of the world's population is unevenly distributed: 1% of the population has 48% global wealth and can pass the midway of 50 percent in 2016 (Oxfam 2015), 19% of the population has 46% of the global wealth, while the remaining 80% of the population account for only 6% of the global wealth (Human Development Report 2015, p. 65).

This discrepancy can be observed in all countries with a high or low population density, as well as with a high or low level of development of human capital. The reducing income inequalities between countries can even rise inequality within a single chosen country.

This can affect poor countries more than rich ones due to their huge population and cause a huge migration in the cities. In the world scale, this inequality will provide a huge contribution to global urbanization.

On the current date, almost 80% of the world's population is concentrated in developing countries, producing 40% of world GDP. The most populated areas are Asian countries—4.463 billion people; Africa—more than 1.216 billion people; China—1.384 billion people; India—1.284 billion people. Where according to projections in the near future, a significant increase in urbanization is expected. Growing inequality in the global distribution of wealth will force the migration process (Earthdata SEDAC 2016).

The highly developed countries have 16% of the world's population and produce more than 50% of world GDP.

The analysis is aimed at the World wealth distribution among different groups of countries and among the population within the exact country. The focus of the research was paid to what extent the development of human capital can have an impact on economic growth.

Different groups of countries (G7, highly developed countries, oil and gas rich countries, etc.) have significantly different GDP and GNI per capita.

Among the significant factors with high impact on the HDI are Gross national income (GNI) per capita, PPP and Trade (Exports and imports) per capita. We can make a 3D analysis of these parameters with the help of the 3D plot (Fig. 3.1).

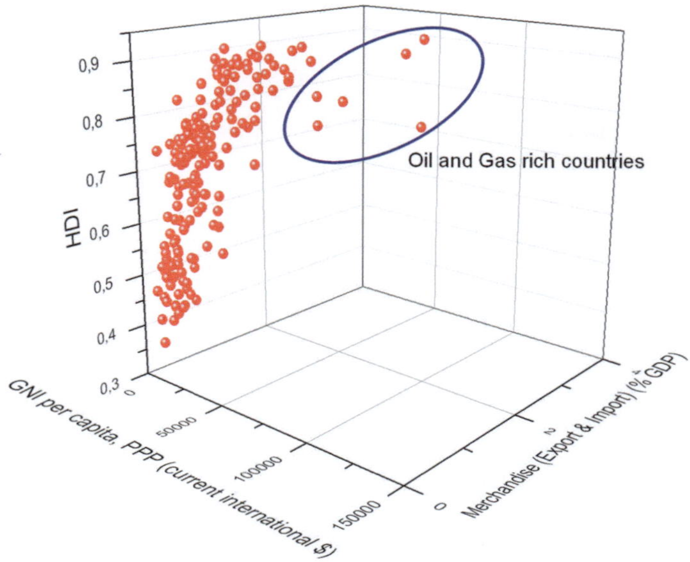

Fig. 3.1 The relation between indicators: Merchandise (Exports and imports) % GDP—OX axis; GNI per capita, PPP—OY axis; Human Development Index (HDI)—OZ axis. *Data source* Human Development Report (2016), World Bank World Development Indicators (2016)

On the 3D Plot (Fig. 3.1), GNI per capita, PPP is significantly dependent on exports and imports of a particular country. We can separate the countries into several groups with significantly different levels of Trade Merchandise (Exports and imports) % GDP and GNI per capita, PPP.

There is a large group of oil-producing countries: Qatar, Kuwait, Brunei Darussalam, UAE, Saudi Arabia, Oman, Russia and Bahrain (Production of Crude Oil including Lease Condensate 2016), with high levels of income from trade and GNI per capita, PPP (Fig. 3.1) and they have high values of HDI ($0.8 < \text{HDI} < 0.9$).

From Fig. 3.2 we can see that Qatar is the leader of oil and gas rich countries so the reserves are more than $ 6 million for each citizen. These huge reserves in a short period provided Qatar with significant economic growth and the top rank between the oil and gas rich countries in terms of GNI per capita, PPP. The second one is Kuwait where for each citizen there are oil and gas reserves for more than $ 4 million. The oil and gas reserves in United Arab Emirates, Turkmenistan, Saudi Arabia are 1.6, 1.5 and 1.1 million USD per capita, respectively (Fig. 3.2).

According to the UN report (Human Development Data 2015) and World Bank (2015, 2016), the percentage of income from the export of oil and gas from the total exports of goods of countries for 2015 is distributed as follows: Brunei—92.98%; Kuwait—89.11%; Qatar—82.77%; Saudi Arabia—78.4%; Russia—63%; Oman—62%; Bahrain—50.35%; United Arab Emirates—42.5%. At the same time for these countries GNI per capita, PPP has a huge difference: Qatar—138480 USD; Kuwait—83150 USD; Brunei—83010 USD; United Arab Emirates—72830 USD;

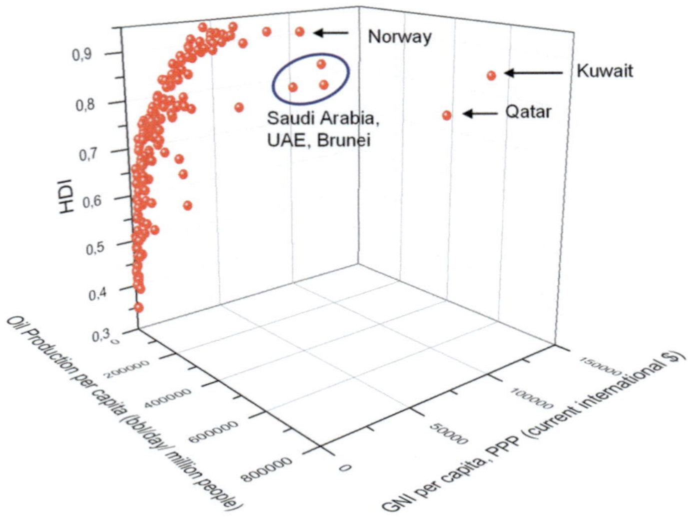

Fig. 3.2 The relation between indicators: GNI per capita, PPP (current international $)—OX axis; Oil Production per capita (bbl/day/million people)—OY axis; Human Development Index (HDI)—OZ axis. *Data source* Human Development Report (2016), World Bank World Development Indicators (2016), Production of Crude Oil including Lease Condensate (2016)

Saudi Arabia—55750 USD; Oman—38650 USD; Russia—24120 USD; Bahrain 22579 USD; Tajikistan—3500 USD. One can see the huge variance in HDI values (Human Development Report 2016) from 0,865 (Brunei) to 0,624 (Tajikistan).

Forcing of natural resources can be a privilege to gain income and export revenue. It looks like that exports of primary commodities in general of oil and gas these countries prefer to invest in the primary sector of the economy and public services within the economy. In order to get some privilege to the future generations they increase the revenue. But if they do not invest in the High-Technology exports and Research & Development (Ravi et al. 2010) then in the nearest future the natural resources (Jain et al. 2010), and obtained capital will run out and these countries will be impoverished.

Another group of countries, for example, Sudan, Nigeria, Eritrea with low income from trade (19% GDP, 30.9% GDP, 37.2% GDP) have both low value of GNI per capita, PPP (4290 USD, 5740 UDS, 1130 USD) and HDI (0.467, 0.514, 0.391). This is an example of poor countries with fewer opportunities (Fig. 3.1).

G7 group of countries has the highest HDI > 0.9 (Fig. 3.1), but a lower GNI per capita, PPP (World Bank 2016) income compared to commodity-dependent countries (oil rich countries): USA—58700 USD; Canada—44020 USD; Germany—49690 USD; United Kingdom—41640 USD; France—42000 USD; Japan—43540 USD; Italy—38460 USD. The high quality of life with the high cost of education, healthcare together with high environmental standards (Global Metrics for the Environment 2016) can explain the reduction of GNI per capita income for this group of countries.

As a result of analysis the countries with high HDI and GNI per capita has a clusterization in different groups due to the role of scientific research and industrialization as in the case with a high volume of trade (exports and imports) per capita (Fig. 3.1) and high urbanization.

Thus, as a result of the study of different groups of countries in terms of their income from trade and the distribution of income among the population, it is established that the division of society is seriously intensifying in the world, the problem of inequality in the distribution of income both between countries and within the country (an example of Tajikistan).

Over the past 50 years, the gap between the richest and poorest countries has doubled. In addition, it was noted that even in developed countries the rich layers of the population are more rich, and the poor—poorest.

That is, the country's economic growth does not automatically ensure the welfare of its people. It was mentioned in the UN report on sustainable development goals, which stated that economic growth is not sufficient to reduce poverty if it is not comprehensive and does not include three aspects of sustainable development—economic, social and environmental.

The UN report for 2015 argues that growing urbanization is also one of the main factors of the country's economic growth, but will the uniform distribution of the good among the population and its development follow from this growth?

The paradox of the Growing Urbanization is the outstanding one and we'll discuss it in the next part.

3.4.2 *Paradox 2. The Growing Urbanization*

Since the end of the nineteenth century, the process of global urbanization was intensified. Faster growth of the urban population became more evident in the middle of the twentieth century.

According to the UN statistical data (World Urbanization Prospects 2014) in 1950 a third of the world's population lived in cities (746 million), in 2000, nearly half were city dwellers (2.9 billion), and in 2050 more than two-thirds (66%) of the world's population (6.3 billion) is projected to be living in urban areas. This rapid increase will take place mainly in developing countries (Human Development Report 2015, p. 67).

"Urbanization has the potential to improve the economic well-being of societies. More than half of the world's people live in cities, but they generate more than 80% of global GDP" (World Bank 2015).

The source for this rapid urbanization is the rural population of China (Hazelhurst 2015), India and developing countries mostly Africa and Asia (UNDP 2013). China has a leading position in urbanization where migrant workers from the countryside are the third of the labour force (Hazelhurst 2015). In Africa the urban population will increase from 40 % to 56 % by 2050, a similar tendency is observed in Asia

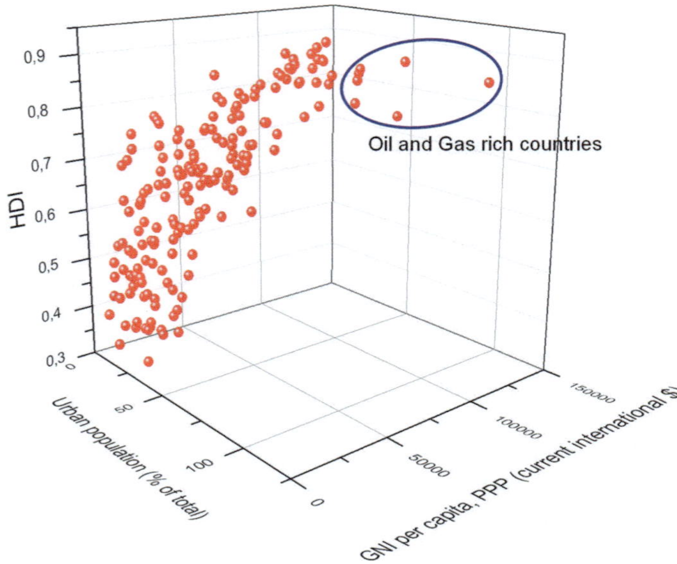

Fig. 3.3 The relation between indicators: HDI from GNI per capita, PPP and the Population Urban (%). *Data source* Human Development Report (2016), World Bank World Development Indicators (2016)

from 48 % to 64 % (UNDP 2013). This trend is observed in the background of the exponential growth of the world population.

From Fig. 3.3 we can see that the level of the urban population has a significant impact on the distribution of GNI per capita for countries with different economic development.

The oil rich countries such as Qatar, Kuwait, Saudi Arabia, UAE, Bahrain, Brunei Darussalam and Oman with a large urban population and huge GNI per capita, PPP, have a rather big HDI value (0.8 < HDI < 0.9).

The percentage of the urban population in these countries is the following: Qatar—99.24%; Kuwait—98.34%; United Arab Emirates—85.54%; Saudi Arabia—83.13%; Brunei—77.2%, Bahrain—88.78%; Oman—77.64%; Russia – 74.014%. Tajikistan, which has a low urban population level (26.78%) and a low HDI, stands out especially from this group of countries.

The group of highly developed countries has a higher level of HDI (HDI > 0.9) and a more evenly population distribution. The high ecological level countries have better transport infrastructure and separate industrial plants from cities that make them more comfortable for the population. This is due to the population distribution all over the country including city, towns, industrial regions, agricultural plants and transport infrastructure (Global Metrics for the Environment 2016).

The highly developed countries rely on a highly developed high-tech sector, which contributes to their economic growth, thereby determining the high level of development of the human capital.

However, high per capita income does not mean the high level of development of the human capital; this is the main difference between these groups of countries.

Thus, in highly developed countries, the urban population actively participates in the economic life of the country and creates material wealth, while in the commodity countries the urban population is the main consumer of income from resources trade.

Countries with low income from trade (export and import) and low HDI have a low level of the urban population.

The group of countries (China, India, Pakistan Bangladesh, etc.) with a huge population (1 393.8; 1 267.4; 185.1; 158.5 million people, accordingly) and growing urbanization has a low level of GNI per capita, PPP (15470 UDS; 6490 UDS; 5560 USD; 3790 USD) and medium levels of HDI (0,727; 0,609; 0,538; 0,57).

This fact illustrates the different roles of growing urbanization in generating the benefits (World Bank 2016) of this particular group of countries and the development of human capital (Human Development Report 2016).

According to the World Bank (2016), growing urbanization can improve society well-being. However, this does not mean the automatic growth of human well-being in a particular country (Paradox 1).

The same factors will influence the development of the country's human capital, economy and society. It is obvious that a knowledge-based economy will manage sustainable development.

A high level of human development and significant expenditure on research and development are necessary, but not sufficient conditions for generation income level by country. We study this fact by considering the following paradox.

3.4.3 Paradox 3. There is no Automatic Link Between Income and Human Development

According to Nelson and Romer (1996) investments add value to technology, and technology to investment, and this vicious circle contributes to economic growth.

In modern conditions, scientific knowledge and information technology are the main sources of economic growth and increasing human welfare (UNESCO 2018). We could not imagine this without scientists and scientific research.

The quality of human capital influence on economic growth; hence, investments in human capital will also contribute to economic growth.

Human development enriches people's lives but unfortunately not every human life (Human Development Report 2016). Let us consider this fact in more detail.

We can refer to the GDP structure, it is a composite statistic of life expectancy, education, and per capita income indicators.

However, in different countries the income structure is specified by different sources. So, the main income of commodity-dependent countries (Qatar, Kuwait and others) formed by resources export. High developed countries (USA, Norway, Japan, etc.) are creating the high-technology markets and increase the human capital.

Thus, the US share in the country's exports of high-tech products is 19.96% of total exports and is 153.187 billion USD at current prices for 2016; Norway—19.49% (3.913 billion USD); Japan—16.22% (92.88 billion USD) (IMF 2016).

China getting privilege from the use of cheap labour force, low environmental standards (Global Metrics for the Environment 2016; Policymakers Summary 2016) and the deployment of foreign enterprises. As well as a large group of countries with the direction of economic development on agriculture, tourism, the banking sector and others.

Among these groups, the level of income may differ significantly, for example, Saudi Arabia and Turkmenistan are oil and gas rich countries but with a significantly different standard of living. For comparison, Turkmenistan and Saudi Arabia rank the 4th and 6th places, respectively, for oil and gas reserves per capita, with huge export of these resources, but the quality of life of these countries differ from each other almost in 15.9 times GNI per capita, PPP (Turkmenistan—3500 USD and Saudi Arabia—55750 USD). At the same time, these group of countries have different possibilities for investment in health, education and science.

Based on this approach it is possible to separate these countries using different groups: with the same HDI value and different income or similar income and different HDI values.

For example, the HDI level of Gabon is 0.684 and GNI per capita, PPP is 7179.34 USD; the level of HDI in Indonesia is 0.684 and GNI per capita, PPP—3570.29 USD. So these countries have just equal HDI levels (Human Development Report 2016) and at the same time almost 2 times difference in GNI per capita, PPP (World Bank 2016).

There are groups of countries with the same GNI per capita, PPP, but different values HDI.

The HDI level of the Russian Federation is 0.804, and the GNI per capita, PPP is 8748.37 USD. The HDI level of Equatorial Guinea is 0.587 and GNI per capita, PPP is 8747.35 USD. One can see that these countries have equal GNI per capita, PPP (World Bank 2016) but HDI levels (Human Development Report 2016) have a significant difference of almost 1.4 times.

This paradox has a good definition in the UN Human Development Report (2015, p. 56): "… there is no automatic link between income and human development. Income enters into the HDI but is just one of its four indicators. Economic growth does not automatically translate into higher human development".

From Fig. 3.4 one can see that the low Research and Development (R&D) expenditure (% of GDP) corresponds to sluggish growth of GDP per capita while increasing investment in R&D the GDP per capita increases substantially. If the R&D expenditure reaches 3–4% of GDP, then GDP per capita increase to 50–70 thousand USD.

This process is accompanied by the growth of human capital. The low HDI (0.5–0.6) corresponds to the very low level of the GDP per capita and with HDI growth from 0.8 to 0.95 the GDP per capita reached the maximum value.

A similar relationship exists between the HDI and R&D expenditure (% of GDP). The growth of the HDI is accompanied by a significant increase in the cost of R&D

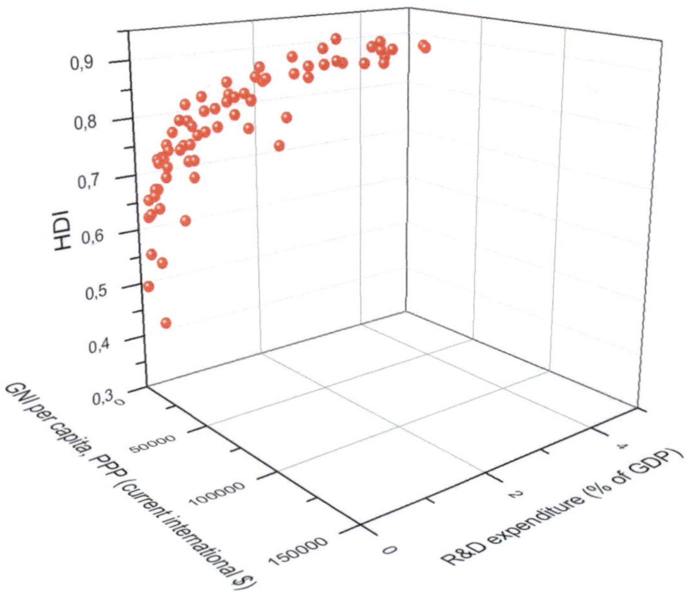

Fig. 3.4 The relation between indicators: Human Development Index (HDI); Research and Development expenditure, % of GDP; GDP per capita (current US$, 2016). *Data source* Human Development Report (2016), World Bank World Development Indicators (2016), IMF GDP data (IMF GDP 2016)

(% of GDP), especially for highly developed countries and countries with innovative economies (Fig. 3.4).

Small investments in R&D do not give growth in GDP per capita (current US$). Significant investments in R&D 4% are possible in countries with high levels of HDI, which can generate high GDP per capita.

However, there are exceptions. Luxembourg GNI per capita is 100738.68 USD with HDI level—0.892, however, it spends only 1.29% of GDP (World Bank 2016) for R&D. The Republic of Korea and Israel are top leaders of high-tech sector, their expenditures were 4.23% and 4.27% of GDP, respectively (World Bank 2016). GNI per the capital of Israel was 37180.53 USD; for Korea—27538.81 USD (World Bank 2016).

We can see the opposite situation with oil and gas rich countries, except Russia. They have low internal R&D expenditures: Qatar—0.46%; Kuwait—0.3%; the United Arab Emirates—0.87%; Saudi Arabia—0.08%; Bahrain—0.1%; Brunei—0,03%, Tajikistan—0,09%, but rather high GNI per capita excluding Tajikistan.

The investment of science should be dependent on the country's aims. A number of countries may not conduct their own research and development, but buying ready-made high technologies. These countries can get into technological dependence from industrially developed countries (USA, Canada, Japan, Germany, France, etc.).

In the case of political disagreements or territorial conflicts, the industrially developed countries can impose their political conditions and implement economic sanctions.

We can refer to the international relations of USA and European Union with Russia and China. India is searching for investments and technologies, including from the USA, but under the risks of losing its technological independence. Consequently, this means a threat to the economic security of India.

3.4.3.1 Research and Development Export

Innovation is the main source of long-term economic growth (Jain et al. 2010), the basis of competitiveness and the solution to many social problems. Human capital becomes one of the determining factors in the effective development of the economy and the achievement of competitive advantages (Martin et al. 2001). Economic growth depends on the technological changes that happen due to the human capital rise.

We can focus on the relations between HDI, high-tech exports and R&D costs (Fig. 3.5).

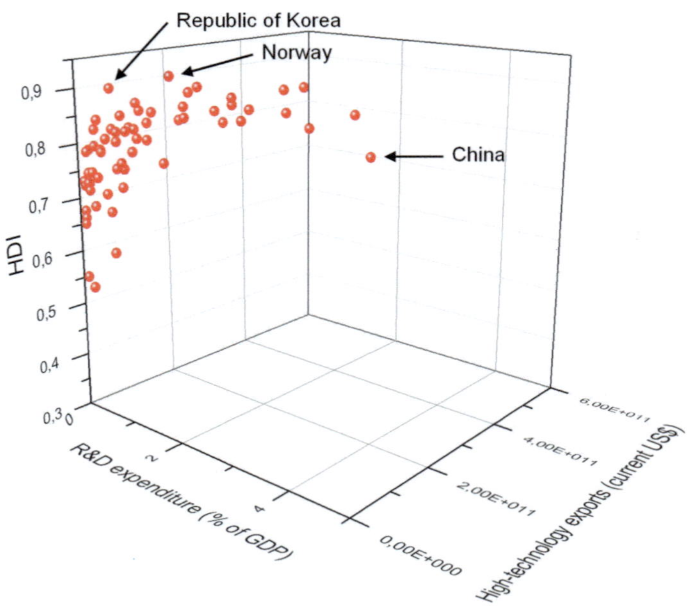

Fig. 3.5 HDI from High-Technology exports (Current US$) and R&D expenditure (% of GDP). *Data source* Human Development Report (2016), World Bank World Development Indicators (2016)

One can see (Fig. 3.5) a big gap between highly developed countries, G7 and countries with high income from the high-tech export (China, the Republic of Korea, etc.).

Figure 3.5 can illustrate the huge difference between the countries with high level HDI as Norway and the countries with huge High-Technology exports as China.

Highly developed countries and G7 with a high level of human capital have a good investment in the development of a high-tech sector, but at the same time have an average level of income from the high-tech products because of their high cost.

Fast-growing economy countries (China, Republic of Korea), but with a lower level of human capital also have high R&D expenditures, but also provide high income from the high-tech production.

We can illustrate this with the example of the above mentioned countries. China: R&D expenditure, 2.07% of GDP; High-technology exports (current US$) 4.96E + 11; 0.727 HDI level. Republic of Korea: R&D expenditure, 4.23% of GDP; High-technology exports (current US$) 1.18E + 11; 0.898 HDI level. Norway: R&D expenditure, 1.93% of GDP; High-technology exports (current US$) 3.93E + 9; 0.944 HDI level.

Norway has the highest HDI and the medium R&D, China is the leader of the High-Technology exports and the Republic of Korea is the leader of R&D expenditure.

The high-technology exports of countries members of G7 (6.6E + 11 USD) or European Union (6.13E + 11 USD) can be compared with China (4.96E + 11 USD). China declared "from the traditional to the innovative", and became the leader in the export of high-tech goods. The G7 countries and the European Union are close to the Chinese economy and should continue to develop high-technology markets and provide cost optimization. Otherwise, it will be very difficult for them to make a competition with the growing Chinese economy.

It is very important for China to improve environmental standards (Policymakers Summary 2016) and conduct its own high-technology research.

For sustainable development, new technologies and investments in education, fundamental science and the high-tech sector are required. It is very important to make investments in social innovations and the digital economy. For the sake of future generations, it is necessary to abandon the short-term benefits and change the priorities and goals of civilization development.

3.5 Conclusion

The economic growth and sustainable development are on the different sides of the swing.

The highly developed countries with higher human development are slowing down their economic growth due to the high level of environmental standards and research & development expenditure.

For some countries, the high rate of economic growth can be explained by a cheap workforce, work discrimination, low-level environmental standards, which can be bad for future generations.

It is possible to maintain the economic growth on the high level providing the innovation economy, knowledge economy, digital revolution, renewable energy and low emission technologies.

As a result of analysis, the countries with high HDI and GNI per capita have a clusterization in different groups due to the role of scientific research and industrialization as in the case with a high volume of trade (exports and imports) per capita and high urbanization.

The process of fast economic growth can lead to serious problems. We can underline two main consequences of economic development: the first one—economic development will increase the human capital; the second—forcing of economic development will inevitably lead to the environmental disaster, the depletion of natural resources and the struggle for their ownership.

During the International Conference on "Spatial Decision Support System For UN Sustainable Development Goals" held in India (Department of Geography, Kalindi College, University of Delhi) in 2017, Dr. Shri Murli Manohar Joshi Member of Parliament and Former HRD Minister of Government of India expressed concern about the rapid economic development, which cannot continue indefinitely. This will inevitably lead to the depletion of all possible resources: raw, human, natural etc. According to Dr. Shri Murli Manohar Joshi, there is no need to aim for fast economic growth, it can be replaced by sustainable consumption and reducing inequalities together with improving quality of life.

The heterogeneity of economic development can lead to serious problems

(1) changes in the structure of the world's population (rapid global urbanization);
(2) polarization of society (the rapid growth of innovative economy and quality of life of some countries and slowing down the economy of poor countries;
(3) enhancement of mutual influence of some countries on the economic situation and quality of life of other Countries;
(4) depletion of natural resources;
(5) the struggle for gaining natural resources.

These can lead to social instability, environmental disasters, depletion of natural resources, to the global crisis and as a consequence the war for gaining the world's resources.

References

Acemoglu D (2009) The solow growth model. Introduction to modern economic growth. Princeton University Press, Princeton, pp 26–76. ISBN 978-0-691-13292-1

Cornell University, INSEAD, WPO (2016) The global innovation index 2016: Winning with global innovation. Dutta S, Lanvin B, Wunsch-Vincent S (eds). Ithaca, Fontainebleau, and Geneva. ISBN 979-10-95870-01-2

Dodds F, Donoghue D, Leiva R (2017) Negotiating the sustainable development goals: A transformational agenda for an insecure world. Routledge, Abingdon, p 215

Earthdata SEDAC (2016) Center for International Earth Science Information Network–CIESIN–Columbia University 2016. Gridded Population of the World, Version 4 (GPWv4): NASA Socioeconomic Data and Application Center (SEDAC). http://dx.doi.org/10.7927/H4NP22DQ. Accessed 15 August 2018

Global Metrics for the Environment (2016) The Environmental Performance Index ranks countries' performance on high-priority environmental issues. http://epi.yale.edu/sites/default/files/2016EPI_Full_Report_opt.pdf. Accessed 25 December 2016

Haines JD, Sharif NM (2006) A framework for managing the sophistication of the components of technology for global competition. Competitiveness Rev Int Bus J Emerald 16(2):106–121. https://doi.org/10.1108/cr.2006.16.2.106

Hazelhurst J (2015) The search for seed capital. FT Wealth: Entrepreneurs, 8 May, pp 40–41

Human Development Report (2015) Work for Human Development. The United Nations Development Programme (UNDP). ISBN 978-92-1-126398-5. http://hdr.undp.org/en/2015-report. Accessed 20 December 2016

Human Development Report (2016) Work for Human Development. UNDP. ISBN 978-92-1-126413-5. http://hdr.undp.org/sites/default/files/2016_human_development_report.pdf. Accessed 25 September 2018

Human Development Data (2015). http://hdr.undp.org/en/data. Accessed 10 September 2018

Human Development Data (2016). http://hdr.undp.org/en/data. Accessed 10 September 2018

IMF Annual Report 2016 (2016) Finding Solutions Together. http://imf.org/external/pubs/ft/ar/2016/eng/pdf/ar16_eng.pdf. Accessed 25 September 2018

IMF GDP data (2016) World Economic and Financial Surveys. World Economic Outlook Database. http://www.imf.org/external/pubs/ft/weo/2015/02/weodata/index.aspx. Accessed 15 September 2018

Jain RK, Triandis HC, Weick CW (2010) Managing research, development, and innovation: managing the unmanageable, 3rd edn. Wiley, p 237. https://doi.org/10.1002/9780470917275.ch11

Kiely R (1998) Industrialisation and development: a comparative analysis. UCL Press, London, p 196 ISBN 1857285441(HB)

Knowledge Economy Index-countries ranking (2016). https://knoema.com/atlas/topics/World-Rankings/Knowledge-Economy-Index/Knowledge-Economy-Index. Accessed 17 December 2016

Martin HP, Schuman H (2001) The global trap: globalization and the assault on prosperity and democracy. Al'pina Pablisher Publ, Moscow, p 335

Nelson RR, Romer PM (1996) Science, economic growth and public policy. Challenge 39(1):9–21. https://doi.org/10.1080/05775132.1996.11471873

Oxfam (2015) Wealth: having it all and wanting more. Issue Briefing. Oxford, UK. www.oxfam.org/sites/www.oxfam.org/files/file_attachments/ib-wealth-having-all-wanting-more-190115-en.pdf. Accessed 13 September 2018

Policymakers Summary (2016) Global Metrics for the Environment. The Environmental Performance Index ranks countries' performance on high-priority environmental issues. http://epi.yale.edu/sites/default/files/2016EPI_Policymakers_Summary.pdf. Accessed 25 December 2016

Production of Crude Oil including Lease Condensate (2016). https://www.eia.gov/beta/international/data/browser/. Accessed 25 September 2018

Research and development expenditure, % of GDP-countries ranking (2016) https://knoema.com/atlas/ranks/Research-and-development-expenditure-percent-of-GDP?origin=knoema.ru. Accessed 10 September 2018

Solow Robert M (1956) A contribution to the theory of economic growth. Q J Econ Oxf J 70(1):65–94. https://doi.org/10.2307/1884513.JSTOR1884513.Pdf

Solow RM (1957) Technical change and the aggregate production function. Rev Econ Stat 39(3):312–320. https://doi.org/10.2307/1926047.jstor1926047.pdf

Swan TW (1956) Economic growth and capital accumulation. Econ Rec 32(2):334–361. https://doi.org/10.1111/j.1475-4932.1956.tb00434.x
UN Sustainable Development Goals, Goal 10: Reduce inequality within and among countries. http://www.un.org/sustainabledevelopment/inequality/. Accessed 20 December 2016
UNESCO (2018) UIS. Stat (Education, Science, Technology and Innovation). http://data.uis.unesco.org. Accessed 25 September 2018
United Nations, Department of Economic and Social Affairs, Population Division (2014). World Urbanization Prospects: The 2014 Revision
UNDP (United Nations Development Programme) (2013) Accelerating Progress: Sustaining Results. New York
World Bank (2015) Overview. www.worldbank.org/en/topic/urbandevelopment/overview. Accessed 25 September 2018
World Bank (2016) World Development Indicators. https://data.worldbank.org/indicator/. Accessed 25 September 2018
World Bank WDL XL (2016) World Development Indicators (Excel). https://datacatalog.worldbank.org/dataset/world-development-indicators. Accessed 25 September 2018
World Rankings-Human Development Index-countries ranking (2016). https://knoema.com/atlas/topics/World-Rankings/World-Rankings/Human-development-index?origin=knoema.ru. Accessed 25 September 2018

Chapter 4
Land Use Land Cover Dynamics Using Remote Sensing and GIS Techniques in Western Doon Valley, Uttarakhand, India

Ajay Kumar Taloor, Vaibhav Kumar, Vivek Kumar Singh, Anil Kumar Singh, Ravindra V. Kale, Rahul Sharma, Varun Khajuria, Girish Raina, Beena Kouser and Naveed Hassan Chowdhary

Abstract Land use land cover (LULC) change analysis emerged as one of the most significant factors which assist decision makers to ensure sustainable development and to understand the dynamics of our changing environment. An integrated approach of remote sensing and GIS has been used to study the land use land cover dynamics of the Western Doon Valley, Uttarakhand. Landsat satellite imageries of two different time periods, i.e., Landsat ETM + data of 2001 and 2010 were acquired and used

---

The original version of this chapter was revised: The incorrect affiliation of authors "A. K. Singh and R. V. Kale" has now been replaced with the correct one. The correction to this chapter is available at https://doi.org/10.1007/978-981-15-2097-6_23

A. K. Taloor (✉) · R. Sharma · V. Khajuria · G. Raina
Department of Remote Sensing and GIS, University of Jammu, Jammu 180006, India
e-mail: ajaytaloor@gmail.com

R. Sharma
e-mail: rahul29453@gmail.com

V. Khajuria
e-mail: varunkhajuria8182@gmail.com

G. Raina
e-mail: girishrainaraina@gmail.com

V. Kumar
Centre for Urban Science and Engineering, Indian Institute of Technology, Bombay, India
e-mail: vaibhav.iirs@gmail.com

V. K. Singh
Centre for Atmospheric Sciences, Indian Institute of Technology, Delhi, India
e-mail: december.keviv@gmail.com

A. K. Singh
Agromet Advisory Services Division, Indian Meteorological Department, New Delhi 110003, India
e-mail: singhanil854@gmail.com

S. Sahdev et al. (eds.), *Geoecology of Landscape Dynamics*,
Advances in Geographical and Environmental Sciences,
https://doi.org/10.1007/978-981-15-2097-6_4

to quantify the land use land cover changes in the study area from 2001 to 2010 over a period of one decade. ERDAS Imagine 10 software has been used to carry out the supervised classification using a maximum likelihood technique. The images of the study area were categorized into five different classes, viz., agricultural land area, settlement area, forest cover area, wasteland area, and water body area. The result indicates that during the decadal period, the agriculture forest and settlement area have increased about 6.22% (i.e., 25.19 km^2), 0.30% (i.e., 2.66 km^2), 2.17% (20.47 km^2), respectively, while area under other land categories such as wasteland and water bodies have decreased about 6.16% (i.e., 22.67 km^2) and 2.52% (i.e., 0.22 km^2), respectively. The Shuttle Radar Topographic Mission (SRTM), digital elevation model (DEM) data have been used for determination of slope analysis and it is found that most of the LULC changes have occurred in the area where slope percentage was in nearly level to gentle categories. The accuracy assessment and Kappa coefficient of both data sets have also been determined and found that in the 2001 accuracy assessment was 85.35% and in 2010 accuracy assessment was 89.59%. The technique used in the study shows the importance of digital data-based change detection techniques for the nature and location of a change in the study area.

Keywords Land use land cover · Change detection · Landsat data · Kappa coefficient · Accuracy assessment

4.1 Introduction

Human beings are one of the most destructive agents of nature who continuously changes and modifying the landscape depends upon its suitability for survival and wellbeing. Since the history of human being the land surface have witnessed the many changes in the form of national boundary barrier, great walls, embankments, urban planning, industrialization, settlement agricultural practice etc. Human alteration of a landscape from natural vegetation to any other use typically results in habitat loss, degradation, and fragmentation, all of which can have a devastating effect on biodiversity. The changes in land use/land cover represent an important part of the global change affecting the environment. These changes occurred by altering (increasing or decreasing) the number, structure, or conditions of the elements

R. V. Kale
Western Himalayan Regional Center, National Institute of Hydrology, (Ministry of Water Resources, River Development and Ganga Rejuvenation, Govt. of India) Irrigation & Flood Control Complex, Opposite Military Hospital, Satwari, Jammu Cantt., Jammu 180003, India
e-mail: ravikale2610@gmail.com

B. Kouser · N. H. Chowdhary
Department of Geology, University of Jammu, Jammu 180006, India
e-mail: beenajucryosphere@gmail.com

N. H. Chowdhary
e-mail: hassannavid4@gmail.com

in the satellite image over various spatial and temporal scales (Stow et al. 1990; Sreenivasulu and Bhaskar 2010). Although, quantifying, monitoring, and evaluating the spatial and temporal dynamics of the land use land cover is quite critical for better understanding many of the Earth's land surface processes (Midekisa et al. 2017). Besides this, to understand these changes allow us to quantify and monitor trends in agriculture (Ramankutty and Foley 2011), freshwater resources (Costa et al. 2003), forest cover (Hansen et al. 2014), and disease transmission (Patz and Norris 2004; Midekisa et al. 2014). Moreover, we are aware that land conversion is the greatest cause of extinction of terrestrial species, of which particular concerns are deforestation, expansion of urban centers, industrial expansions, major roads, and railways network corridors have really created a great impact on the ecology and survival of many species that previously existed (Tripathy et al. 1996).

Large number of researchers around the world are monitoring these changes of land use which is a product of interactions between a society's cultural background, state, and its physical needs on the one hand and the natural potential of land on the other, so that better understanding can be made among man, nature, and natural resources (Balak and Kolarkar 1993; Chaurasia et al. 1996; Agarwal et al. 2002; Jasrotia et al. 2012; Jasrotia et al. 2013; Taloor et al. 2018). Researchers around the world have started to monitor land use land cover changes by involving traditional surveys and inventories from the nineteenth century. With the passage of time and an enhancement in the technology, remote sensing and GIS are quite advantageous as it is economically billable and time saving for micro to macro scale LULC changes with geographic spatial information (William et al. 1994; Yuan et al. 2005; Xiao et al. 2006; Shalaby and Tateishi 2007; Noor et al. 2008; Prakasam 2010; Friedl et al. 2010; Dong et al. 2012; Giri et al. 2013; Yan and Roy 2015; Xiong et al. 2017). The classification of the image is not completed until its accuracy assessment is not assessed although, the applications of LULC classification is increasing day by day with the enhancement in remote sensing technology (Congalton and Green 2008; Martellozzo and Clarke 2011).

In recent years, there has been tremendous increase in the availability of high performance cloud computing such as the NASA Earth Exchange (NEX) platform which allows the processing and analysis of NASA earth observation data (Nemani 2011), Amazon Web Service (AWS) also now provides access to the Landsat data archive, enabling analysis of this dataset on the cloud. In the recent times, Google Earth Engine (GEE) has enhanced the scientific capability to explore and analyze as it is a new high performance computing platform which gives access to a vast and growing amount of earth observation data. In the recent times, Google Earth Engine (GEE) has enhanced the scientific ability to explore and analyses of the earth surface, as it is a new high-performance computing platform which gives access to a vast and growing amount of earth observation data and also the processing power to analyze these data at planetary as well as micro-scale (Midekisa et al. 2017).

The main objectives of the present study are to examine the land use/land cover temporal changes during 2001–2010, determination of accuracy assessment, kappa coefficient, and role of slope in land use land cover change dynamics. The study also highlights the importance of digital change detection techniques for the nature and location of change in the Western Doon valley.

4.2 Study Area

The Western Doon valley lies between latitude 30° 14′1″ to 30° 30′51″ and longitude 77° 38′ 05′ to 78° 05′ 50″ covers the total area of 898.33 km^2 (Fig. 4.1). The Western Doon valley is an intermountain valley that lies between two intermittent ranges of the Himalayas. It is bounded on all sides by mountains, with one range running from the west to the east in a semi-circular arc; and one running at the south from Paonta Sahib to Haridwar. The valley also forms a watershed between the Yamuna and Bindal River in the systems. Doon or Dun is a local word for valley, particularly an open valley in between the Siwaliks and higher Himalayan foothills. The average annual rainfall is 2200 mm out of which 1700 mm is monsoonal. Geologically, Western Doon valley is an asymmetrically, longitudinal structurally synclinal valley formed of Siwalik rocks of sedimentary origin having the trend of the northwest to southeast of Upper Tertiary Age (Jasrotia et al. 2018).

4.3 Materials and Methods

The present study was carried out using the various primary and secondary data. These include Survey of India (SoI) topographic sheet of 1:50000 scale. Landsat ETM + satellite images of Western Doon Valley were acquired for 2001 and 2010, respectively, with the spatial resolution of 30 m. These datasets were obtained from the Global Land Cover Facility (GLCF) an earth science data interface. To find out the changes, Landsat ETM + data of 2001 and 2010 were geo-referenced and supervised classification was used to determine the change detection analysis by using the maximum likelihood algorithm in ERDAS Imagine 10 software. The supervised classification depends on the accuracy of the user, techniques, experience, and accuracy of his optical capability to define and detect the different signatures among the various patterns in the satellite images. Spectral information represented by the one spectral band is used to classify each individual pixel. The Arc GIS 10 software was used for the integration of spatial data and the preparation of thematic maps. Adequate field checks have been made before finalizing of thematic maps. Slope map was prepared from SRTM, DEM data to envisage the role of slope in landscape change dynamics. The approach used in the present study is shown in Fig. 4.2.

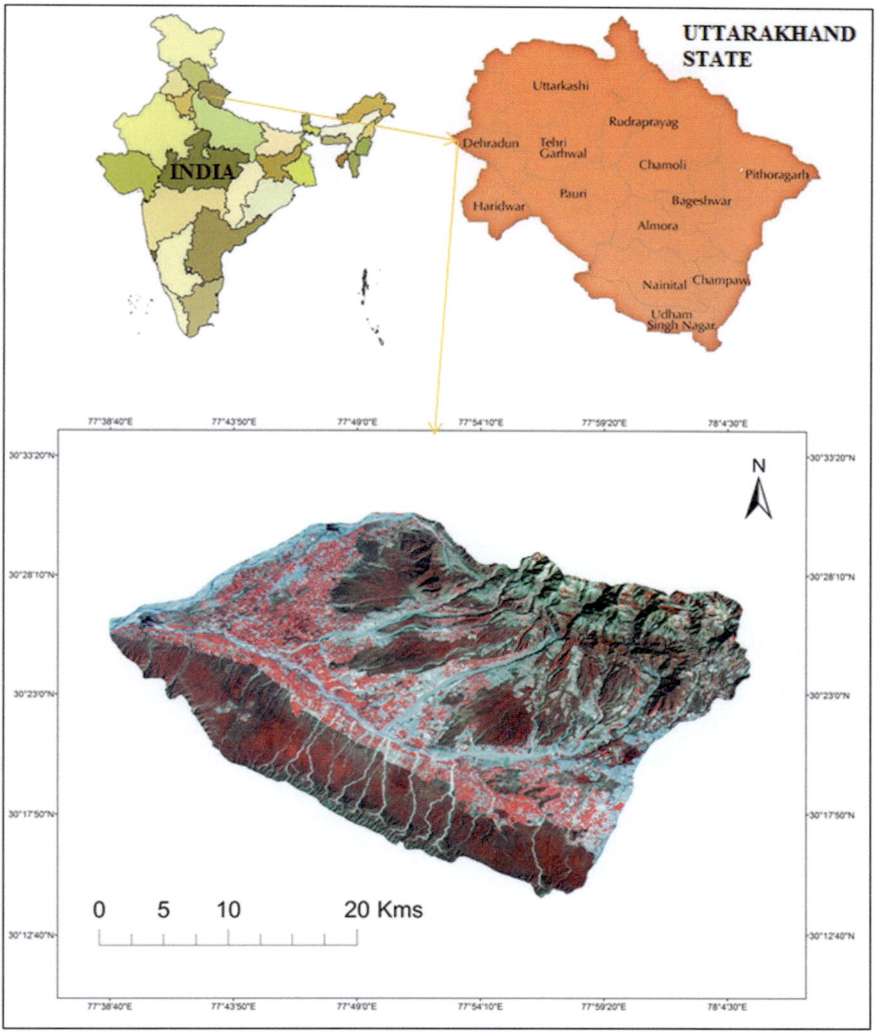

Fig. 4.1 Location map of the study area (*Source* Landsat-7, ETM+)

4.4 Results and Discussions

4.4.1 Slope Map

The slope is a measure of the steepness of a line, or a section of a line, connecting two points and is also one of the indicators of human development in many cases. Level and gentle slope areas are mostly developed with agricultural activities or human

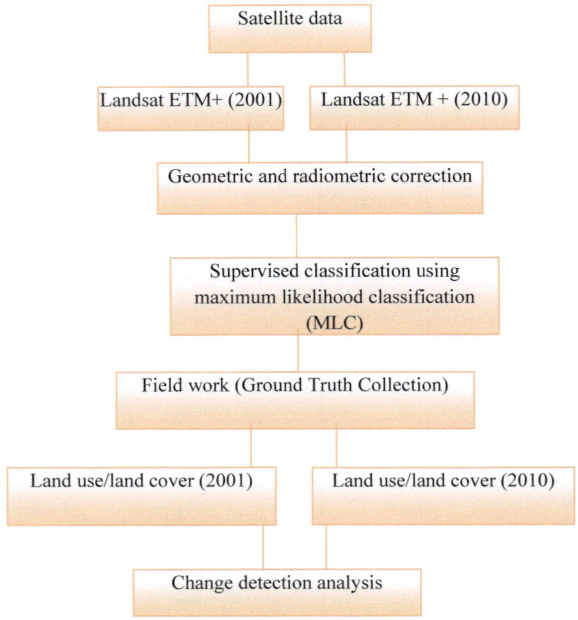

Fig. 4.2 Methodology adopted in the present study

settlements compared to moderate and steep slopes. The Shuttle Radar Topographic Mission (SRTM), Digital elevation model (DEM) data were used to prepare the slope map of the study area. The derived slope map was classified into seven categories (Taloor et al. 2017;) such as nearly level (0–1%), very gentle (1–3%), gentle (3–5%), moderate (5–10%), steep (10–15%), moderately steep (15–35%), and very steep (>35%) (Fig. 4.3). It is found in the study by comparing the slope map with change detection map that most of the changes were made in the area which has a level to gentle slope due to human activities which suggest that anthropogenic activities play a vital role in changing the landscape surface in the Western Doon Valley.

4.4.2 Land Use/Cover Status

The study area is classified into five major classes from Landsat TM satellite images of 2001 and 2010 are shown in Fig. 4.4 and Fig. 4.5, respectively. The different classes analyzed from the satellite data are shown in Table 4.1. The land use land cover study depicts that there is a positive growth in agriculture, settlement, forest cover; negative growth in water bodies and wasteland (Fig. 4.6). The detail description of the different classes is given in the following subheading.

Settlement area: Settlement included the area under residential, commercial, industrial, parking and transportation facilities. In the satellite imagery, the class was identified by blocky appearance, light bluish colored, fine to medium texture

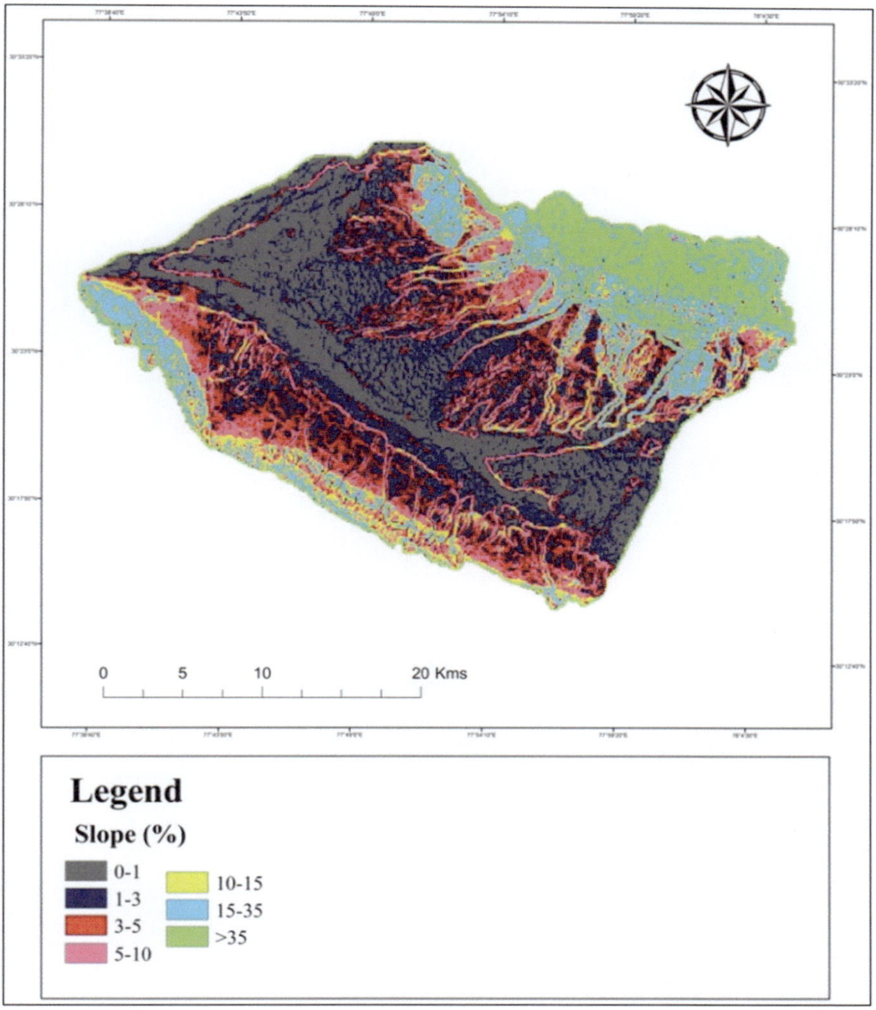

Fig. 4.3 Slope map of the study area (*Source* SRTM, DEM)

with regular shape and varying size. An increase in the settlement area means the expansion of mankind which has positive, as well as negative impact on the land it surges. In the 2001 thematic layer, the area covered by settlement class is 175.07 km^2 (19.49%) and increased 2.17% of the total area in 2010 as 194.54 km^2 (21.66%). In the study area, it is found that most of the expansion in the settlement is in the fringes of the earlier built up area and generally in the area with level to the gentle slope.

Agriculture land area: Agriculture appears light pink in the FCC image characterized by the shades of red color and textural variability including the areas cultivated with various cultures of corn, wheat, barley, oat, potatoes, tea plantation etc. In the

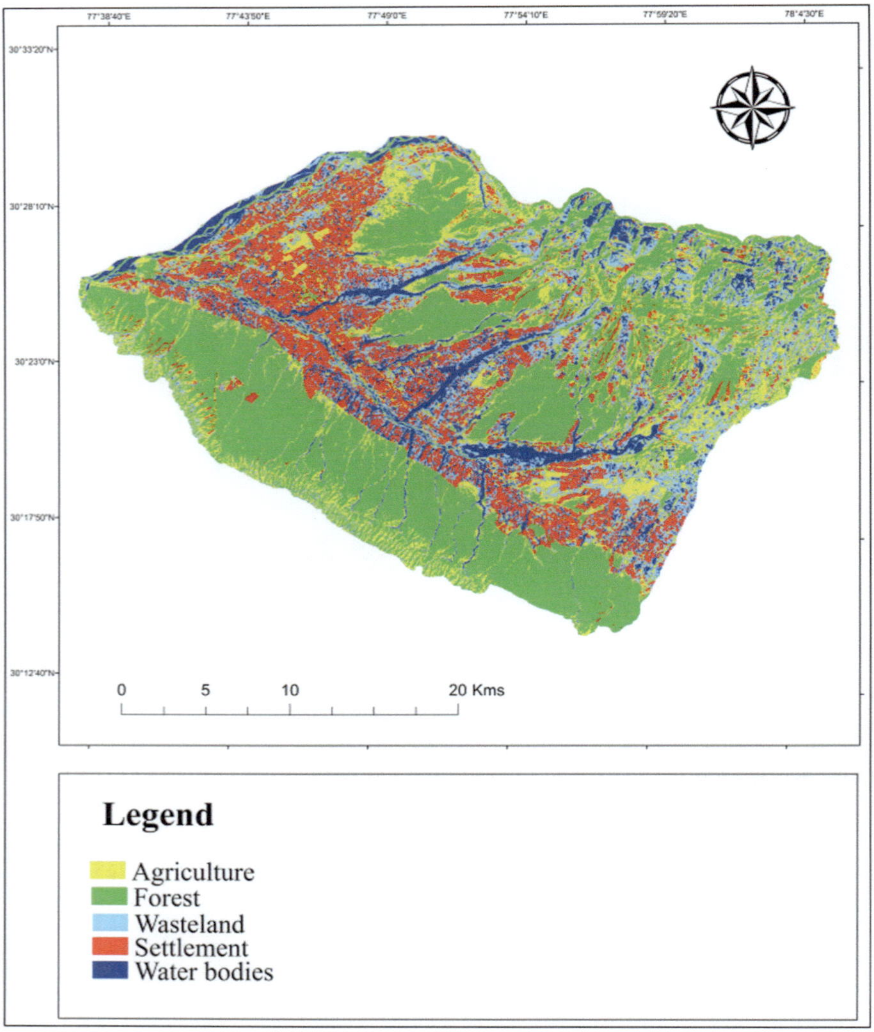

Fig. 4.4 Land use land cover map 2001 (*Source* Landsat-7, ETM+)

land use classes of 2001, the agriculture land covers area covers 131.31 km^2 (14.62%) of the total area whereas in 2010 this agricultural land covers 187.19 km^2 (20.84%) of total area with an increase of 6.22%. The increase in agriculture due to population pressure and availability of a large amount of fallow land in the Western Doon Valley. A certain portion of the forest land is also converted into the agricultural land by making the reckless cutting of the trees in the area adjoining to the water bodies.

Forest cover area: Forest cover includes the evergreen forests, deciduous forests, mixed forests, shrubs (hazelnuts, willow trees) open forest in the study area Open

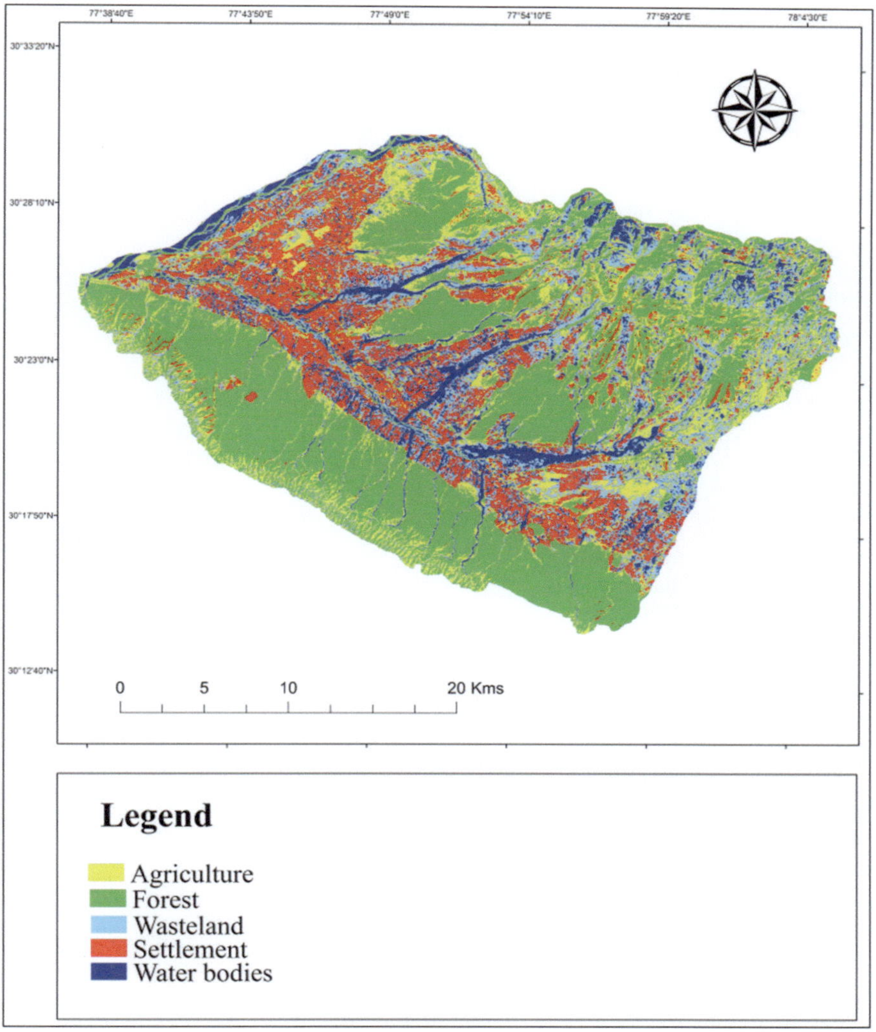

Fig. 4.5 Land use land cover map 2010 (*Source* Landsat-7, ETM+)

forest is identified by dull red-greenish color in false color composite (FCC), the dense forest bright red color, deciduous forest shows light gray color in the image. A complete stretch from southwest to southeast covered by the forest cover and there major patches of forest are lying in the central parts of the study area. In 2001, LULC the area covered by the forest cover was 89.56 km^2 (9.97%) and in 2010 it increases to 92.22 km^2. It is also a well-established fact that despite the increase in population pressure and an increase in the agriculture growth in the Western Doon Valley forest cover has a positive growth.

Table 4.1 Statistical information of land use land cover of the study area

Classes	Description	Area (2001)		Area (2010)		Growth rate (%)
		Km^2	%	Km^2	%	
Settlement	Residential, commercial, industrial, parking, transportation, and facilities	175.07	19.49	194.54	21.66	2.17
Agricultural land	Areas cultivated with various cultures of corn, wheat, barley, oat, potatoes, tea plantation	131.31	14.62	187.19	20.84	6.22
Forest cover	Evergreen forests, deciduous forests, mixed forests, shrubs (hazelnuts, willow trees)	89.56	9.97	92.22	10.27	0.30
Wasteland	Uncultivated agricultural lands pasture and consisting of arid land with short vegetations or no vegetation cover	169.03	18.82	113.67	12.65	−6.16
Water bodies	Rivers, lakes and other water bodies	333.37	37.11	310.70	34.59	−2.52
Total		898.33	100	898.33	100	

Source Landsat-7, ETM+

Wasteland area: The wasteland appears light white in FCC and fine to medium texture covers including the uncultivated agricultural lands, fallow land, pasture, arid land with short vegetations, stony and rocky land with no vegetation cover. The wasteland in the study area has been decreased over the period of 2001 to 2010 by 6.6% which is a positive trend in human development. In the Western Doon valley, the wasteland area was mixed with agriculture and settlement and it maybe further reduced with temporal changes in the future course of time. In the study area, the wasteland has been converted into agriculture land, settlement, and forest covers. In 2001 the area cover by this class was 169.03 (18.82) which decreases in 2010 as 113.67 (12.65%) of the total study area with a negative growth of 6.16%.

Water bodies area: The water bodies appear cyan in color and light dark in deep water conditions. The Yamuna and the Bindal are the two major rivers fallows in the

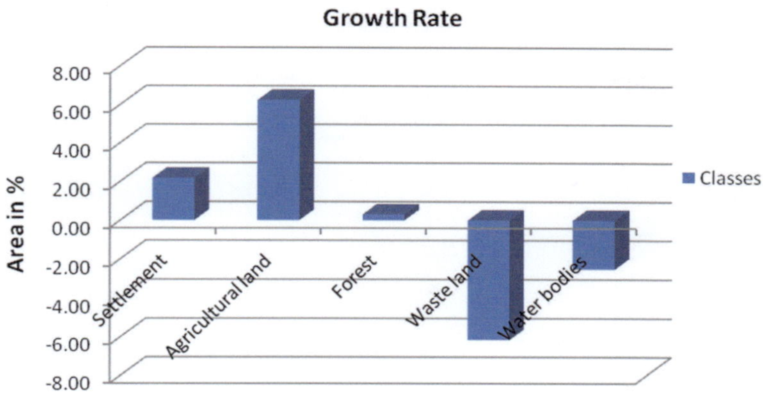

Fig. 4.6 Growth rate between the periods of 2001–2010 (*Source* Landsat-7, ETM+)

Western Doon Valley with a large number of seasonal tributaries that joins them from all over the study area. The Yamuna flows in the western side of the study area as northeast to the southwest whereas Bindal flows from northeast to west. In the land use land cover maps, the area covered by water bodies was 333.37 km^2 (37.11%) in 2001 and 310.70 km^2 (34.59%) in 2010 showing a negative growth of 2.52% over the period of 2001 to 2010.

4.4.3 Accuracy Assessment

Accuracy assessment has become vital with the passage of time as remote sensing techniques emerged as one of the most powerful tools in the classification of land use land cover. This process defines the degree of coherence of the classified image with the ground truth of an image classification of samples reference images used for analysis. The accuracy assessment usually evaluates the effectiveness of classifiers with the help of statistical significance computation of overall accuracies. A considerable number of references (pixels) are taken from the classified image and made a field check visit to evaluate the correctness of the classification process. The kappa coefficient ranges from 0 to 1; values higher than 0.7 is considered acceptable, while those equal to or lower than 0.4 identify a very low correlation between the classified image and the ground truth as a reference available images and maps of the respective time period. This process was supplemented with previous knowledge and ground checks. In the present study, the overall accuracy of the different classes was achieved 85.35% and kappa coefficient 0.88 for 2001 dataset whereas for the data set of 2010 the accuracy was 89.59% and Kappa coefficient was 0.91 (Table 5.2).

Table 4.2 Accuracy assessment and kappa coefficient

Time period	(2001 data)		(2010 data)	
Classes	Total accuracy (%)	Kappa coefficient	Total accuracy (%)	Kappa coefficient
Settlement	85.96	0.93	88.22	0.95
Agricultural land	89.95	0.94	93.45	0.92
Forest cover	88.12	0.88	93.56	0.93
Wasteland	82.76	0.83	84.76	0.89
Water bodies	79.98	0.81	87.98	0.87
Total	85.35	0.88	89.59	0.91

Source Landsat-7, ETM+

4.4.4 Change Detection

Based on the post-classification comparison (PCC) method was applied to change detection analysis, which is recognized as the most accurate change detection technique, detects LULC changes by comparing independently produced classifications of images from different data sets. In PCC each date of rectified imagery is independently classified to fit a common land type schema (equal number and type of land cover classes). The resulting land cover maps are then overlaid and compared on a pixel-by-pixel basis. The change detection analysis was performed by using a simple pixel-by-pixel mathematical combination of images for two different time periods. The change map produced by overlaying the two classified images assisted in locating the changes occurring in LULC classes (Fig. 4.7).

The formula used for the caluclation of rate of change has been derived from the formula (Puyravaud et al. 2003)

$$r = \frac{1}{\mathrm{t2} - \mathrm{t1}} \times \mathrm{In}\frac{\mathrm{At2}}{\mathrm{At1}}$$

where, r is the rate of land cover change, and At1 and At2 are the forest cover at time t1 and t2 respectively, In is the logarithm.

4.5 Conclusion

The study conducted in one of the most important and vital regions of India located in the Lesser Himalayas of the Uttarakhand State. The study reveals that the major land use in Western Doon Valley is the built-up area. During one decade, the area under built-up land has been increased by 2.17% (19.47 km^2) due to the construction of new buildings on fallow land and wasteland and in the area adjoining to the river

Fig. 4.7 Change detection map (*Source* Landsat-7, ETM+)

Table 4.3 Change detection percentages

Classes	Change detection (%)
Settlement	11.12
Agricultural land	42.56
Forest cover	2.96
Barren land	−32.75
Water bodies	−6.80

Source Landsat-7, ETM+

beds which was earlier a part of water bodies. The agricultural and vegetation land have been increased by 6.22% (55.88 km^2) tremendously due to population pressure and high inflation rate during the period of (2001–2010) in the Western Doon Valley and it is also observed that most of these changes have occurred in the area which is flat wasteland and having slope very level to gentle. Another significant fact of the study is that the water bodies have been decreased by 2.52% (2.67 km^2) which is one of the major concerns for ecology and environment of the Western Doon Valley where more than 2 lakhs migratory birds visit annually. Although, the forest cover has been also increased by 2.62 km^2 due to the effective and efficient policies of the administration, which is a positive sign for the growth of ecology and habitat. The results of the present study clearly demonstrated the potential of remote sensing and remote sensing techniques in deciphering the changing pattern of land use/cover in a study area.

Acknowledgements The authors are grateful to NASA for making the Landsat and SRTM, DEM datasets freely available under the umbrella of USGS web server. The authors are highly thankful to the Head, Department of Remote Sensing and GIS, University of Jammu, Jammu for providing the facility to carry out the research work timely.

References

Agarwal C, Green GM, Grove JM, Evans TP, Schweik CM (2002) A review and assessment of land-use change models: dynamics of space, time, and human choice, technical report, US Department of Agriculture, Forest Service, Northeastern Research Station, Pennsylvania, USA

Balak R, Kolarkar AS (1993) Remote sensing application in monitoring land use changes in arid Rajasthan. Int J Remote Sens 14(17):3191–3200

Chaurasia R, Closhali DC, Dhaliwal SS, Sharma PK, Kudrat M, Tiwari AK (1996) Landuse change analysis for agricultural management-a case study of Tehsil Talwandi Sabo. Punjab. J Indian Soc Remote Sens 24(2):115–123

Congalton RG, Green K (2008) Assessing the accuracy of remotely sensed data: principles and practices. CRS press

Costa MH, Botta A, Cardille JA (2003) Effects of large-scale changes in land cover on the discharge of the Tocantins river Southeastern Amazonia. J Hydrol 283(1–4):206–217

Dong JW, Xiao XM, Sheldon S, Biradar C, Duong ND, Hazarika MA (2012) Comparison of forest cover maps in Mainland Southeast Asia from multiple sources: PALSAR, MERIS, MODIS and FRA. Remote Sens Environ 127:60–73

Friedl MA, Sulla-Menashe D, Tan B, Schneider A, Ramankutty N, Sibley A, Huang X (2010) MODIS Collection 5 global land cover: algorithm refinements and characterization of new datasets. Remote Sens Environ 114(1):168–182. https://doi.org/10.1016/j.rse.2009.08.016

Giri C, Pengra B, Long J, Loveland TR (2013) Next generation of global land cover characterization, mapping, and monitoring. Int J Appl Earth Obs 25:30–37. https://doi.org/10.1016/j.jag.2013.03.005

Hansen MC, Egorov A, Potapov PV, Stehman SV, Tyukavina A, Turubanova SA, Roy DP, Goetz SJ, Loveland TR, Ju J, Kommareddy A (2014) Monitoring conterminous United States (CONUS) land cover change with web-enabled landsat data (WELD). Remote Sens Environ 140:466–484

Jasrotia AS, Bhagat BD, Kumar A, Kumar R (2013) Remote Sensing and GIS approach for delineation of groundwater potential and groundwater quality zones of Western Doon valley, Uttarakhand. India. J Indian Soc Remote Sens 41(2):365–377

Jasrotia AS, Kumar A, Bhat A (2012) Morphometric analysis and hydrogeomorphology for delineating groundwater potential zones of Western Doon Valley, Uttarakhand, India. Int J Geomat Geosci 2(4):1078–1098

Jasrotia AS, Taloor AK, Andotra U, Bhagat BD (2018) Geoinformatics based groundwater quality assessment for domestic and irrigation uses of the Western Doon valley, Uttarakhand, India. Groundwater Sust Develop 6:200–212

Martellozzo F, Clarke KC (2011) Measuring urban sprawl, coalescence, and dispersal: a case study of Pordenone Italy. Environ Plan B Plan Des 38(6):1085–1104

Midekisa A, Holl F, Savory DJ, Andrade-Pacheco R, Gething PW, Bennett A, Sturrock HJ (2017) Mapping land cover change over continental Africa using landsat and google earth engine cloud computing. PLoS ONE 12(9):e0184926. https://doi.org/doi.org/10.1371/journal.pone.0184926

Midekisa A, Senay GB, Wimberly MC (2014) Multisensor earth observations to characterize wetlands and malaria epidemiology in Ethiopia. Water Resour Res 50(11):8791–8806

Nemani R (2011) Nasa earth exchange: Next generation earth science collaborative. Int Arch Photogramm. 38–8(W20):17

Noor AM, Alegana VA, Gething PW, Tatem AJ, Snow RW (2008) Using remotely sensed night-time light as a proxy for poverty in Africa. Popul Health Metr 6:5. https://doi.org/10.1186/1478-7954-6-5PMID:18939972;PubMedCentralPMCID:PMC2577623

Patz JA, Norris DE (2004) Land use change and human health. Geoph Monog Series 153:159–167

Prakasam C (2010) Land use and land cover change detection through remote sensing approach: a case study of Kodaikanal Taluk, Tamil Nadu. Int J Geo Geosc 1(4):1150–1158

Puyravaud JP (2003) Standardizing the calculation of the annual rate of deforestation. For Ecol Manag 177(1–3):593–596

Shalaby A, Tateishi R (2007) Remote sensing and GIS for mapping and monitoring land cover and land-use changes in the Northwestern coastal zone of Egypt. Appl Geogr 27(1):28–41

Sreenivasulu V, Bhaskar PU (2010) Change detection in land use and land cover using remote sensing and GIS techniques. Int J Engineer Sci Techno 2(12):7758–7762

Stow DA, Collins D, McKinsey D (1990) Land use change detection based on multi date imagery from different satellite sensor systems. Geocarto Int 5(3):1–12

Taloor AK, Ray PK, Jasrotia AS, Kotlia BS, Alam A, Kumar SG, Kumar R, Kumar V, Roy S (2017) Active tectonic deformation along reactivated faults in Binta basin in Kumaun Himalaya of north India: inferences from tectono-geomorphic evaluation. Z Geomorphol 61(2):159–180

Taloor AK, Kotlia BS, Jasrotia AS, kumar A, Alam A, Ali S, Kouser B, garg PK, kumar R, Singh B, Jasrotia R (2018) Tectono-climatic influence on landscape changes in the glaciated Durung Drung basin, Zanskar Himalaya, India: A geospatial approach. Quaternary International https://doi.org/10.1016/j.quaint.2018.09.030

Tripathy GK, Ghosh TK, Shah SD (1996) Monitoring of desertification process in Karnataka state of India using multi-temporal remote sensing and ancillary information using GIS. Int J Remote Sens 17(12):2243–2258

William ER, William BM, Turner BL (1994) Modeling land use and land cover as part of global environmental change. Clim Change 28(1–2):45–64

Xiao J, Shen Y, Ge J, Tateishi R, Tang C, Liang Y, Huang Z (2006) Evaluating urban expansion and land use change in Shijiazhuang, China, by using GIS and remote sensing. Landsc Urban Plann 75(1–2):69–80

Xiong J, Thenkabail PS, Gumma MK, Teluguntla P, Poehnelt J, Congalton RG, Yadav K, Thau D (2017) Automated cropland mapping of continental Africa using google earth engine cloud computing. ISPRS J Photogramm Remote Sens 126:225–244. https://doi.org/10.1016/j.isprsjprs.2017.01.019

Yan L, Roy DP (2015) Improved time series land cover classification by missing-observation-adaptive nonlinear dimensionality reduction. Remote Sens Environ 158:478–491. https://doi.org/10.1016/j.rse.2014.11.024

Yuan F, Sawaya KE, Loeffelholz BC, Bauer ME (2005) Land cover classification and change analysis of the twin cities (Minnesota) Metropolitan Area by multitemporal Landsat remote sensing. Remote Sens Environ 98(2):317–328

Chapter 5
Dying and Dwindling of Non-glacial Fed Rivers Under Climate Change (A Case Study from the Upper Kosi Watershed, Central Himalaya, India)

J. S. Rawat and Geeta Rawat

Abstract A large number of the Himalayan watersheds which are densely populated are drained by the network of non-glacial fed rivers. Due to anthropogenically accelerated climate change, the tributary rivers of these non-glacial watersheds are dying and their master rivers are dwindling steadily causing socio-economic and environmental disruption in the region imposing a threat on the sustenance of mountain biodiversity, ecosystems and human civilization. The present paper demonstrates this fact by presenting results of historical geohydrometeorological records (1992–2016) of one of the non-glacial fed representative watersheds of the Central Lesser Himalayan region, viz., the Upper Kosi watershed in Uttarakhand state. Study reveals that despite of increase in annual rainfall and vegetation cover during the last two and half decades, about 82% network of the tributary streams of the major non-glacial fed rivers have been transformed from their perennial to non-perennial nature, therefore, the lean flow of their master river has been dropped down about sixteen times low during the last two and half decades, hence, the non-glacial master rivers are dwindling and dying steadily struggling for their existence and the inhabitants are threatened due to crises of water for drinking (both for people and cattle) and irrigation. The study suggests that there is an urgent need to develop a new water policy for rejuvenation of the non-glacial fed rivers by the Federal and concerned State Governments by fixing accountability of river rejuvenation works through constituting River Rejuvenation Authority, because it is so important that in near future, there may be need of a separate Department/Ministry of River Rejuvenation to save human civilization in the Himalaya.

J. S. Rawat (✉)
Uttarakhand Centre on Climate Change, Department of Geography, Kumaun University, SSJ Campus Almora, Uttarakhand 263601, India
e-mail: jsrawat1955@gmail.com

G. Rawat
Centre of Excellence for NRDMS in Uttarakhand, Department of Geography, Kumaun University, SSJ Campus, Almora 263601, Uttarakhand, India
e-mail: drgeeta2406@gmail.com

S. Sahdev et al. (eds.), *Geoecology of Landscape Dynamics*,
Advances in Geographical and Environmental Sciences,
https://doi.org/10.1007/978-981-15-2097-6_5

Keywords Climate change · Perturbation in hydrological cycle · Hydrometeorology · Kosi river rejuvenation · Dying non-glacial fed rivers · River rejuvenation

5.1 Introduction

A change in the state of the climate that can be identified by changes in the mean and/or variability of its properties, and that persists for an extended period typically decades or longer is known as climate change. It refers to any change in climate over time, whether due to natural variability or as a result of human activities (IPCC 2007). Studies reveal that mountains are among the most fragile ecosystems on the earth. These are highly sensitive to climate change and are being affected at a faster rate than the terrestrial ecosystem (Beniston 2003; Xu et al. 2009; Thakur et al. 2016, Beniston and Rebetez 1996; Messerli and Ives 1997; Diaz and Bradley 1997; Ceppi et al. 2010; Rangwala and Miller 2012). Several studies have revealed that mountainous systems warm at a faster pace in comparison of low lying areas, often with a greater increase in daily minimum temperatures than daily maximum temperatures (Beniston and Rebetez 1996; Diaz and Bradley 1997; Rangwala et al. 2009; Pederson et al. 2010).

The Himalayan watersheds are among the most vulnerable regions to climate change owing to their higher altitudes and abundance of freshwater resources in the form of lakes, rivers, snow and glaciers, similar to other mountainous ecosystems (Negi 2012; Zomer et al. 2014; Hwang et al. 2014; Rocca et al. 2014; Kalaninova et al. 2014). Studies reveal that climate change in mountainous regions affects many of the ecosystem services such as regional hydrology (Rawat 2007; Rawat et al. 2012, 2016; Rawat and Rawat 2017), carbon cycle (Pandey 2011; Semwal et al. 2013), agriculture (Shweta et al. 2013; Rawat 2014; Tiwari and Joshi 2012), animal husbandry (Bardhan et al. 2010) and biodiversity (Rana et al. 2011). The Himalayan regions are also subjected to higher anthropogenic pressures (Tiwari and Joshi 2012), as this mountain system houses one of the most densely populated regions. However, research on the impact of climate change on hydrometeorological and ecohydrological processes in the Himalayan regions is still in its infancy, compared to other ecosystems of the world. Studies have shown that many climate change induced ecohydrological phenomena in the northwestern Himalaya are already active (Tambe et al. 2012; Pandey et al. 2014). Most of these phenomena are directly or indirectly related to perturbations in the local hydrological cycle (Tambe et al. 2012; Rawat et al. 2012). Recent studies include increasing monsoon rainfall (Srivastava et al. 2013), increasing temperature (Murty et al. 2008; Singh et al. 2013; Liuy and Chen 2000), changes in rainfall pattern (Basistha et al. 2009), increased frequency of extreme events (Tambe et al. 2012; Rao et al. 2014; Ghosh et al. 2012), steady recession of glaciers (Raj et al. 2014; Bahuguna et al. 2014; Nainwal et al. 2008), rapid depletion of tributary glaciers (Mehta et al. 2013), shrinking of snow cover (Srivastava et al. 2013), decreased contribution of glacier-melt water to river discharge (Mehta et al. 2013), shifting of natural vegetation and fruit belts to higher elevation (Kumar 2014),

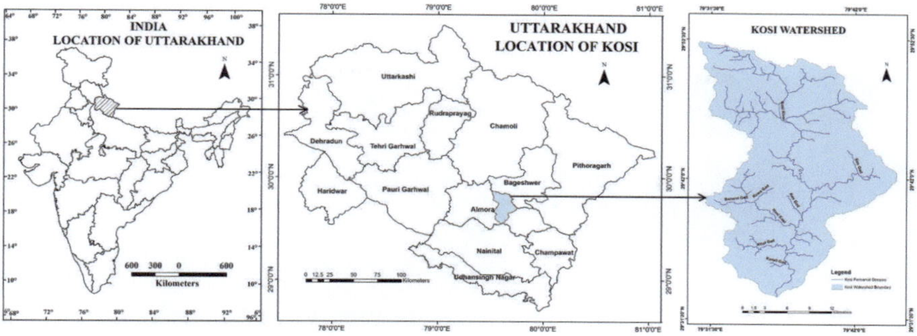

Fig. 5.1 Location map of the experimental Upper Kosi watershed (District Almora, Uttarakhand, India)

changes in agricultural patterns (Rawat 2013, 2014; Chauhan et al. 2014), emergence of new pests (Priyanka and Joshi 2013), decrease in the length of perennial river network and summer runoff due to depletion of groundwater (Malik and Umesh 2011), etc.

The objectives of this paper are to demonstrate that how non-glacial fed tributary streams are dying by the transformation of their perennial nature to non-perennial nature and their major rivers are dwindling steadily due to climate change in the non-glacial fed rivers in the Lesser Himalayan domain which is densely populated and to draw an attention of the local (Himalayan) and global leadership towards the fate of the water resources and sustenance of human civilization in the lofty, young, dynamically active, ecologically fragile, environmentally very sensitive domain of the earth, viz., the Himalaya. To achieve above objectives, a non-glacial fed representative experimental watershed, viz., the Upper Kosi river watershed from the Central Himalayan domain has been employed as a natural laboratory (Fig. 5.1) which is being hydrometeorologically monitored since the last two and half decades, i.e., 1992 onwards (Valdiya et al. 1993; Rai 1993; Rawat and Geeta 1999, 2017; Rawat 2005, 2012, 2017; Rawat et al. 2016).

5.2 About the Representative Watershed

The Upper Kosi Watershed lies in district Almora of the Uttarakhand State, India (Fig. 5.1). Encompassing an area of 462.61 km^2, the watershed extends in between 29° 33′47″ N to 29° 52′20″ N Latitudes and 79° 33′12″ E to 79° 48′11″ E Longitudes in the Lesser Himalayan terrain in Central Himalaya. Topographically (Fig. 5.2), the watershed has mature hill topography predominantly of convex hill crest, wide multitier terraced valleys composed of alluvium and the mid-crest are characterized by concave slopes with pluvial cones and fans. Attitudinally, the height of the watershed varies between 1000 and 2676 m. Climatically, the watershed is complex. A large part

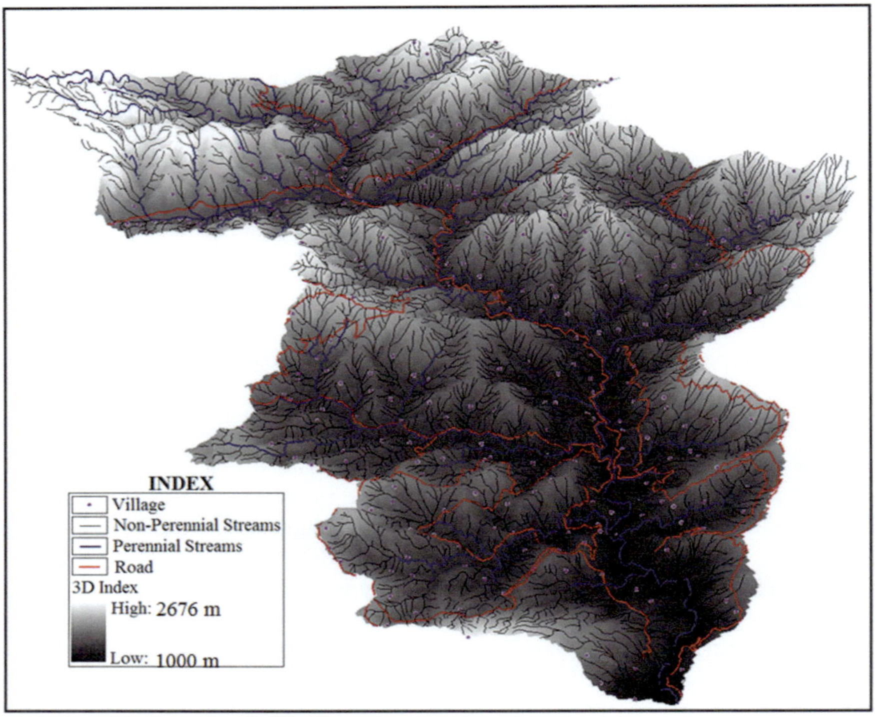

Fig. 5.2 Topographic map of the Upper Kosi watershed depicting relief pattern and geographical distribution of villages and road networks. *Source Based on Cartosat-1 data*

of the watershed in between 1200–1800 m enjoys cool temperate climatic conditions while a small part near the mouth of the watershed having elevation <1200 m falls under sub-tropical climatic condition. The remaining part of the watershed having elevation >1800 m has cold climatic conditions.

Geologically, the rocks of the Kosi watershed are composed of four tectonic units and seven different lithological units of the Lesser Himalayan region. These are (i) Almora Nappe constituted of the rocks of Saryu Formation (quartzites and mica schists), Almora Group (gneisses) and Gumalikhet Formation (schist, slate/phyllite) and (ii) Ramgarh Nappe constituted of rock of Devguru Formation (quart porphyry) and Sedimentary Belt made up of the rocks of Berinag Quartzitess (Valdiya 1980). The rocks are dipping towards north-east mainly 15 to 20° but upstream the Kosi river near Takula, the rock dip is south-west. As per Strahler's (1964) stream orders, the Kosi is a seventh order river having a network of as many as 9969 streams of different orders, i.e., 7805 first order (total length 3588.63 km), 1750 s order (total length 729.83 km), 315 third order (total length 254.77 km), 76 fourth order (total length 119.64 km), 16 fifth order (total length 65.32 km), 6 sixth order (length 39.73 km) and 1 seventh order (length 35.75 km). The total length of non-perennial and perennial stream network in the watershed is about 4833.49 km having a density of 10.43 km/km^2.

Out of the total watershed, about 52.22% area is under forest land, 30.24% area under barren land, 13.74% is under agricultural land and the remaining 0.08% area is under urban area. The Kosi is one of the thickly populated non-glacial watersheds of the Himalaya. Within this watershed, there 343 villages (Fig. 5.2) having a density of 1.34 village/km^2. Besides are 343 villages and the Almora Town, there are as many as 12 rapidly growing sub-urban centres, viz., Someshwer, Majkhali, Hawalbagh, Kosi, Chanoda, Kathpuria, Sitlakhet, Daulaghat, Takula,Kathpuria, Kosi and Gewapani which are located in different parts of the watershed. For drinking and irrigation purposes, there are more than 100 gravity flow drinking water schemes, 5 drinking water lift schemes, 181 hand pumps, 15 irrigation lift schemes, 22 hydrum schemes and 50 irrigation canals in the Kosi watershed.

5.3 Database and Methodology

5.3.1 Hydrometeorological Database

For studying hydrometeorological processes under climate change, the first monitoring station was set up in the Central Himalayan region in 1987 (Haigh et al. 1988). In 1991, a network of six other micro-watersheds having diverse ecohydrological characteristics was identified and hydrometeorologically instrumented in 1991 to monitor a set of hydrometeorological variables (Valdiya et al. 1993). The operational monitoring of these watersheds was carried out under different sponsored research projects (Rai 1993; Rawat and Geeta, 1999, 2017; Rawat 2005, 2012; Rawat et al. 2016). Presently, these instrumented micro-watersheds are being monitored on a daily basis under a research project funded by the Uttarakhand Science and Education Centre, Dehradun (Rawat 2017). For present study, the long-term historical (1992–2016) meteorological data of 5 stations of the Upper Kosi watershed, viz., Kausani, Sitlakhet, Salla Rautela, Khunt and Deolikhan were used to find the average annual rainfall and evaporation loss; and the water discharge records recorded near the mouth of the Kosi watershed at Kosi were used to define the lean water discharge flow and runoff (Chow 1964) pattern and trend of the Kosi River. Using rainfall, runoff and evaporation records of the watershed, the pattern and trend of water balance (Schendel 1975) was computed to examine the status of water balance for groundwater recharge.

The mean monthly data of two and half decades (i.e., 1992–2016) have been used to define the monthly pattern while the annual data have been subjected to trend analysis of the hydrometeorologic parameters. The main objective of the trend analysis is to assess whether values of hydrometeorological parameters are increasing, decreasing or trendless over time. To define the relationship between hydrometeorological parameters, i.e., rainfall, evaporation, runoff, water balance with time, correlation

Table 5.1 Statistics of the hydrometeological parameters of the Upper Kosi Watershed

Parameters	Annual mean	"r" value	Regression equation
Rainfall	1012 mm	0.481[a]	$Y = 829.5893 + 11.8036x$
Rainy days	99 days	−0.514[b]	$Y = 102.6619 - 0.554x$
Rainfall intensity	7.88 mm/day	0.605[b]	$Y = 5.2102 + 0.1698x$
Discharge	3.64 m^3/s	0.56[b]	$Y = 0.938 + 0.0656x$
Runoff	502 mm	0.52[b]	$Y = 94.0618 + 8.1107x$
Evaporation	772 mm	0.192	$Y = 736.3553 - 0.3155x$
Water balance	−19 mm	−0.41[a]	$Y = -138.293 + 0.3101x$
Lean flow	295.96 l/s	−0.82[a]	$Y = 662.5 - 28.19x$

[a]Significant at 99% confidence level
[b]Significant at 95% confidence level
Source Collected in the field by the authors

analysis and to define the trend of these parameters, regression analysis was conducted using log term (1992–2016) hydrometeorological data. The statistical results of the hydrometeorological parameters are presented in Table 5.1.

5.3.2 *Land Use/Land Cover Data*

The LANDSAT is a scientific program which is operated by NASA and USGS, which offers the longest global record of Earth's surface. The LANDSAT-5 imagery is of February, 1990 and LANDSAT imagery is of February, 2016 with the resolution of 30 m nominal and panchromatic resolution 15 m. Besides these, land use pattern based on Survey of India Topographic sheets was also computed for the year 1965. The LANDSAT images were processed in the ERDAS Imagine 9.3 software The results of land use/land cover of three different years are presented in Table 5.2.

5.4 Geohydrometeorological Results

5.4.1 *Land Use/Land Cover Dynamics*

Land use/land cover study of the study area was carried out for three different years, i.e., 1965 based on Survey of India Topographic sheets, and 1990 and 2016 based on LANDSAT satellite images. The results are presented in Table 5.2. and Fig. 5.3 depicts a spatial distribution of land use/land cover pattern in the Upper Kosi watershed in 1965 and 2016. The results of land use/land cover dynamics reveal that the forest cover, barren land and urban area have been increased about 4.41% (in

Table 5.2 Area under different land use categories in the Upper Kosi watershed in district Almora, Uttarakhand

Land use land cover categories	1965[a]		1990[b]		2016[b]	
	Area (km^2)	Area (%)	Area (km^2)	Area (%)	Area (km^2)	Area (%)
Forest	235.165	50.81	255.6295	55.23	255.6011	55.22
Barren land	104.595	22.59	130.2653	28.14	139.9678	30.24
Agriculture land	120.984	26.14	74.6176	16.12	63.7404	13.74
Urban area	2.066	0.46	2.2996	0.51	3.5027	0.80
Total	462.81	100	462.812	100	462.812	100

Source Based on Survey of India Toposheets and LANDSAT Data

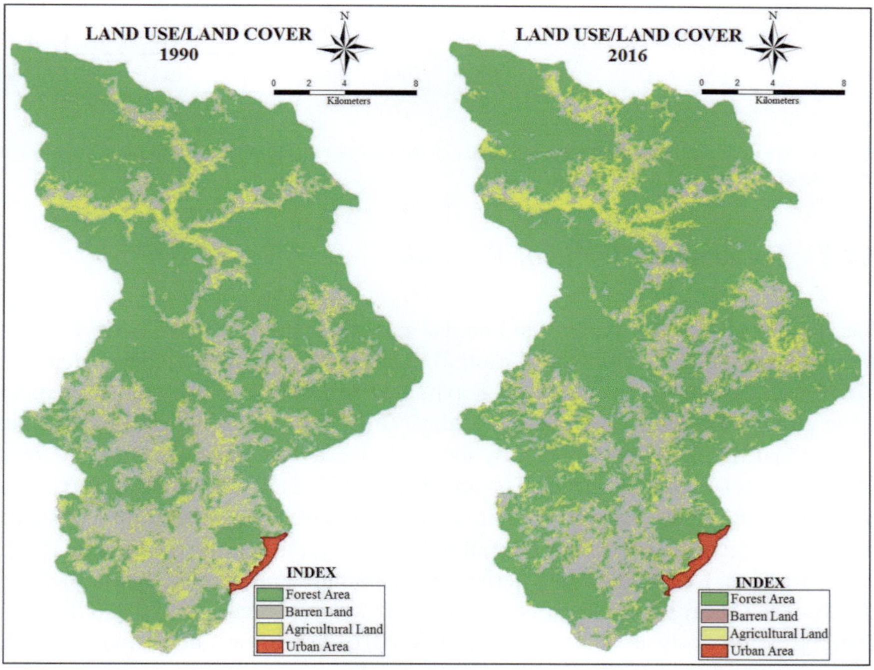

Fig. 5.3 Spatial distribution of land use/land cover in the Upper Kosi watershed during 1990 and 2016. *Source* based on LANDSAT data

20.42 km^2 area), 7.67% (in 35.10 km^2 area) and 0.34% (in 1.44 km^2 area), respectively; while the agricultural land has decreased about 12.4% (or 57.24 km^2) area during the last five decades in the study area (Table 5.2).

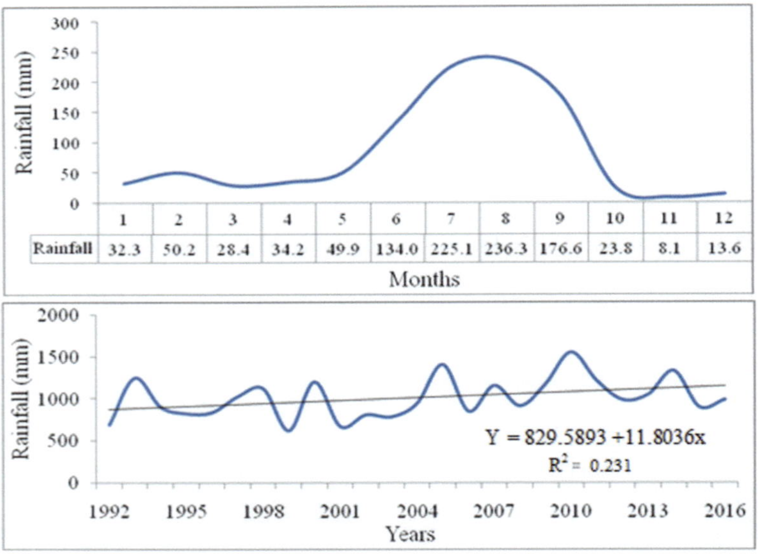

Fig. 5.4 Mean monthly rainfall pattern (1992–2016) (above) and annual rainfall trend (below) in Upper Kosi Watershed

5.4.2 *Pattern and Trend of Rainfall*

The long-term monthly pattern and annual rainfall trend are presented in Fig. 5.4 which reveal that the mean annual rainfall in the representative watershed stands at ~1012 mm. August is the month of maximum rainfall (~236 mm) and November is the month of minimum rainfall (~8.2 mm) (Fig. 5.4). Approximately 65% of the total annual rainfall is contributed by the south-western monsoon and ~8% occurs during the post-monsoon season. The remaining ~15% and ~12% of the influxes are contributed by winter and summer rainfall, respectively. The rainfall time series data (Fig. 5.4) depicts the amount of rainfall in different years and its trend by the linear regression line. During the study period, the maximum rainfall occurred in the year 2010 (~1507 mm) while the minimum rainfall was recorded in 1999 (~628 mm). The regression analysis of long-term (1992–2016) data shows that annual rainfall has rising trend ($r = 0.481$, p-value 0.05) significant at the 95% confidence level (Fig. 5.4).

5.4.3 *Pattern and Trend of Rainy Days*

The rainy days' records reveal that on an average during a year, the watershed receives rainfall in ~99 days only. July is the month of maximum rainy days (~21 days) while

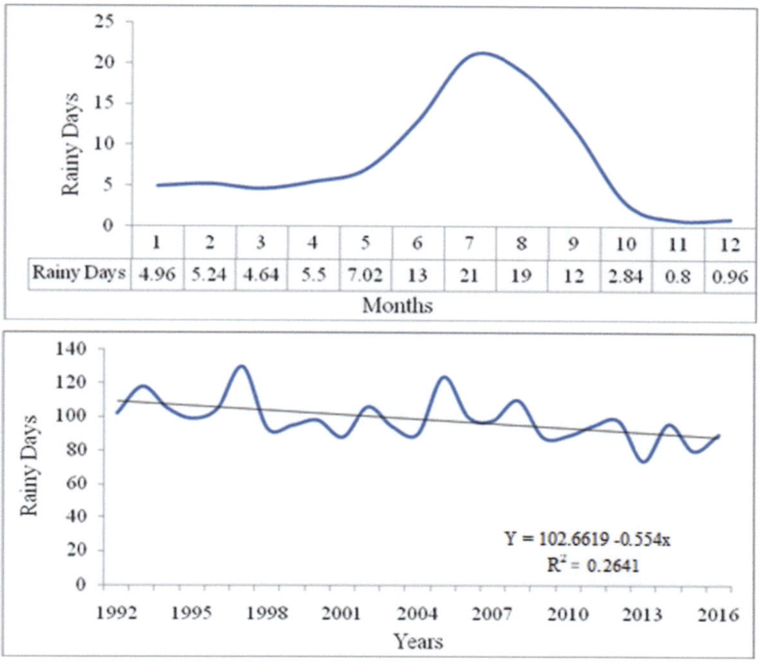

Fig. 5.5 Mean monthly rainy days (1992–2016) (above), and annual rainy days trend (below) in the Upper Kosi Watershed. *Source* Rawat and Geeta Rawat (2016)

November is the month of minimum rainy days (~0.8 days) (Fig. 5.5). Approximately 59% of the total annual rainy days is contributed by the south-western monsoon and ~21% rainy days occurs during the summer season. The remaining ~13% and ~7% of the rainy days are contributed by winter and post-monsoon seasons, respectively. The rainy days time series data (Fig. 5.5) depicts the distribution of rainy days in different years and its trend by the linear regression line. The long-term analysis of total annual rainy days shows a decreasing trend ($r = -514$, p-value 0.01) significant at the 99% confidence level (Fig. 5.5).

5.4.4 Pattern and Trend of Rainfall Intensity

The long-term mean annual rainfall intensity stands at ~7.88 mm/day. August is the month of maximum rainfall intensity (~14 mm/day) and November is the month of minimum rainfall intensity (~3 mm/day) (Fig. 5.6). In the monsoon season, mean rainfall intensity stands at ~12.38 mm/day while in the post-monsoon season, mean rainfall intensity stands at ~3.54 mm/day. The mean rainfall intensity during summer and winter seasons stands at ~6.10 mm/day and ~6.54 mm/day, respectively. The time

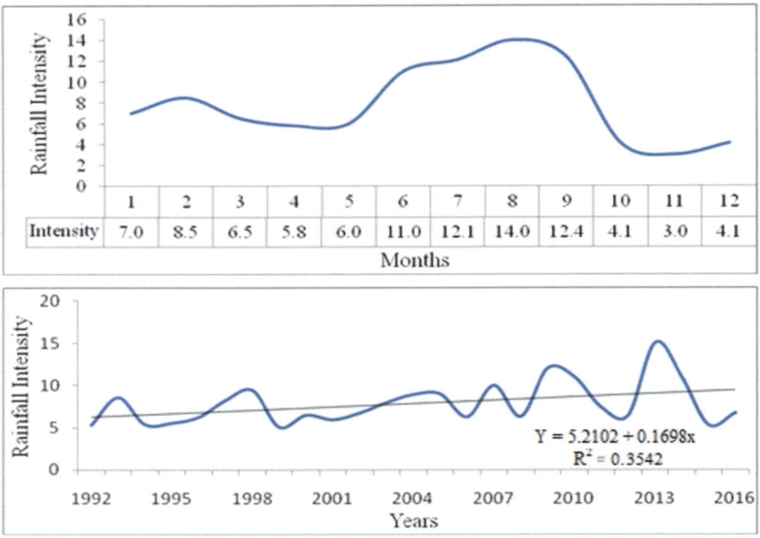

Fig. 5.6 Mean monthly (1992–2016) rainfall intensity (above) and annual rainfall intensity trend (below) in the Upper Kosi Watershed. *Source* Rawat and Geeta Rawat (2016)

series data (Fig. 5.6) depicts the mean rainfall intensity in different years and its trend by the linear regression line. The long-term linear trend line of the rainfall intensity has high increasing trend ($r = 0.605$, p-value 0.01) significant at the 99% confidence level (Fig. 5.6).

5.4.5 *Pattern and Trend of Water Discharge*

The long-term monthly pattern and annual water discharge trend of the Kosi watershed are presented in Fig. 5.7 which reveals that the mean annual water discharge of the Kosi river stands at ~3.64 m^3/s. August is the month of maximum discharge (~10.60 m^3/s) and May is the month of minimum discharge (~0.80 m^3/s) (Fig. 5.7). Approximately 53% of the total annual water discharge is contributed by the southwestern monsoon and ~6% occurs during the summer season. The remaining ~27% and ~14% of the influxes are contributed by post-monsoon and winter discharge, respectively. The water discharge time series data (Fig. 5.7) depicts the rate of water discharge in different years and its trend by the linear regression line. During the study period, the maximum water discharge occurred in the year of 2010 (~8 m^3/s) while the minimum water discharge was recorded in 1999 (~1.88 m^3/s). The regression analysis of long-term (1992–2016) water discharge data shows that annual discharge of the Kosi river has increasing trend ($r = 0.56$, p-value 0.01) significant at the 99% confidence level (Fig. 5.7).

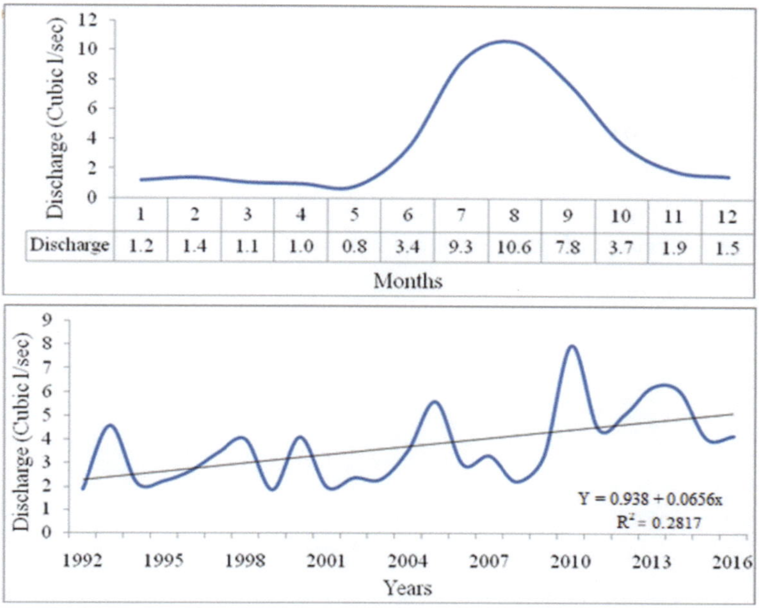

Fig. 5.7 Mean monthly (1992–2016) water discharge (above) and annual water discharge trend (below) in the Upper Kosi Watershed. *Source* Rawat and Geeta Rawat (2016)

5.4.6 *Pattern and Trend of Runoff*

The monthly pattern and trend of annual runoff depth in the Kosi watershed are presented in Fig. 5.8 which reveals that the mean annual runoff depth in the of the Kosi river stands at ~502 mm. August is the month of maximum runoff (~135 mm) and May is the month of minimum runoff (~9 mm) (Fig. 5.8). The runoff depth time series data (Fig. 5.8) depicts the magnitude of runoff depth in different years and its trend by the linear regression line. During the study period the maximum runoff depth occurred in the year 2010 (~820 mm) while the minimum runoff was recorded in 1999 (~250 mm). The regression analysis of long-term data shows that annual runoff depth in the Kosi river has increasing trend ($r = 0.52$, p-value 0.01) significant at the 99% confidence level (Fig. 5.8).

5.4.7 *Pattern and Trend of Evaporation*

The monthly pattern and trend of annual evaporation in the Kosi watershed are presented in Fig. 5.9 which reveals that the mean annual evaporation in the of the Kosi watershed stands at ~772 mm. May is the month of maximum evaporation (~108 mm) and August is the month of minimum evaporation (~42 mm) (Fig. 5.9).

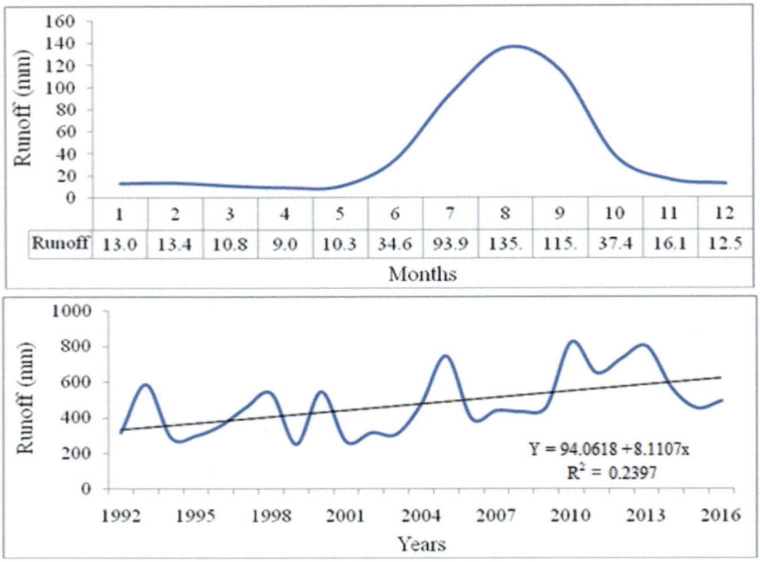

Fig. 5.8 Mean monthly (1992–2016) runoff depth (above), and annual runoff trend (below) in the Upper Kosi Watershed. *Source* Rawat and Geeta Rawat (2016)

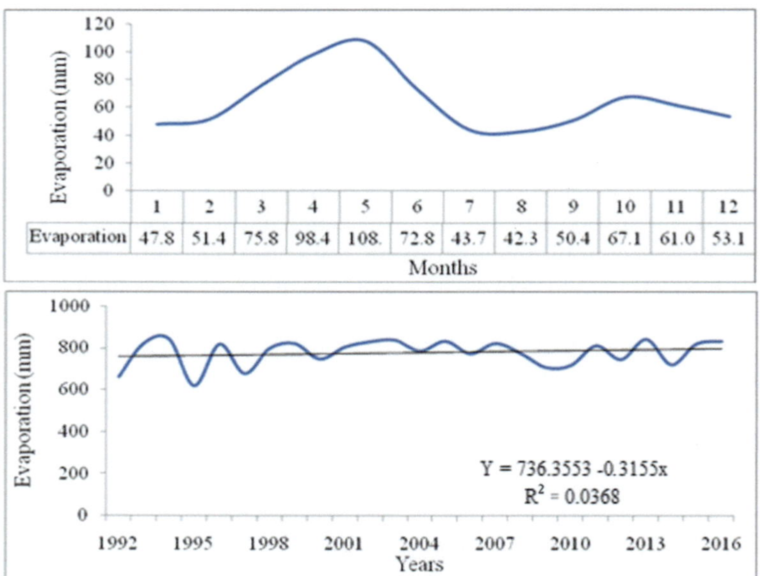

Fig. 5.9 Mean monthly(1992–2016) evaporation loss (above) and annual evaporation trend (below) in the Upper Kosi Watershed. *Source* Rawat and Geeta Rawat (2016)

The evaporation time series data depicts the regression line of long-term data which shows that annual evaporation has a slightly increasing trend (r = 0.195, p-value 0.09) which is insignificant (Fig. 5.9).

5.4.8 *Pattern and Trend of Water BalancePattern and Trend of Water Balance*

The annual water balance (Schendel 1975) of the watershed is deficit which stands at −19 mm. August is the month of maximum water balance (+91 mm) and May is the month of minimum water balance (−74 mm) (Fig. 5.10). During south-western monsoon months of July, August and September, the water balance is surplus while in all other 9 months of the year, the water balance was found always deficit (Fig. 5.10).

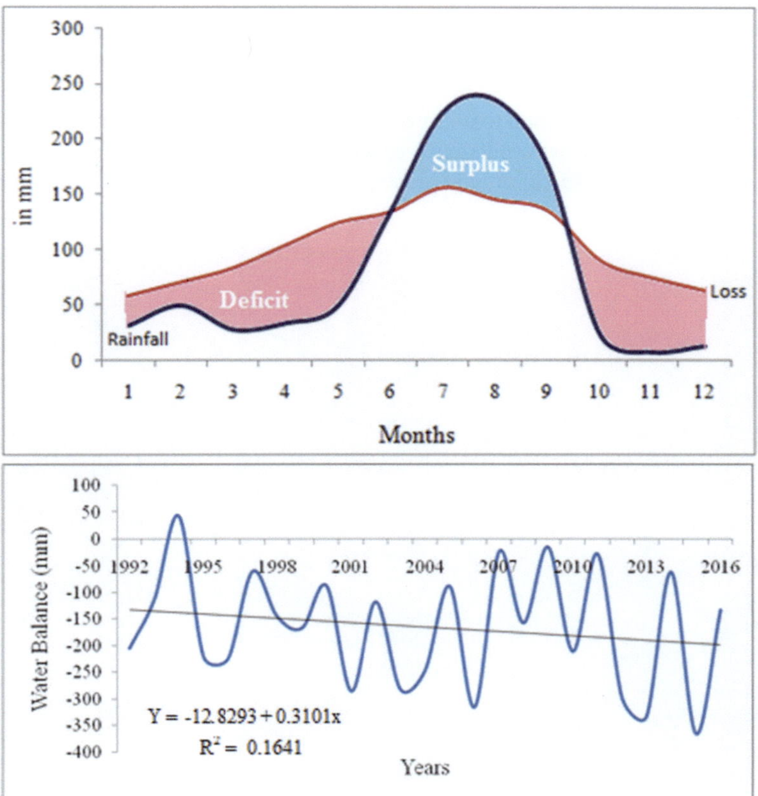

Fig. 5.10 Mean monthly (1992–2016) water balance (above) and annual water balance trend (below) in the Upper Kosi Watershed. *Source* Rawat and Geeta Rawat (2016)

The water balance time series data (Fig. 5.10) depicts the amount of water balance in different years and its trend by the linear regression line. The long-term linear trend line indicates that water balance has decreasing trend ($r = -0.41$, p-value 0.05) significant at the 95% confidence level (Fig. 5.10).

5.4.9 Dying of Tributary Streams

In mountains, the non-glacial fed streams originate from perched aquifers which are recharged each year by rainfall. In hills, normally there is no continuous water table and the stream flow is maintained by draining such areas of gravity water in the perched layers (viz., aquifer) or in rock interstices (Chow 1964). If there is no recharge of aquifers during rains, the rocks of aquifer become dry and there is no water flow for streams. The study reveals that on an average, the annual water balance of the Kosi watershed is negative which suggests that the perched aquifers of the watersheds are not being recharged by rainwater, therefore, during non-monsoon days, there is no water supply for these streams from their aquifers. In this way, the minor and major tributary streams are dying from the headwater region by their transformation from perennial to non-perennial nature (Fig. 5.11). The total length of perennial rivers in the Kosi watershed was about 225.86 km during about five decade back which is only 41.49 km only at present (Fig. 5.12 and Table 5.3) and remaining

Fig. 5.11 Dying of tributary streams/rivers the headwater regions due to their transformations from perennial system to non-perennial system: Kosi river upstream Chaunoda, at Chaunoda, at Someshwar and Kosi. *Source* Rawat and Geeta Rawat (2016)

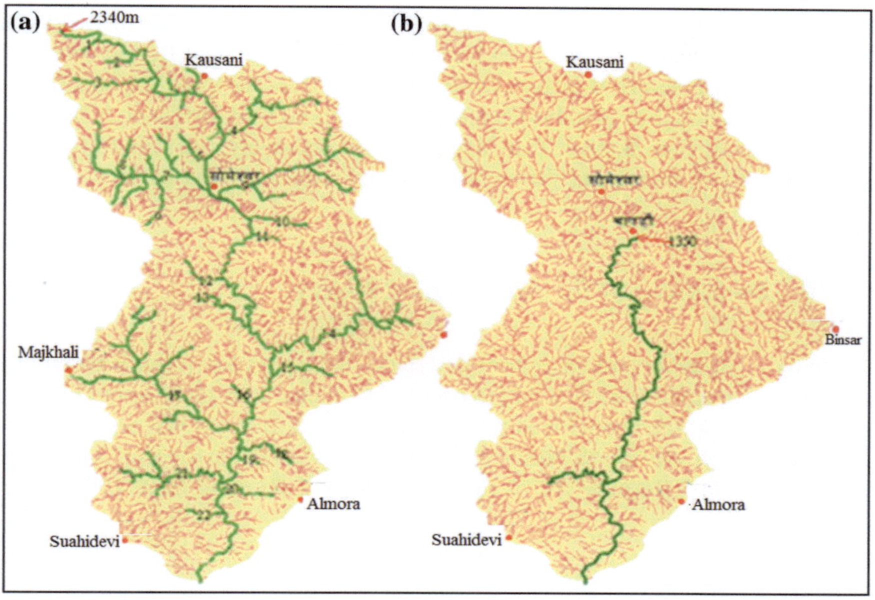

Fig. 5.12 Distribution of perennial streams in the Kosi watershed during 1965 (**a**) *Source* Based on SoI toposheet and at present (**b**) *Source* Based on field survey (Rawat 2007)

Table 5.3 Length of major tributary rivers and their Master River, viz., Upper Kosi During 1965 and at present

Name of tributary streams	Length in km		
	1965[a]	At present[b]	Dead during summer
1. Koshalya Ganga	13.785	2.737	11.138
2. Dev Gad	5.47	0	5.47
3. Menal Gad	1.73	0	1.73
4. Ben Gad	3.028	0	3.028
5. Janeli Gad	5.967	0	5.967
6. Rengad	2.231	0	2.231
7. Kurali Gad	2.73	0	2.73
8. Sim Gad	19.761	8.123	11.638
9. Nani Kosi	29.031	9.450	19.581
10. Bagania Gad	17.971	0	17.971
11. Jamthara Gad	4.131	0	4.131
12. Khul Gad	14.043	4.759	9.284
13. Other	34.88	0	34.88
14. Kosi Master River	52.889	16.43	36.459
Total length	225.850 (100%)	41.49 (18%)	184.8 (82%)

184.87 km which constitutes about 82% of the total perennial river networks of the Upper Kosi watershed has been transformed from perennial to non-perennial (intermittent/ephemeral) in nature. During this last 5 decades, the major tributary rivers of the Kosi such as Koshlya Ganga, Dev gad, Menal Gad, Ben Gad, Jaineli Gad Ren Gad, Kurali Gad, Bagania Gad, Jamthara Gad and many other minor tributary streams have been transformed from perennial to non-perennial streams.

5.4.10 Dwindling of the Master Kosi River

Due to dying of minor and major tributary rivers, the Upper Kosi river is dwindling steadily during summer season due to decreasing annual lean flow. About two and half decade back, the annual lean flow of the Upper Kosi river was about 792 l/s (Fig. 5.13) which was found 180 l/s in 2010 (Fig. 5.14a) and at present, it has been dropped down to 50 l/s only (Fig. 5.13 and Fig. 5.14b). The long-term (1992–2016)

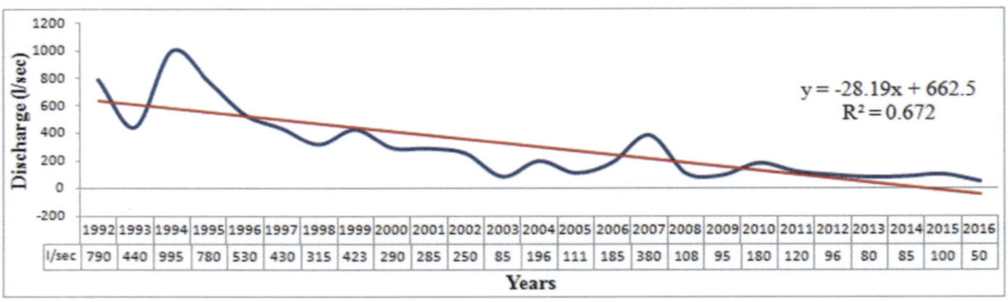

	1992	1993	1994	1995	1996	1997	1998	1999	2000	2001	2002	2003	2004	2005	2006	2007	2008	2009	2010	2011	2012	2013	2014	2015	2016
l/sec	790	440	995	780	530	430	315	423	290	285	250	85	196	111	185	380	108	95	180	120	96	80	85	100	50

Fig. 5.13 Hydrograph of the Kosi river showing lean annual flow magnitude and its trend recorded near Almora Pumping House at Kosi. *Source* Rawat and Geeta Rawat (2016)

Fig. 5.14 Dwindling of the so-called mighty Kosi river at Kosi, i.e., 200 m upstream of the Almora Drinking water Pump House: capacity of ground water flow 180 l/s in 2010 (**a**) and 50 l/s in 2016 (**b**). *Source* Photographs taken by the authors

lean annual flow or the annual minimum discharge of the Kosi river is presented in Fig. 5.13 which illustrates that the lean flow of the Kosi river has a decreasing trend ($r = -0.82$, p-Value 0.01) significant at 99% confidence level. These lean flow records reveal that during the last two and half decade, the lean flow of the Upper Kosi river has been dropped down about 16 times low, i.e., 50 l/s in 2016 compared to 1992 when it was about 792 l/s.

5.5 Discussion

The above results of historical geohydrometeorological records reveal that during the last two and half decade, there has been no marked decrease in the amount of annual rainfall in the region, rather it has slightly increasing trend, on the one hand; and there has been an increase of 4.41% (in 20.43 km^2) area in vegetation cover in the experimental watershed. Despite the increase in annual amount of rainfall and in vegetation cover, the study reveals that minor and major tributary rivers are dying and the Master Upper Kosi river is dwindling steadily. It seems ironical. But the fact behind is that there is a drastic reduction in the total number of annual rainy days and alarmingly increasing significant trend of rainfall intensity due to anthropogenically accelerated climate change. In addition to these, during the last five decades, about 57.2 km^2 agricultural land (which controls overland flow and increases rainwater infiltration due to frequent tillage practices), has been converted mainly into barren land due to migration of local inhabitants.

The study suggests that due to (i) decreasing trend of rainy days, (ii) very high increasing trend of rainfall intensity and (iii) increasing area under barren land, the magnitude of the overland flow on the hill slopes is being accelerated alarmingly hence the runoff is increasing which results in drastic reduction in the hydrologic processes, i.e., infiltration and percolation of rainwater into the perched aquifers. In this way, the perched aquifers which give rise to the perennial streams are drying from the headwater region and dwindling the major rivers in the downstream valley region.

Examining the water balance, it was found that due to evaporation and very high loss of rainfall water through runoff, the average annual water balance of the experimental watershed is deficit (i.e., −19 mm). Month wise status of water balance reveals that only during the south-western monsoon months of July, August and September, the water budget is in surplus while in remaining all the nine months, the water budget is in deficit state in this non-glacial fed experimental watershed causing acute drinking/irrigation water crises. We are monitoring the lean flow of the Upper Kosi river since the last 25 years. Due to deficit water balance of the watershed, the groundwater storage in the form of perched aquifers are depleting and the Kosi River is dwindling steadily. The lean flow records of the Kosi river reveals that it has significant decreasing trend which has been dropped down 16 times low, i.e., 50 l/s in 2016 compared to 1992 when the lean flow was about 792 l/s. Thus, just like slow poisoning, the Kosi river is dwindling or dying steadily due to climate change.

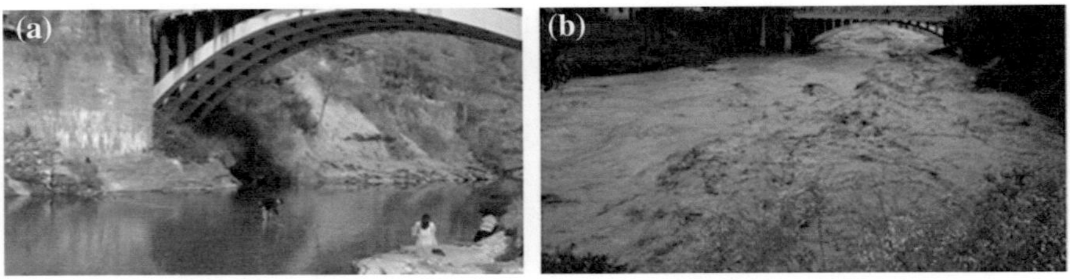

Fig. 5.15 Very low flow during summer (50 l/s in May 2016) (**a**); and very high monsoon flow (capacity 618.10 m^3/s on 18th September 2010) (**b**) in the Kosi river at Kosi. *Source* Photographs taken by the authors

A large part of the Himalayan watersheds are drained by the non-glacial fed rivers like the Upper Kosi river. In the Central Himalayan State, viz., the Uttarakhand, about 40% geographical area is drained by the non-glacial fed rivers. The condition of all the non-glacial fed rivers in the Himalaya is similar to the Kosi river. The spot surveys reveal that the lean flow of the Western Ranganga river at Marchula has been dropped down from 2282 l/s in 2005 to 1384 l/s in 2016; of Saryu river at its confluence with Panar River has been dropped down from 3324 l/s in 2005 to 1580 l/s in 2016; and of Panar river at confluence of Panar and Saryu, the lean flow has been dropped down from 251 l/s in 2005 to 158 l/s in 2016. Similarly, the other non-glacial fed Himalayan rivers are dwindling and dying steadily. In all these rivers, now the flow pattern has become too low during summer and too high during monsoon rain (e.g. see Fig. 5.15 in case of the Upper Kosi river).

5.6 Conclusions

Under the anthropogenically accelerated climate change, the water resources are under deep stress in the non-glacial Himalayan regime; consequently, the hydrologic cycle has been perturbed alarmingly leading towards the process of desertification. The sharp hydrologic indicators of the beginning of desertification in non-glacial fed river watersheds in Himalaya as advocated through this study are rapid transformation of perennial tributary rivers into non-perennial (ephemeral/intermittent) rivers and steady dwindling of Major Rivers. The representative watershed study presented in this paper demonstrates that during the last two and half decade, about 82% network of the tributary streams of the major non-glacial fed rivers have been transformed from their perennial to non-perennial nature, therefore, the lean flow of their Master Rivers has been dropped down about sixteen times low particularly during the last two and half decade (i.e., 1992–2016), hence, all the Master Rivers of the non-glacial fed river watersheds of the Central Himalaya are dwindling steadily and struggling for their existence. These hydrologic characteristics are sharp indicators that the processes of desertification in Himalaya have been started and at present, the desertification in this young mountain chain is in incipient stage. If no river rejuvenation measures were taken, consequences would be very adverse in the near future on water availability for both out-of-stream and in-stream uses (Rawat et al. 2012).

In view of this serious problem, there is an urgent need to start river rejuvenation mechanical and biological measures massively to save the dwindling and dying rivers of the Non-glacial fed river of the Himalaya. Although it is a herculean task but not impossible, if there is a will like "Bhagirath Prayas". But this Bhagirat Prayas cannot be done alone by one or two departments/NGOs or local people. To save the dying and dwindling rivers, there is an urgent need to develop new water policy and to constitute "River Rejuvenation Authority" (RRA) by which all stakeholders could launch and monitor regularly a massive river rejuvenation program throughout the year continuously under one umbrella, viz., RRA. Water is life and it is depleting very fast in Himalaya, therefore, it is so important that to cope with this problem in near future, there may be need of a separate "Department of River Rejuvenation" or even may be a separate "Ministry of River Regeneration".

Acknowledgements This work is based on field data collected through a number of sponsored research projects, viz., (i) Response of hydrologic cycle to environmental degradation, funded by the Ministry of Environment and Forests, Government of India; (ii) Environmental hydrology of the Almora region; (iii) Watershed monitoring and modeling in Central Himalaya; (iv) Geohydrometeorological data at micro and macro scale watersheds; (v) Application of geohydrological database in groundwater augmentation, funded by the Department of Science and Technology, New Delhi and (vi) Impact of climate change in hydrology under varied ecological conditions in the Kumaun region, funded by the Uttarakhand State Council for Science and Technology, and Impact of Climate Change in Hydrometreorology; A case study from the Upper Kosi Watershed, Uttarakhand, funded by the Uttarakhand Science, Education and Research Centre, Dehradun.We thank Prof. K. S. Valdiya and Prof. Martin Haigh for their encouragement since the inception of this work in 1992. Thanks are due to Mr. Bhupal Singh (Khunt) and Mr. David (Laxmi Ashram Kausani) with their Team of 11 Data Recorders for monitoring hydrometeorological data since 1992 from different hydrometeorological stations of the Upper Kosi watershed.

References

Bahuguna IM et al (2014) Are the Himalayan glaciers retreating? Curr Sci 106:1008–1013

Bardhan D et al (2010) Livestock in Uttarakhand: growth patterns and determinants of composition and intensity. Indian J Anim Sci 80:584–589

Basistha A et al (2009) Analysis of historical changes in rainfall in the Indian Himalayas. Int J Climatol 29:555–572

Beniston M (2003) Climate change in mountain regions: a review of possible impacts. Clim Change 59:5–31

Beniston M, Rebetez M (1996) Regional behaviour of minimum temperature in Switzerland for the period 1979–1993. Theor Appl Climatol 53:231–243

Ceppi P, Scherrer S, Fischer A, Appenzeller C (2010) Revisiting Swiss temperature trends 1959–2008. Int J Climatol 32(2):203–213

Chauhan BS et al (2014) In advances in agronomy. In: Spark DL (eds) vol 123, pp 65–121

Chow VT (1964) Runoff. In: Chow VT (eds) Handbook of applied hydrology. McGraw Hill, Newyork, pp 14.1–14.54

Diaz HF, Bradley RS (1997) Temperature variations during the last century at high elevation sites. Clim Change 36:253–279
Ghosh S et al (2012) Lack of uniform trends but increasing spatial variability in observed Indian rainfall extremes. Nat Clim Change 2:86–91
Haigh MJ, Rawat JS, Bisht HS (1988) Hydrology of mountainous areas. In: Proceedings of the Štrbské Pleso Workshop, Czechoslovakia, International Association of Hydrological Sciences (IAHS), vol 90, pp 419–423
Hwang T et al (2014) Divergent phenological response to hydroclimate variability in forested mountain watersheds. Glob Change Biol 20:2580–2595
IPCC (2007) Synthesis Report. Contribution of Working Groups I, II and III to the Fourth Assessment Report of the Intergovernmental Panel on Climate Change (IPCC), Genewa, Switzerland, p 104
Kalaninova D et al (2014) Caddisflies (Trichoptera) as good indicators of environmental stress in mountain lotic ecosystems. Biologia 69:1030–1045
Kumar R (2014) Effect of climate change and climate variable conditions on litchi (Litchi chinensis Sonn.) productivity and quality. Acta Hortic 1029:145–154. https://doi.org/10.17660/actahortic..1029.16
Liuy X, Chen B (2000) Climate warming in the Tibetan Plateau during recent decades. Int J Climatol 20:1729–1742
Malik DS, Umesh B (2011) Ecological characteristics of Sahastradhara stream at Dehradun (Uttarakhand). Environ Conserv J 12:69–74
Mehta M et al (2013) Glacier changes in Upper Tons River basin, Garhwal Himalaya, Uttarakhand. India Z Geomorphol 57:225–244
Messerli B, Ives JD (1997) Mountains of the world: a global priority. Parthenon Publishing Group, p 495
Murty NS et al (2008) Climate and its variability over the Western Himalaya. J Agrometeorol 10:296–299
Nainwal HC et al (2008) Temporal changes in rate of recession: evidences from Satopanth and Bhagirath Kharak glaciers, Uttarakhand, using total station survey. Curr Sci 94:653–660
Negi PS (2012) Climate change, alpine treeline dynamics and associated terminology: focus on northwestern Indian Himalaya. Trop Ecol 53:371–374
Pandey AK et al (2014) Climate footprints in the Late Quaternary-Holocene landforms of Dun Valley, NW Himalaya. India Curr Sci 106:245–253
Pandey R (2011) Forest biomass extraction for livestock feed and ass ociated carbon analysis in lower Himalayas. India Mitigat Adapt Strat Global Change 16:879–888
Pederson GT, Graumlich LJ, Fagre DB, Kipfer T, Muhlfeld CC (2010) A century of climate and ecosystem change in Western Montana: what do temperatures pretend? Clim Change 98:133–154
Priyanka N, Joshi PK (2013) Effects of climate change on invasion potential distribution of Lantana camara. J Earth Sci Clim Change 2011:4. https://doi.org/10.4172/2157-7617.1000164
Rangwala I, Miller JR (2012) Climate change in mountains: a review of elevation dependent warming and its possible causes. Clim Change 114(3–4):527–547
Rangwala I, Miller J, Xu M (2009) Warming in the Tibetan Plateau: possible influence of the changes in surface water vapour. Geophys Res Lett 36:1–16
Rai SP (1993) Hydrological and geomorphological studies of the Syahidevi–Binsar area, district Almora, Kumaun Himalaya. PhD thesis, Kumaun University
Raj KBG et al (2014) Remote sensing based assessment of glacial lake growth on Milam glacier, Goriganga basin, Kumaon Himalaya. J Geol Soc India 83:385–392
Rana CS et al (2011) New distributional record of Gentiana tetrasepala Biswas (Gentianales: Gentianaceae) from the Valley of Flowers National Park, Garhwal Himalaya. J Threat Taxa 3:2100–2103
Rao KK et al (2014) Projected changes in mean and extreme precip itation indices over India using PRECIS. Glob Planet Change 113:77–90

Rawat JS, Rawat G (1999) Environmental hydrology of the Almora region. Technical Report of the DST Project, Department of Geography, Kumaun University, Almora, pp 1–392
Rawat JS (2005) Geo-hydrometeorological Data at micro and macro scale watersheds for the Stakeholders of the Local Level Planning, Technical Report of DST Project, Department of Geography, Kumaun University Almora, pp 1–125
Rawat JS (2007) Khoo Na Jaai Kaheen Kosi. Gyanodaya Prakashan, Nainital, pp 1–54
Rawat JS (2012) Application of Geo-hydrometeorological Database in Groundwater Augmentation, Technical Report of DST Project, Department of Geography, Kumaun University Almora, pp 1–87
Rawat JS, Rawat G (2016): Hydrometeorological records under varied ecological conditions in the Kumaun Lesser Himalayan Terrain, USERC, Dehradun Project Report, Uttarakhand Center on Climate change, Kumaun University SSJ Campus Almora-263601, Uttarakhand (India), pp 1–261
Rawat JS (2017) Impact of climate change in hydrometeorology; a case study from the Upper Kosi Watershed, Uttarakhand, funded by the USERC Dehradun, January 2017 onwards
Rawat J, Rawat G (2017) Saving the Himalayan dwindling rivers through river rejuvenation measures. In: Gupta S, Gupta S (eds) Water for Future, Ankit Prakashan Haldwani, 39.54
Rawat JS, Dobhal R, Arora VPS (2012) Climate change. In: Dobhal R (eds) Uttarakhand SOAR 2012, Uttarakhand State Council of Science and Technology, Dehtadun, pp 244–279
Rawat JS, Govind A, Rawat G, Joshi M, Rai SP, Gahlot N (2016) Perennial to ephemeral transformation of a Lesser Himalayan watershed. Curr Sci III(4):686–693
Rawat PK (2013) Observed changes in extreme wet and dry spells during the South Asian summer monsoon season. Nat Clim Change 4:456–461
Rawat PK (2014) GIS development to monitor climate change and its geohydrological consequences on non-monsoon crop pattern in Himalaya. Comput Geosci 70:80–95
Rocca ME et al (2014) Climate change impacts on fire regimes and key ecosystem services in Rocky Mountain forests. For Ecol Manage 327:290–305
Schendel V(1975) The world's water resources and water balance, natural resources and development, vol 1, Hannover, W Germany, pp 8–14
Semwal RL et al (2013) Growth and carbon stocks of multipurpose tree species plantations in d egraded lands in Central Himalaya, India. For Ecol Manage 310:450–459
Shweta G et al (2013) Climatic water balance and moisture adequacy index in Tarai region of Uttarakhand. Pantnagar J Res 11:317–321
Singh O et al (2013) On rising temperature trends at Dehradun in Doon valley of Uttarakhand, India. J Earth Syst Sci 122:613–622
Srivastava P et al (2013) Early Holocene monsoonal fluctuations in the Garhwal higher Himalaya as inferred from multi-proxy data from the Malari paleolake. Quat Res 80:447–458
Strahler AN (1964) Quantitative geomorphology of drainage basins and channel networks. In: Chow VT (eds), Handbook of applied hydrology, Mc Graw Hill, USA
Tambe S et al (2012) Reviving dying springs: climate change adaptation experiments from the Sikkim Himalaya. Mt Res Dev 32:62–72
Thakur SN, Kahlon S, Bhutani S (2016) Assessment of climate change through temperature and rainfall trend analysis in Chamba and Kangra Districts of Himanchal Pradesh, Panjab Geographer, 12, pp 21–42. ISSN 0973-3485
Tiwari PC, Joshi B (2012) Environmental changes and sustainable development of water resources in the Himalayan headwaters of India. Wat Resour Manage 26:883–907
Valdiya KS (1980) Geology of kumaun lesser Himalaya. Wadia Institute of Himalayan Geology, Dehradun, pp 1–291
Valdiya KS, Rawat JS, Rai SP (1993) Response of hydrologic cycle to environmental degradation. Central Himalayan Environment Association, Nainital, pp 1–56
Xu J, Grumbine RE, Shreshta A, Eriksson M, Yang X, Wang YUN, Wikes A (2009) The melting Himlayas: cascading effects of climate change on water, biodiversity and livelihood. Conserv Biol 23(3):520–530

Zomer RJ et al (2014) Projected climate change impacts on spatial distribution of bioclimatic zones and ecoregions within the Kailash Sacred Landscape of China, India, Nepal. Clim Change 125:445–460

Chapter 6
Robust and Reliable Technique of Automatic Building Extraction from High Resolution Imagery

Arvind Pandey, Mriganka Shekhar Sarkar, Gajendra Singh, Sarita Palni, Nisha Chand and Manish Kumar

Abstract The automation in man-made object extraction such as building habitation from urban area imagery has become a challenging task for photogrammetry, computer vision, and remote sensing. This study aims to automatically extract building of an urban area using high resolution intensity data and fuzzy membership logic to classify the image object by using *e-Cognition* software. The object oriented method was implemented and high resolution Quick-Bird imagery was used for automatic building extraction in Dehradun city of Uttarakhand district, India. We have further evaluated the performance of this automated extracted building feature by using accuracy completeness (89.74%), correctness (94.50%), and the quality (85.29%). The study however, clearly shows that the segmentation-based classification is much superior to the traditional per-pixel methods mainly used on high resolution images. It also shows that high spatial resolution satellite data and appropriate data processing play not only an important role in modern urban planning but also reduce the cost of manpower and saves time.

Keywords e-cognition · Automatic building extraction · Fuzzy logic · Photogrammetry

A. Pandey (✉) · M. S. Sarkar · G. Singh
Uttarakhand Space Application Center, 131/2 Vasant Vihar, Dehradun 248006, Uttarakhand, India
e-mail: pandeyarvind02@gmail.com

M. S. Sarkar
e-mail: mriganka.bio@gmail.com

G. Singh
e-mail: gajendrawat@yahoo.com

S. Palni · N. Chand
Kumaun University, Nainital, Uttarakhand, India
e-mail: saritapalni206@gmail.com

N. Chand
e-mail: chandnisha72@gmail.com

M. Kumar
Department of Geography, Kalindi College, University of Delhi, New Delhi, India
e-mail: manish.ks1@gmail.com

S. Sahdev et al. (eds.), *Geoecology of Landscape Dynamics*,
Advances in Geographical and Environmental Sciences,
https://doi.org/10.1007/978-981-15-2097-6_6

6.1 Introduction

The automatic delineation of man-made objects such as buildings from high resolution imagery taken in human habitation areas has become a challenging task for photogrammetry, computer vision, and remote sensing. Studies on this area have been initiated during the late 1980s where several sources of remotely sensed imagery were used such as "*color images*", "*single intensity images*", "*stereo*", "*laser range images*", and "*multi band images*" (Peng et al. 2005). Some important applications of this method involve automatic extraction of information from imagery and apprising geographic information system (GIS) databases. Especially in the research field of urban planning, the method has been utilized intensively, wherein the manual interpretation is very hectic and time taking process. Several research attempts have been recently made to automate or semi-automate the procedure (Zhang et al. 2005; Hu et al. 2004; Eidenbenz et al. 2000). A vast range of methods such as mathematical and statistical algorithms have been implemented to automatically construct two dimensional (2D) or three dimensional (3D) building models using satellite and aerial borne imagery. However, Dash et al. (2004) have developed a method relying on the standard deviation of the data to separate woody tree structures and building structures by considering the height variation at the edge of an object. Samadzadegan et al. (2005) have proposed a novel viewpoint for object re-Cognition by implying fuzzy reasoning algorithm that distinguishes structural, textural, and spectral information. After extracting and connecting the edge pixels of an object, Hongjian and Shiqiang (2006) has delineated the height of buildings from a sparse laser-based data and reconstructed the 3D structure for each building. Sohn and Dowman (2002) have delineated the building structures automatically from a hybrid image of the IKONOS having pan-sharpened multispectral bands and the low-sampled airborne laser scanning image. Lafarge et al. (2010) have proposed an object oriented approach that automatically extracts buildings from digital elevation models. In this method, rough approximations of the building footprints are first recognized by a technique based on marked point processes. Further, these polygons are regularized by the connection between the neighboring rectangles and finally, find the roof height discontinuities between them. Another very new method namely the snake model has brought recent attention to building extraction. The model was implemented by Kass et al. (1988). Without any prior information about the remotely sensed image, the technique considers global information about the image contour to receive a closed or open curve like structures. Due to its robustness, the model is intensively used in several fields of image-processing, such as "*image segmentation*", "*image tracking*", and "*3D modeling*" (Lam and Yan 1994). The model is an "*energy-minimizing spline*" guided by outer constraint forces and affected by image forces that pull toward the lines and edges features (Shih and Zhang 2004).

Recent advancements in remote sensing techniques can produce very high resolution images such as WorldView, Quick-Bird, GeoEye-1, IKONOS, and Pléiades. Research on extracting information from those imagery is still in the infantry stage. Some mathematical and statistical algorithm exists to extract information from those high resolution imagery that take minimum computational time and perhaps more accurate than others. High resolution satellite images are gaining popularity in extracting information features. Automatic feature extraction using such images has become the main objective to save time in updating data and reduces 20–90% labor cost, as a result of this people are changing progressively to new procedures.

Therefore, the main aim of this research is to automatically extract building information of human habitation area using high resolution intensity data range and implying fuzzy membership logic for classifying the image object by using *e-Cognition* software. In this research work, object oriented approach and high resolution Quick-Bird data are used for automatic building extraction. In this project, three categories of sites are used for extraction buildings. Further, accuracy for each site is calculated and compared with visually interpreted features or references.

6.2 Study Area

Dehradun is situated in the northwest part of India and it is the capital city of the Uttarakhand district (Fig. 6.1). The image of the study area falls between latitude 30.32°N and longitude 78.04°E. The city is situated in the "*Doon Valley*" on the foothills of the Himalayas, fall between two of India's prominent rivers—the Ganges on the eastern side and the Yamuna on the western side. The city is famous for its picturesque landscape and pleasant climate and this provides a gateway to all pilgrimage and tourist intense regions. The city is well connected by road networks and its geographical location situated within the proximity of popular Himalayan tourist destinations such as the prominent Hindu holy cities of Haridwar and Rishikesh. The city is also connected to Nainital, Mussoorie, Auli, and the Himalayan pilgrimage of "*Char Dham*".

6.3 Materials and Methods

6.3.1 Spatial Data

The Quick-Bird scene was procured from Space Imaging LLC that covers the full extent of the Doon valley, Uttarakhand. This image was acquired on September 1st, 2010 at 3:44 pm. The entire scene includes a negligible cloud cover. Quick-Bird image encompasses 11 bit radiometric resolution and 4 multispectral bands (blue, green, red, and near-infrared bands) each band have 2.4 m spatial resolution. In addition, it also contains a panchromatic band of 0.6 m spatial resolution. False color composite (FCC) map was prepared for the purposes of discriminating buildings from other features of the scene by combining bands four, three, and two. This was further sharpened by merging the 0.6 m resolution separate panchromatic band.

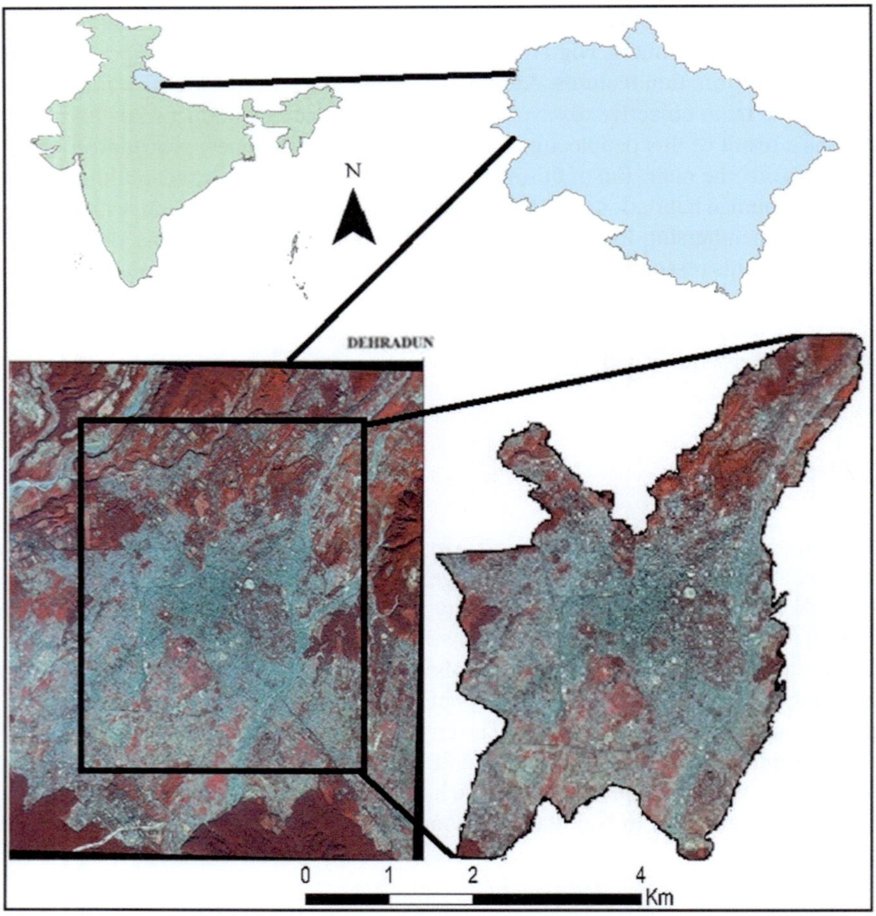

Fig. 6.1 Map showing the geographical location of the study area—Dehradun city, Uttarakhand, India

6.3.2 Unsupervised Classification Method

An unsupervised classification was carried out on the Quick-Bird FCC image specifying initial 50 classes with 20 iterations, a 0.99 confidence interval, using the principal axis, and 2 standard deviations (Long and Srihann 2004). This classification was aimed to identify possible building features from other land cover classes in Dehradun City (Fig. 6.1). The Kappa Coefficient method was used to assess the accuracy of the final classified map (Congalton 1991).

6.3.3 *Object Based Classification Method*

Usually, in human habitation areas, building structures are often characterized by heterogeneity and different feature classes. The intensity values of building structures are not constant values because cars, trees, shadows, and roads produce noise and rapidly change the image intensity. Therefore, the efficient segmentation statistical algorithms are necessary to detect building features before imposing any feature extraction algorithm. We use two-step method to get the result of building structure extraction, first one is to use multi-resolution image and spectral difference segmentation for getting the shape and sharpness of object then the second one is the image classification technique using object oriented approach and fuzzy logic by using e-Cognition professional 9.0 software. The entire methodological flow chart is displayed in Fig. 6.2.

6.4 Results

Automatic feature extraction method has provided the approximate area of feature in 0.12 km^2, using the e-Cognition automatic feature extraction method (algorithm). The feature output was created on shape file. The total boundary area of 0.12 km^2 was extracted whereas the ward area is 0.33 km^2 (Fig. 6.3).

On the contrary, the unsupervised classification is hectic and time-consuming and gives 6% low accuracy in comparison to the e-Cognition (Table 6.1). While analyzing the unsupervised classification, approximate building area was 0.11 km^2. While comparing the both results, we obtain the difference of 0.01 km^2 area. This confirmes that e-Cognition gives 100% accuracy where as, ERDAS gives 94% accuracy.

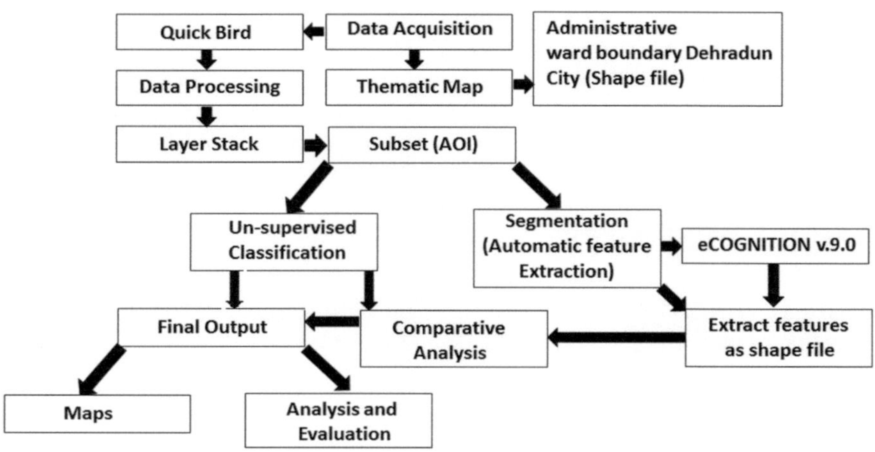

Fig. 6.2 Flow chart of building extraction process by two parallel methods

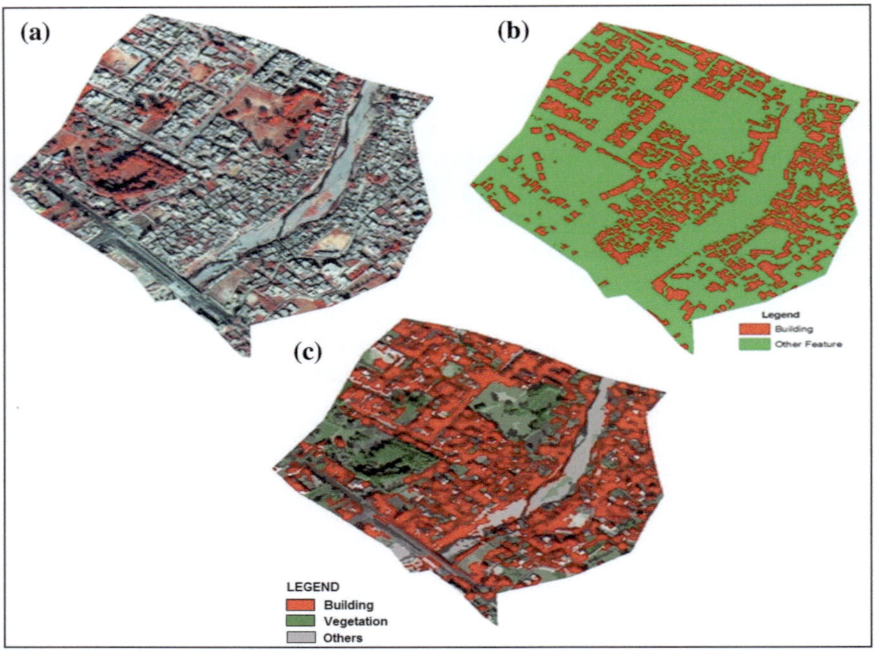

Fig. 6.3 **a** Area showing study area and false color composite map of Dehradun city, **b** unsupervised classified map showing building features, and **c** Automated extracted building using segmentation method implemented in e-Cognition software. *Source* Quick-Bird satellite data

Table 6.1 Extracted building area statistics of two different methods and their accuracy using kappa statistics

Methods	Building area (km^2)	Others area (km^2)	Total area (Km^2)	Kappa (overall)
Unsupervised classification method	0.11	0.22	0.33	0.72
Automatic classification method	0.12	0.21	0.33	0.81

Source Quick-Bird satellite data

When the results of both softwares were compared the accuracy of Automatic feature extraction of the e-Cognition was 94.01% in area of intrest.

6.4.1 Discussion and Conclusion

Automatic information extraction of urban features like buildings by using object based approach and fuzzy logic on high resolution Quick-bird imagery has proven to be a very useful and robust method for urban planning. This is also very helpful for e-Cognition to detect building features in urban areas, due to the availability of high resolution informative features. Even though the variety of the type of building shape, structure, and building roofs together influence the building extraction process much complicated, but the required acceptable level of accuracy for the extracted buildings was generated by the automated building extraction method. The major advantage of this new technique is independent of any other additional information, such as building height information that had been usually used for former building extraction methods. Moreover, our method provides GIS-ready data. We conclude that our approach generates acceptable results. Automated building extraction process has also proven to be better than conventional unsupervised image classification method for building extraction. However, the output image obtained through the automated process may not produce the exact shape of the building structures. Some motorable roads and pavements had created noise, which is also sensed by the algorithm as building features. Further, more research is still required to improve the algorithms to reduce these errors to eliminate linear features such as motorable roads and pavements.

Acknowledgements The authors would like to thank Dr. Durgesh Panth, Director of Uttarakhand Space Application Center (USAC), Uttarakhand for providing funds to carry out this short-term research. We thank Mr. Shashank Lingwal, Scientist, USAC for providing support during this study.

References

Congalton RG (1991) A review of assessing the accuracy of classifications of remotely sensed data. Remote Sens Environ 37(1):35–46

Dash J, Steinle E, Singh RP, Bähr HP (2004) Automatic building extraction from laser scanning data: an input tool for disaster manage. Adv Space Res 33(3):317–322

Eidenbenz C, Kaeser C, Baltsavias E (2000) ATOMI–automated reconstruction of topographic objects from aerial images using vectorized map information. Int Arch Photogr Remote Sens 33(B3):462–471

Hu X, Zhang Z, Tao CV (2004) A robust method for semi-automatic extraction of road centerlines using a piecewise parabolic model and least square template matching. Photogr Eng Remote Sens 70(12):1393–1398

Hongjian Y, Shiqiang Z (2006) 3D building reconstruction from aerial CCD image and sparse laser sample data. Opt Lasers Eng 44(6):555–566

Kass M, Witkin A, Terzopoulos D (1988) Snakes: active contour models. Int J Comput Vis 1(4):321–331

Lafarge F, Descombes X, Zerubia J, Pierrot-Deseilligny M (2010) Structural approach for building reconstruction from a single DSM. IEEE Trans Pattern Anal Mach Intell 32(1):135–147

Lam KM, Yan H (1994) Fast greedy algorithm for active contours. Electron Lett 30(1):21–23

Long W, Srihann S (2004) Land cover classification of SSC image: unsupervised and supervised classification using ERDAS imagine. In: IEEE international proceedings of geosci and remote sensor symposium, 2004. IGARSS'04, vol 4. IEEE, pp 2707–2712

Peng J, Zhang D, Liu Y (2005) An improved snake model for building detection from urban aerial images. Pattern Recogn Lett 26(5):587–595

Samadzadegan F, Azizi A, Hahn M, Lucas C (2005) Automatic 3D objects recognition and reconstruction based on neuro-fuzzy modeling. ISPRS J Photogr Remote Sens 59(5):255–277

Shih FY, Zhang K (2004) Efficient contour detection based on improved snake model. Int J Pattern Recogn Artif Intell 18(02):197–209

Sohn G, Dowman IJ (2002) Terrain surface reconstruction by the use of tetrahedron model with the MDL criterion. Int Arch Photogr Remote Sens Spat Info Sci 34(3/A):336–344

Zhang Q, Pavlic G, Chen W, Fraser R, Leblanc S, Cihlar J (2005) A semi-automatic segmentation procedure for feature extraction in remotely sensed imagery. Comput Geosci 31(3):289–296

Chapter 7
Soil Resource Inventory for Meeting Challenges of Land Degradation: A Remote Sensing Approach

Dinesh Kumar Tripathi

Abstract In this study an attempt has been made to delineate, map out, and generate database on soil resources for meeting challenges of land degradation in irrigated agro-ecosystem using geospatial tools of remote sensing (RS) and geographic information system (GIS). Gauriganj block, Amethi district (lies between 26° 7′5″ N to 26° 19′5″ N latitudes and 81° 36′45″ E to 81° 45′18″ E longitudes), Uttar Pradesh was selected for study. The space born multispectral Landsat 7 ETM+data of year, 2014 and corresponding survey of India Topographical sheets numbered 63 F/11, 63 F/12, and 63 F/16 were applied for soil survey. The satellite image of the study area was processed using standard visual image interpretation approach incorporating field check and attribute data in ERDAS imagine 9.1 and ARC view 3.2a software. Digital image processing techniques were also applied for generation ad-on-data for visual image interpretation. On the basis of satellite image analysis and information regarding soil surveys conducted earlier under Sharda Sahayak C.A.D project (1988) Lucknow (U.P.), entire study area was classified into 83 soil interpretation units. The database on both units was generated in GIS environment considering USDA soils classification system. The soils of the study area were grouped into two orders, four sub-orders, five great groups, six subgroups, five families, and seven series. The study reveals that the RS and GIS techniques can be used in an effective manner in soil resource investigation and mapping. This study may prove a better input in proper management of degraded lands in the study area.

Keywords Remote sensing · Geographic information system · Landsat 7 ETM+data · ERDAS imagine 9.1 · Arc view 3.2a · Field check · Soil interpretation units

D. K. Tripathi (✉)
Department of Geography, Kamla Nehru Institute of Physical and Social Sciences, Sultanpur, UP, India
e-mail: tripathidk.geoinformatics@gmail.com

S. Sahdev et al. (eds.), *Geoecology of Landscape Dynamics*,
Advances in Geographical and Environmental Sciences,
https://doi.org/10.1007/978-981-15-2097-6_7

7.1 Introduction

Soils are fundamental to the well-being and productivity of agricultural and natural ecosystems (Singer and Ewing 2000). It is finite, fragile, and nonrenewable natural resource (Lal 1995). The continuous deterioration of their quality due to land degradation processes in agro-ecosystem has been an issue of global concern because it poses a serious threat to human well-being. In agriculture-based countries like India, it is a great challenge before scientists, researchers, and decision-makers. Over exploitation of soil resources without understanding its sustainable limit has caused extensive soil degradation and causing serious threat to present and future agricultural growth and sustainability. In a developing country like India where agriculture is a main stay of the economy, soil degradation has emerged as a serious threat. As per the reports of the Department of Agriculture and Co-operation (DAC 1994), all the various categories of degraded land were spread over about 107 million hectares area. It is estimated that six billion tons of soils are eroded from crop land each year (Narayan and Babu 1983). In some studies, it is accounted that about sixty percent cropped soil in India is affected from degradation problems (Sehgal and Abrol 1994; Mandal and Mandal 1996; Biswas et al. 1999) and has become incapable to produce the adequate food for sustainable livelihood of people distributed on it. About 40% of the total degraded soils of the country are still under cultivation which is an indication of environmental ignorance as well as farming compulsion of the peoples who are engaged in a perpetual war of friction with land resources. On the other hand today the population has exceeded one billion and by 2025 at the current growth rate of 1.6%, it would be 1.37 billion. Four hundred million tons of food grain would be needed to feed this population (Patil 2003). It would therefore be necessary to plan the proper management of land resource considering sustainability measures. The reliable and up-to-date information on spatial extent, property, and limitations of soil resources is a prerequisite for soil resource conservation and land degradation management in any region. Proper inventory and mapping of soils serve to gain spatial information on the soil resources and primarily help in solving agro-ecological and land resource management problems for any region.

The conventional soil survey and mapping techniques are expensive, tedious, and time-consuming task. It needs a number of in situ measurements to locate soil boundaries. In the recent years, as both RS and GIS are cost effective and technologically sound geospatial tools, they have emerged as popular viable substitutes. Rawashdeh and Saleh (2006) offer permanent and authentic record of spatial patterns (Prakash and Gupta 1998). These techniques have been proved to be most efficient, economical, and reliable for comprehensive inventory of soil resourses and land use pattern. RS data helps detection of soil boundaries admirably because of variations in spectral response of the different soils, attributable to their varied physical make-up and chemical composition (Karale 1992). The spectral reflectance of soil is governed by its properties such as color, texture, organic matter, and minerals. The collection of information on these characteristics differences by remote sensing techniques reduces fieldwork, overcomes errors associated with subjectivity and is able to generate soil map of inaccessible areas.

In India, several studies were carried out on soil survey in deferent regions using aerial photographs (Karale et al. 1970). During early 1980s satellite RS techniques which were used in soil mapping, attracted the attention of scholars and researchers. Initially, works on soil resource mapping using satellite data were carried out by Mirajkar and Srinivasan (1975), NRSA (1976, 1978, 1979, 1981) Venkataratnam and Rao (1977) and Venkataratnam (1980). Several scholars used digital image processing techniques in soil surveys and demonstrated its potential (Venkatratnam 1980), Kudrat et al. (1990), Karale (1992), Ravisankar and Thamappa (2004), Rao et al. (2004) and Milind et al. (2011). Simultaneously, several GIS modeling techniques were also used by scholars to draw reliable and useful informations from soil maps (Kudrat et al. 1990; Saha et al. 1991; Kudrat et al. 1995, 1997). Keeping these in view, a micro level soil inventory was carried out in the Gauriganj block, Amethi district, Uttar Pradesh (India), to generate database on soil resources using modern geospatial tools of RS and GIS. Present study is aimed to identify and delineate the soil units using Landsat 7 ETM+image (2014) and collateral data, map out the soil resources in ERDAS Imagine 9.1, and Arc GIS 9.3 software adopting USDA soils classification system, generate village level database on soil types for land degradation management and analyze spatial pattern of soil types in the study area.

7.2 Study Area

The study area has been undertaken in Gauriganj block (falls between latitude 26° 7′5″ to 26° 10′5″ N and longitude 81° 36′45″ to 81° 45′18″ E) of Amethi district, Uttar Pradesh, India (Fig. 7.1) which lies in the middle Ganga plain. It covers an area of 207.56 km^2, characterized by an even and featureless plain, composed of deep and fertile alluvial deposits. The area falls under typical tropical, semiarid, monsoonal type of climate. The hot and dry summer, hot rainy season, warm autumn and cool winter is its characteristics (Mishra and Sharma 2003). The area receives 977 mm average annual rainfall mainly in rainy season between the months of July and September (Sharma et al. 2001) whereas the winter receives irregular and scanty rainfall. The average annual minimum and maximum temperature in the study area is recorded as 4.10 °C and 47.5 °C, respectively. Soil Survey Staff (1994) has classified soils of this area as "Aquic Petrocalcic Natrustalf" and accumulation of salts above the soil surface is the main feature in a large area. The block is economically underprivileged and majority of the population (about 80%) earns livelihood through agricultural and allied activities. Land degradation is a major environmental issue in this area.

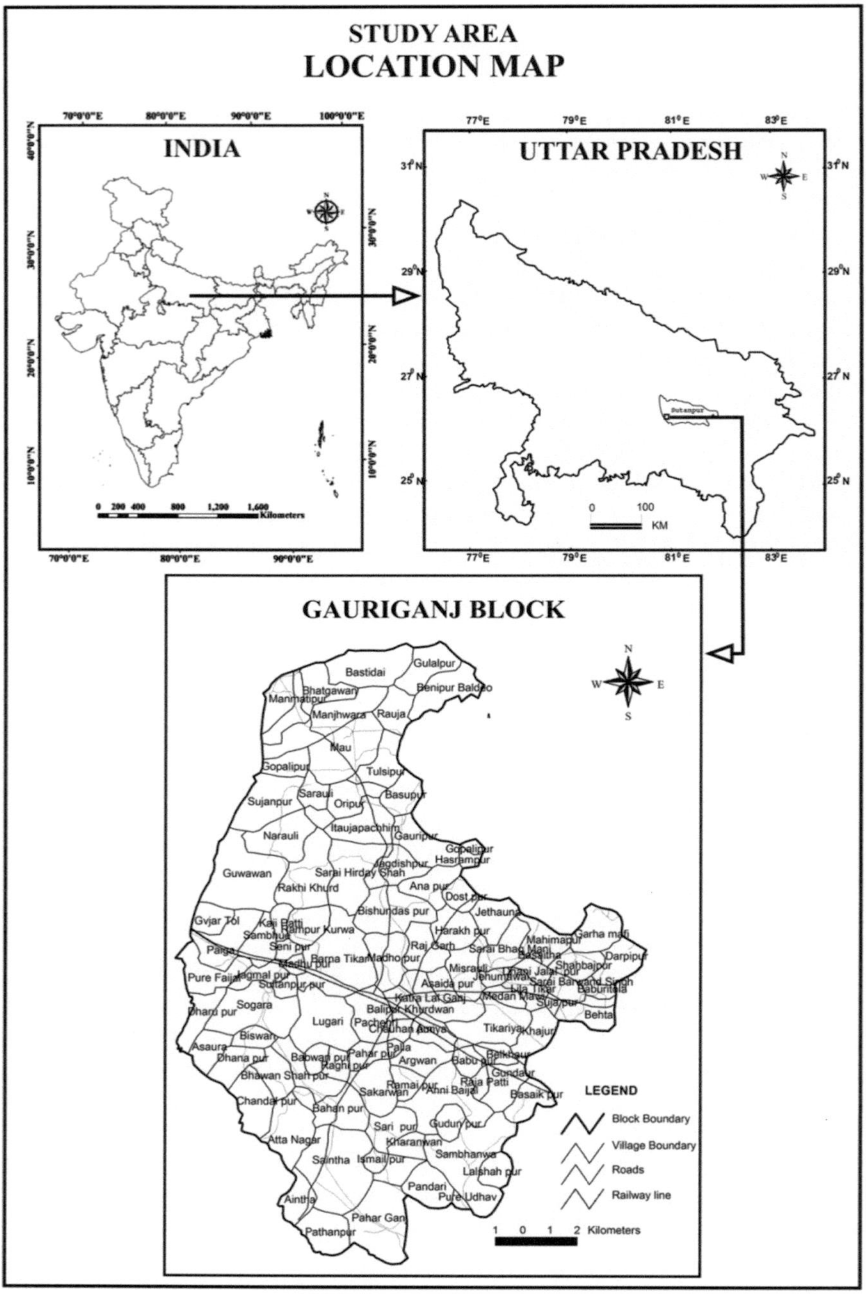

Fig. 7.1 Study area map

Table 7.1 Characteristics of the Lansat-7 ETM+data

Sl. no.	Parameters	Characteristics
1	Spectral range (mm)	0.4–2.4
2	Spatial resolution (m)	30
3	Swath width (km)	185
4	Spectral resolution	Variable
5	Spectral coverage	Discrete
6	Number of bands	7
7	Spectral bands used in this analysis WL (nm)	Band-1:450–520 (nm) Band-2:530–610 (nm) Band-3:630–690 (nm) Band-4:780–900 (nm) Band-5:1550–1750 (nm) Band-7:2090–2350 (nm)

Source Landsat 7 science data users handbook (2004)

7.3 Materials and Methods

7.3.1 Data Used

In this study data used and their sources are: (i) Landsat-7 ETM+multi-spectral images (30 m resolution) acquired in the year 2014 (source: http://glcf.umiacs.umd.edu). Table 7.1 depicts the Characteristics of the Lansat-7 ETM+data used in this analysis, (ii) Google Earth Images (source: http://www.googleearth.com), (iii) Village boundary map prepared by National Natural Resource Database Management System (NNRDMS), Sultanpur (U.P.), (iv) Survey of India (SOI) Topographical sheets numbered 63 F (scale 1:250000), 63 F/11, 63 F/12, 63 F/16 at scale 1:50000, (v) Training/ground truth data collected through selective field survey with GPS handset (Garmin GPS map 76 Cx) in the month of May, 2014, (vi) Soil Survey Report (source Sharda Sahayak C.A.D. Project 1988, Lucknow), (vii) Information regarding land use/land cover collected through the local people informal interview, (viii) GIS/RS packages of Arc GIS 9.3 (ESRI) and ERDAS Imagine 9.1 (Leica Geosystems, Atlana, U.S.A.).

7.3.2 Database Preparation

In order to investigate the soil resources Landsat 7, ETM+satellite image for the years 2014 was downloaded through Global Land Cover Facilities Network (GLCF). The Landsat-7 ETM+image provided by GLCF Network was ortho-rectified (UTM/WGS 84 projection) and radiometrically corrected. The sub-setting of satellite image was performed in Arc GIS 9.3 software to extract study area from the entire image using geo-referenced out line boundary map of Gauriganj block. Subsequently, the data normalization was performed for reducing spatial variation in reflectance caused by sun elevation differences and radiometric gain setting. Primarily, radiance values were calculated in ERDAS Imagine modeler using DN values of the image. The formula given by Markham and Barker in 1986 (Landsat 7 Science Data Users Handbook 2004) was used in this process.

$$L* = (L_{max} - L_{min})/Qcal_{max} * Qcal + L_{min}$$

where L^* = spectral radiance at the sensors aperture $W/(m^2.sr.um)$, Qcal = Calibrated Digital Number, $Qcal_{max}$ = maximum possible DN value (255), L_{max} & L_{min} = maximum/minimum scaled spectral radiance value for a given band (provided in the header file). The radiance values further converted into reflectance using the following formula-

$$\rho P = (\pi * L\lambda * d2)/(\mathrm{ESUN}\lambda * \cos\theta s)$$

where ρP = effect of planetary reflectance, $L\lambda$ = band radiance ($w/m^2/ster/\mu m$), d = distance of earth from Sun (in astronomical units, $d = 0.997052$ for this case), ESUNλ = mean solar exo-atmospheric irradiances for given wavelength in watts/$m^2/\mu m$/ster, θs = solar zenith angle in degrees.

To know the condition of vegetation cover on the soil the most frequently used Normalized Difference Vegetation Index (NDVI) was applied on reflectance image. The NDVI, is the ratio, respects the absorption of photosynthetic active radiation and hence it directly relates to the photosynthetic capacity of plants and energy absorption (Sellers1985; Myneni et al. 1995). The band 3 (Red) and band 4 (Near Infrared) of Landsat-7 ETM+data were used to obtain the NDVI using following formula:

$$\mathrm{NDVI} = (\mathrm{DNIR} - \mathrm{DNR}/\mathrm{DNIR} + \mathrm{DNR})$$

where DNIR = digital numbers of Infrared band, DNR = digital numbers of Red band.

By design, the NDVI varies between -1.0 and $+1.0$, where NDVI ranging between 0.1 and 0.7 typically represents vegetation cover. Higher levels of healthy vegetation cover in any region are associated with higher NDVI values, while NDVI values near zero indicate the less green vegetation. The prepared NDVI image was used during image processing for soil mapping.

7.3.3 Visual Image Interpretation

The major outcome of this study is the mapping and evaluation of soil resources of the study area. Various methods to delineate soil boundaries in remote sensing image data are in vogue in which visual interpretation method was used in most of the cases (Karale et al. 1981; Biswas 1987). However, computer-based digital image processing methods have also been used in soil mapping (Epema 1986; Kudrat et al. 1990; Korolyuk and Sheherbenko 1994) and recommended as a potential tool (Lee et al. 1988; Kudrat et al. 1992). In the present study, on-screen standard visual image interpretation method was employed. Before the image interpretation, a preliminary general traversing of study area was undertaken and some observations were recorded at few places. A legend was formed to identify the tonal behavior of soils and land use/land cover classes on the image. During this field visit, training data were also collected for digital image analysis. Garmin GPS map 76 Cx handset was used to locate training data collection sites. The visual interpretation of image was performed in Arc GIS 9.3 considering image elements (such as tone, texture, shape, size, pattern, site, and association), author's background knowledge, and collateral data and terrain conditions. Eighty-three soil interpretation units were identified and delineated on False Color Composite (FCC) of the satellite data. To improve the image contrast for better delineation of soil boundaries, spectral enhancement and band combination techniques were used. NDVI and classified (maximum likelihood method) images were used as add-on data set to supplement the existing onscreen interpretation on False Color Composite of Landsat 7 ETM+imagery. The information pertaining to soil profile and their physical, chemical, and biological properties collected earlier in Sharda Sahayak C.A.D. Project 1988, Lucknow (U.P.) were incorporated during the entire soil mapping process.

7.3.4 Ground Truth Collection

In order to correlate the image elements and existence of soil units, a field visit was made again and ground truth was collected. After selecting sample sites, the pockets of land which were mapped as specific soil units were precisely located on the ground with the help of topographical sheets and observation was made regarding geotechnical elements.

7.3.5 Post-field Interpretation

According to field observations during ground truth, the preliminary interpreted soil units were finalized and soil maps were prepared. The units having similar soil characteristics were merged and seven soil series were proposed. The soils' series were further classified up to family level following USDA Soil Taxonomy system. Further, entire soil mapping and area estimation were performed in Arc GIS 9.3 software.

7.3.6 *Accuracy Assessment*

Accuracy assessment of soil maps was carried out through analyzing 250 randomly selected sample points on the reference image. The analysis was performed in Accuracy Assessment Option of ERADAS IMAGINE software. The ground truth data and Google earth high-resolution images were used for comparing mapped thematic layers and an error matrix was generated. The quantitative assessment of maps accuracy was performed by computing overall accuracy and Kappa Coefficient (Biahop et al. 1975).

7.4 Results and Discussion

7.4.1 *Soil Mapping Units*

The area under study was estimated to be 20,791 ha in Arc GIS 9.3 software. Visual image interpretation method incorporating field check/training collection and digital processing techniques like image enhancement, band combination, NDVI, maximum likelihood provide useful method for soil mapping. Delineation of soil boundaries on satellite image basically involves their characterization through on-screen visual interpretation in terms of image elements like color/tone, texture, shape, size, association, etc. In this study, visual interpretation of Landsat 7 ETM+data (2014) was performed in Arc GIS 9.1 for soil mapping. On the image of the study area, 83 distinct mapping units were delineated on the basis of their spectral responses (Fig. 7.2). Detailed information pertaining soil profiles and their physical, chemical, and biological properties of each mapping unit were collected from soil survey report of Gauriganj block, Sharda Sahayak Command Area Development Project, 1988, Lucknow (U.P.). The information on soil composition in the interpreted sample strips was extrapolated to unsampled area. The mapping units having similar soil composition were merged together and made as a single unit. In this mapping process seven soil series were recognized. The soils' units were further processed and mapped up to family level following USDA Soil classification scheme (Soil Survey Staff 2004, 2009, 2010). The entire soil mapping and area estimation were performed in Arc GIS 9.3 software.

7.4.2 *Soil Orders and Suborders*

The results of the soil mapping reveal that the study area has been classified into two soil orders, viz., Alfisols and Inceptisols (Fig. 7.3a and Table 7.2). The differentiation in orders is based on highly generalized criteria. It is generally based on soils' genesis related properties. Alfisols are the relatively high fertile soil which form in semiarid

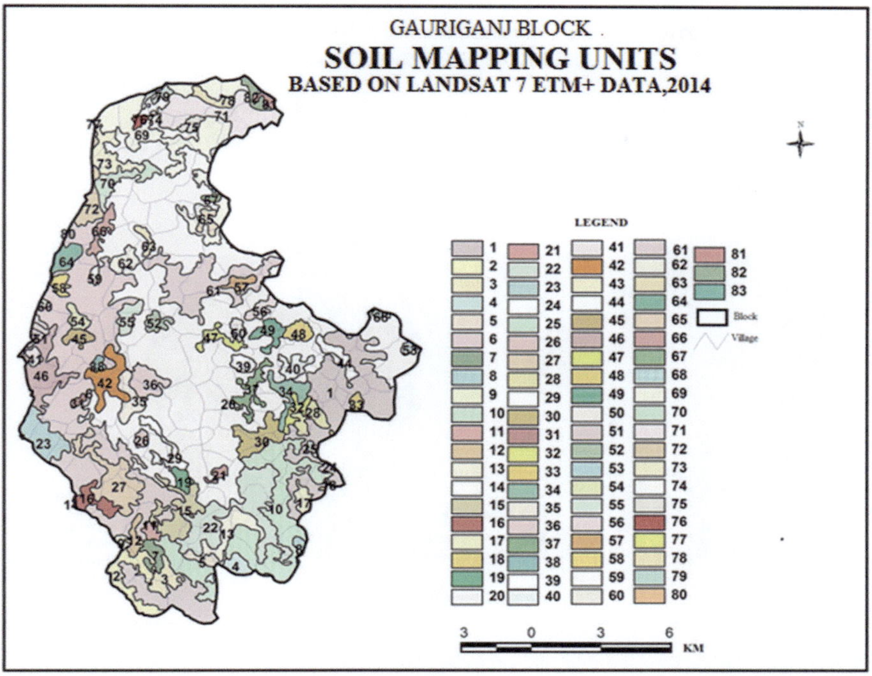

Fig. 7.2 Soil mapping units based on visual interpretation *Source* Landsat 7 data, 2014

to humid conditions. It contains clay-enriched horizon and native fertility. It contains aluminum and iron minerals but relatively low organic matter. It represents an area of 13,663.84 ha (65.72% of the total area) and can be observed in cultivated and low lying areas. The soils of Alfisole order were further classified into Aqualfs and Ustalfs, at suborder level based on soil moisture and temperature.

Aqualfs form in aquic situations of fluctuating water table. During the considerable part of the year, groundwater table is found near the surface. These soils cover about 47.22% area of the block (Fig. 7.3b). Ustalfs soils occur in subhumid to semiarid conditions. These soils accumulate the carbonates in or below the subsoil. These soils occur on 18.50% area of the block. Inceptisols are also mineral soils that have developed over subhumid and semiarid environments characterized by accumulation of clays, gypsum, and salt of translocated alluvium These soils represent an area of 6908.85 ha (33.23%). This soil order was classified into Ochrepts and Ustepts suborders. Ochrepts are characterized by an ochric epipedon (too little organic matter in upper surface, light color), a warm soil temperature regime and an ustic soil moisture regime found on about 3.78% of area. Ustepts (29.45%) are mainly drained freely and have an ustic moisture regime.

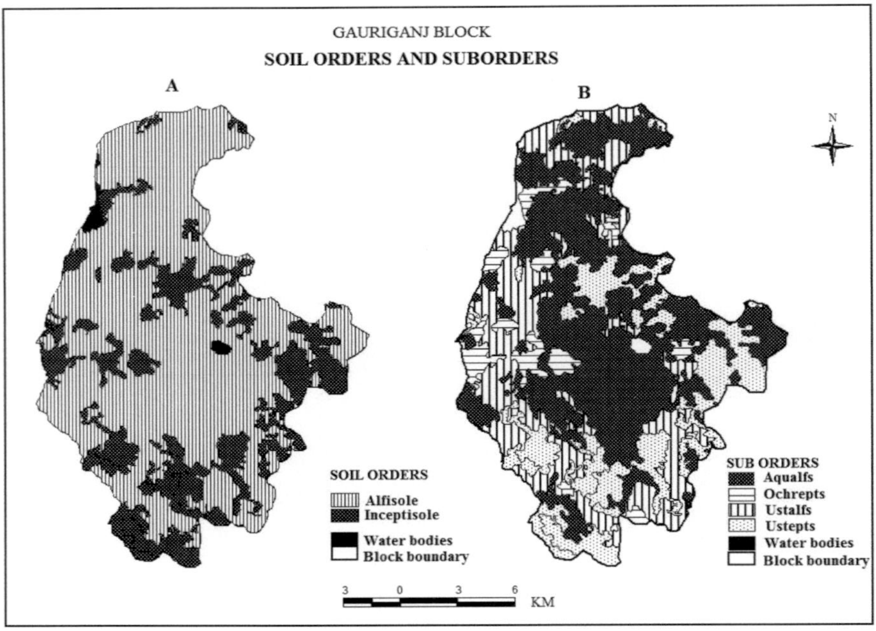

Fig. 7.3 Distribution of soil orders and suborders *Source* Based on Landsat 7 data, 2014

Table 7.2 Soil orders and Suborders in Gauriganj block, Amethi District (U.P)

Sl. no.	Soil order	Suborder	Area in ha	Area in %
1	Alfisols	Aqualfs	9817.51	47.22
		Ustalfs	3846.33	18.50
2	Inceptisols	Ustepts	6122.94	29.45
		Ochrepts	785.89	3.78
3	Waterbody		218.30	1.05
Total			20,791	100

Source Derived from satellite data analysis in GIS

7.4.2.1 Soil Great Groups and Subgroups

On the basis of the kind and sequence of soil horizons, soil suborders of the study area were further divided into great groups. Five great groups are recognized and mapped in the study area, namely, Epiaqualfs, Haplustalfs, Haplustepts, Natraqualfs, and Ustochrepts (Fig. 7.4a, and Table 7.3). **Epiaqualfs** are the Aqualfs that have an epipedon that rests on an argillic horizon without an abrupt textural change if the argillic horizon has low saturated hydraulic conductivity. These soils do not have a kandic horizon, a natric horizon, a fragipan, or a duripan (Soil Survey Staff 1999).

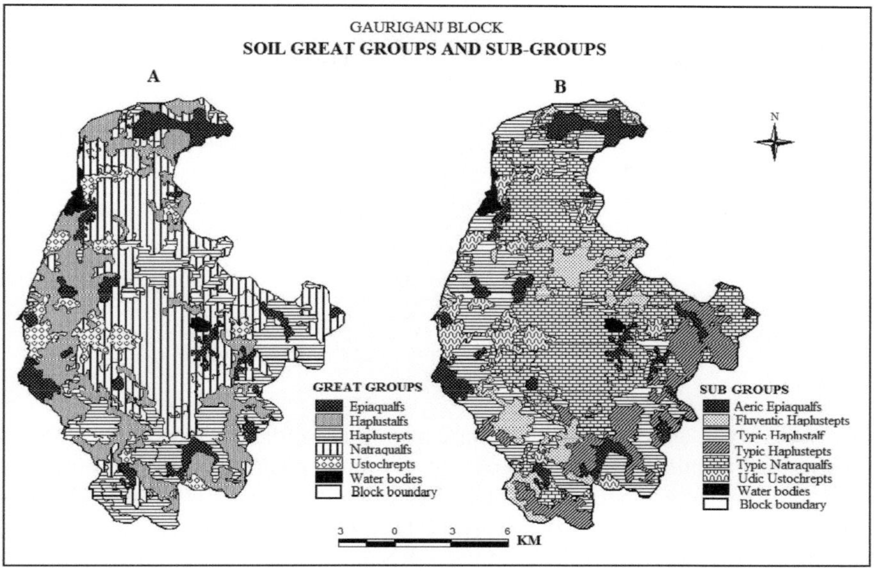

Fig. 7.4 Distribution of soil great groups and subgroups *Source* Based on Landsat 7 data, 2014

Table 7.3 Soil great groups in Gauriganj block, Amethi District (U.P)

Sl. no.	Great groups	Area in ha	Area in %
1	Epiaqualfs	1137.26	5.47
2	Haplustalfs	3846.33	18.50
3	Haplustepts	6122.94	29.45
4	Natraqualfs	8680.24	41.75
5	Ustochrepts	769.26	3.70
6	Water body	218.30	1.05
Total		20,791	100

Source Derived from satellite data analysis in GIS

These soils cover about 1137.26 ha (5.47%) area of the block. Haplustalfs are the Ustalfs that have an argillic horizon. Horizons, do not have a duripan that has its upper boundary within 100 cm of the surface, do not have a petrocalcic horizon within 150 cm of the surface, and do not have much plinthite. (Soil Survey Staff 1999). These soils occupy about 3846.33 ha (18.50%) area in the block.

Haplustepts which is freely drained Ustepts are calcareous at some depth or having high base status. These soils are found on 6122.94 ha (29.45%) area in Gauriganj block. Natraqualfs (Aqualfs that have a natric horizon and warmer temperature regime) and Ustochrepts cover 41.75 and 3.70% area of the block, respectively. On the basis of basic system of Soil Classification System prepared by Soil Survey Staff (1999), the soil great group is further a categorized into four subgroups, i.e., Aeric,

Table 7.4 Soil subgroups in Gauriganj block, Amethi District (U.P)

Sl. no.	Great groups	Area in ha	Area in %
1	Aeric Epiaqualfs	1137.26	5.47
2	Fluventic Haplustepts	1505.26	7.24
3	Typic Haplustalfs	3846.33	18.50
4	Typic Haplustepts	4617.68	22.21
5	Typic Natraqualfs	8680.24	41.75
6	Udic Ustochrepts	785.89	3.78
7	Water Body	218.30	1.05
Total		20,791	100

Source Derived from satellite data analysis in GIS

Fluventic, Typic, and Udic. Aeric subgroups represent the drier conditions of soil whereas the Fluvents soils distributed mainly on flood plain formed by recent water-deposited sediments. It contains an appreciable amount of organic carbon. The soils that do not have the characteristics defined for the other subgroups are kept into Typic subgroups. Udic soil is common in humid regions where moisture is sufficiently high throughout the year to meet plant requirements. In the study area, six subgroups are identified and mapped namely Aeric Epiaqualfs, Fluventic Haplustepts, Typic Haplustalfs, Typic Haplustepts, Typic Natraqualfs, and Udic Ustouchrepts. The spatial extent and distribution are depicted by Fig. 7.4b and Table 7.4.

7.4.3 Soil Families and Soil Series

On the basis of mineralogy, texture, and temperature, the subgroups are further classified into soil families which distinguish between clayey, loamy, and sandy soils. The soils of the Gauriganj block have been classified into five soil texture families, (Fig. 7.5a, Table 7.5) namely, Loamy, Coarse loamy, Fine, Fine loamy, and Fine silty. Loamy soils are highly fertile in nature which consist mainly an equal mixture of sand (30–50%) and clay (less than 30%) together with silt (30–50%) and humus. It can retain some amount of moisture and plant food even under adverse weather and climate conditions. It extends over a 22.21% (4617.68 ha) of the geographical area. This soil is well exposed in southeastern and eastern part of the block. There are some exposure around village Attanagar and Chandanpur and southwestern part of the study area. The coarse loamy is the coarse-textured soil which contains sand (50–70%) and silt (0–50%) with less than 20% clay. These are low water holding capacity soil and good for horticulture, legumes, groundnut, and Bajra. About 7.24% (1505.26 ha) of the total area is under course loamy soil. This soil is exposed in northeast and southwest parts of the study area. Fine-textured soils (Clay) cover an area 45.27% (9412.08 ha) of the total area of the block. These soils are found in middle and eastern parts of the block. About 3.78% (785.89 ha) area is under fine

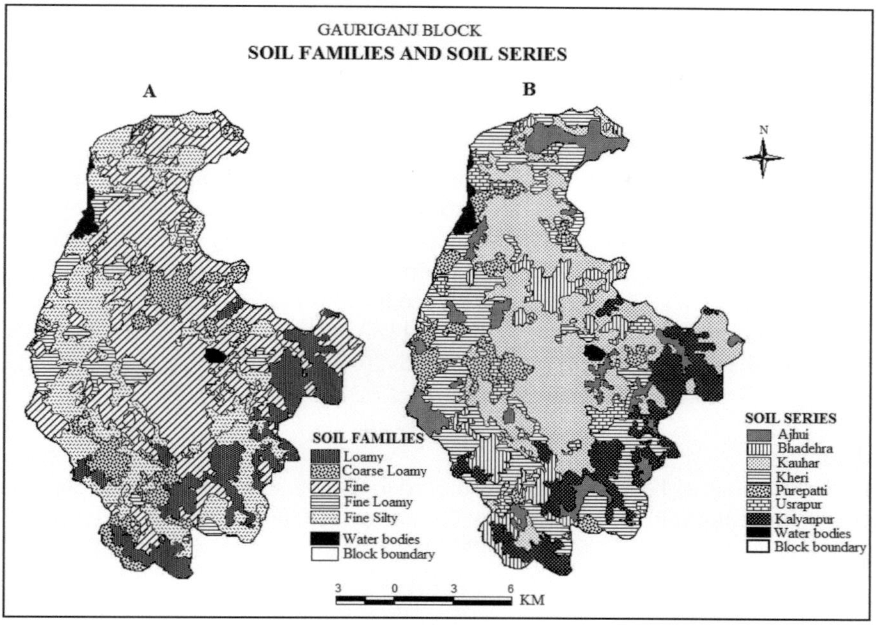

Fig. 7.5 Distribution of soil families and soil series *Source* Based on Landsat 7 data, 2014

Table 7.5 Soil families in Gauriganj block, Amethi District (U.P)

Sl. no.	Great Groups	Area in ha	Area in %
1	Loamy	4617.68	22.21
2	Coarse loamy	1505.26	7.24
3	Fine	9412.08	45.27
4	Fine loamy	785.89	3.78
5	Fine silty	4251.75	20.45
6	Water body	218.30	1.05
Total		20,791	100

Source Derived from satellite data analysis in GIS

loamy soils exposed in Sarai Bhagmani, Jehmawai, Barna Tikar, Lugari, Pure Faizal Dharupur, Jagmal Pur Madhupur, Sultanpur, Sogara, Basupur, Guwawan, Narauli, Pandari, Mau, and Gopalipur villages. The fine silty soils containing sand (0–20%) and silt (40–60%) with 40–60% clay exposed mostly in western and southern parts of the Gauriganj block. It amounts to 20.45% (4251.75 ha) of the total area.

The narrowest category soil in soil taxonomy is called soil series (Soil Survey Staff 1975). It consists of pedons (soil individual) that have similar pedogenesis, soil chemistry, and physical properties. Each series consists of pedons that have horizons similar in soil color, texture, structure, pH, consistence, mineral and chemical composition, and arrangement in the soil. In the study area, seven soil series were identified and mapped, viz., Ajhuri, Bhaderha, Kauhar, Kheri, Purepatti, Usrapur, Kalyanpur (Fig. 7.5b, Table 7.6).

Ajhuri

Ajhuri soil series belongs to Alfisole order, Aqualfs suborder, Epiaqalfs great group, and Aeric subgroup. It represents fine textured high fertile soil characterized by clay-enriched horizon, aluminum and iron minerals, relatively low organic matter, thermic or warmer temperature regime, and fluctuating water table. This soil is suitable for wheat, paddy, pulses, gram, peas, and oil seeds crops like mustered linseed, etc. It exposed in thirty villages covering an area of 1137.26 ha (5.47%) in Gauriganj block. Mapping units numbered 7, 13, 17, 18, 23, 25, 26, 31, 32, 37, 41, 43, 44, 53, 54, 55, 66, 67, 71, 77 represent the Ajhuri series (Fig. 7.2 and Table 7.7).

Bhaderha

Bhaderha soil series belongs to Inceptisols order, Ustepts suborder, Haplustepts great groups, Fluventic aplustepts subgroups, Coarse loamy family (mixed and Hyperthermic). It occupies on 1505.26 ha (7.24%) area over thirty-seven villages (Table 7.6) in the study area. Mapping units numbered 2, 5, 15, 27, 49, 50, 51, 52, 59, 61, 79, 80, and 81 represent the Bhaderha soil series. The soil of this series is coarse textured and consists of sand (50–70%), silt (0–50%) and clay (less than 20%) and suitable for horticulture, legumes, groundnut, and Bajra crops. The soil is characterized by medium fertility and low water holding capacity.

Kauhar

Kauhar soil series belongs to Alfisols order, Aqualfs suborder, Natraqualfs great group, Typic Natraqualfs subgroup and Fine textured soil family (mixed, Hyperthermic). Mapping units numbered 3, 9, 20, 28, 33, 38, 57, 60, 74, 78, 82, and 83 represent the Kauhar soil series. It covers an area of 8274.18 ha (39.80%) and occupies in forty-seven villages. It accounts for 3846.33 ha (18.50%) and illustrated by mapping units number of 6, 10, 14, 24, 29, 34, 47, 63, and 69.

Purepatti

Purepatti soil series is included under Inceptisols order, Ochrepts suborder, Ustochrepts great group, Udic Ustochrepts subgroup and Fine loamy family (mixed, Hyperthermic). This soil series occupies on 785.89 ha (3.78%) lands. Mapping units numbered 4, 11, 36, 40, 42, 45, 46, 62, 64, 65, and 70 represent the Purepatti soil series. This soil series was found in twenty-two villages.

Usrapur

Usrapur soil series belongs to Alfisols order, Aqualfs suborder, Natraqualfs great group, Typic Natraqualfs subgroup, and Fine Silty (mixed, Hyperthermic) family. It accounts for only 405.42 ha (1.95%) area. Mapping units numbered 8, 12, 21, 30, 35, 48, 58, 73, 75, and 76 represent the Usrapur soil series. This soil series is distributed over fifteen villages.

Table 7.6 Spatial Distribution of Soil Series in Gauriganj Block, Amethi District (U.P)

Sl. no.	Soil series	Area in ha	Area in %	Villages
1	Ajhuri	1137.26	5.47	Sogara, Tikariya, Sarai Barwand Singh, Sahbaj Pur, Mahimapur, Pure Fajil, Misrauli, Asaidapur, Katra Lal Ganj, Amiya, Darpipur, Saintha, Rauja, Benipur Baldeo, Gulalpur, Bastidai, Majhwara, Bhatgawan, Paiga, Guwawan, Sujapur, Narauli, Rampur Kurwa, Rohshi Khurd, Chhitepur, Banwari Pur, Pandari, Gundur, Belkhaur, Basupur, Tulsipur, Kharnwan, Samhanwa, Basaikpur, Dhanupur, Asaura, Dharupur and Manmatipur
2	Bhaderha	1505.26	7.24	Benipur Baldeo, Gulalpur, Bastidai, Gopalipur, Hasrampur, Gauripur, Anapur, Jagdishpur, Bishundaspur, Sarai Hirday Shah, Barna Tikar, Harakh Pur, Gvjar Tol, Dhanapur, Bhawan Shah Pur, Atta Nagar, Saripur, Saintha, Aintha, Pathanpur, Pahar Ganj. Pandari, Ismailpur, Bahanpur, Sakrawan, Bahanpur, Shah Pur, Ronhsi Buzurg, Chandaipur, Misrauli, Rajgarh, Paiga, Guwawan, Rohsi Khurd, Bhatgawan and Sujanpur
3	Kauhar	8274.18	39.80	Gulalpur, Bastidai, Manjhawara, Mautulsipur, Oripur, Sarauli, Itaujapachhim, Narauli, Sujanpur, Gauripur, Sarai Hirday Shah, Rohni Khurd, Barna Tikar, Madhopur, Raj Garh, Jethauna, Sarai Bagmani, Madhopur, Pachehri, Bali Pur Khurdawan, Katralal Ganj, Amiya, Argwan, Ramaipur, Anni Baijal, Saripur, Sakarwara, Paharpur, Lugri, Raghipur, Ronhsi Buzurg, Sogara, Garha Mafi, Darpipur, Saintha, Pathanpur, Aintha, Attanagar, Bahanpur, Paharganj, Palia, Chauhanpur, Rampur Kurwa, Bishundaspur, Mahimapur, Bhehta, Tikariya and Khajuri
4	Kheri	3846.33	18.50	Gulalpur, Bastidai, Bhatgawan, Manmatipur, Manjhwara, Rauja, Mau, Basupur, Sujanpur, Guwawan, Rohni Khurd, Kajipatti, Sembhue, Paiga, Gvjar Tol, Sogara, Biswan, Dhanapur, Chandaipur, Atta Nagar, Saintha, Bahanpur, Paharpur, Pandari, Samhanwa, Lal Shahpur, Raja Patti, Babupur, Anni Baijal, Gundaur, Basaikpur, Belkhaur, Medan Mawi, Raj Garh, Madhopur, Narauli, Pure Fajil, Senipur, Jagmalpur, Dharupur, Asura, Bhawan Shah Pur, Tikaria, Itaujapachhim, Tulsipur, Benipur Baldeo and Mahanpur

(continued)

Table 7.6 (continued)

Sl. no.	Soil series	Area in ha	Area in %	Villages
5	Purepatti	785.89	3.78	Mau, Gopalipur, Sujanpur, Guwawan, Narauli, Rohshi Khurd, Basupur, Pure Faijal, Paiga, Sembhue, Dharupur, Jagmalpur, Sultanpur, Barna Tikar, Sogara, Lugari, Saintha, Pandari. Jehumawi, Sarai Bhagmani, Madhu Pur and Senipur
6	Usrapur	405.42	1.95	Manjhwara, Bhatgawan, Manmattipur, Guwawan, Lugari, Jethauna, Argawan, Tikaria, Ramaipur, Lal Shah Pur, Saintha, Babupur, Anni Baijal, Barna Tikar and Sarai Bhagmani
7	Kalyanpur	4617.68	22.21	Garhamafi, Mahimapur, Sahbaj Pur, Sarai Barwand Singh, Baburitola, Behta, Sujapur, Lila Tikar, Khajuri, Belkhaur, Gundur, Basaik Pur, Samhanwa, Gudunpur, Anni Baijal, Saripur, Kharanwan, Chhitepur, Paharganj, Pathanpur, Aintha, Atta Nagar, Chandal Pur and Sakarwan

Source Derived from satellite data analysis in GIS

Table 7.7 Soil series and their association with soil constraints in Gauriganj block, Amethi District (U.P)

Sl. no.	Mapping units	Soil series	Soil classification	Texture
1	7, 13, 17, 18,23, 25, 26,31, 32, 37,41, 43, 44,53, 54, 55,66, 67, 71, 77	Ajhuri	Alfisols, Aqualfs, Epiaqalfs, Aeric Epiaqualf	Fine
2	2, 5, 15, 27, 49,50, 51, 52,59, 61, 79,80, 81	Bhaderha	Inceptisols, Ustepts, Haplustepts, Fluventic Haplustepts	Coarse loamy
3	3, 9, 20, 28, 33,38, 57, 60,74, 78, 82, 83	Kauhar	Alfisols, Aqualfs, Natraqualfs, Typic Natraqualfs	Fine
4	6, 10, 14, 24, 29,34, 47, 63, 69	Kheri	Alfisols, Ustepts, Haplustalfs, Typic Haplustalf	Fine silty
5	4, 11, 36, 40, 42,45, 46, 62,64, 65, 70	Purepatti	Inceptisols, Ochrepts, Ustochrepts, Udic Ustochrepts	Fine loamy
6	8, 12, 21, 30, 35,48, 58, 73, 75,76	Usrapur	Alfisols, Aqualfs, Natraqualfs, Typic Natraqualfs	Fine silty
7	1,16, 19, 22, 56, 68	Kalyanpur	Inceptisols, Ustepts, Haplustepts, Typic Haplustepts	Loamy

Kalyanpur

Kalyanpur soil series is classified under Inceptisols order, Ustepts suborder, Haplustepts great group, Typic Haplustepts subgroup, and Loamy (mixed, Hyperthermic) soil family. Mapping units numbered 1, 16, 19, 22, 56, and 68 represent the Kalyanpur soil series. It is distributed on 4617.68 ha (22.21%) and found in twenty-four villages.

7.5 Conclusion

In this study, Landsat 7 ETM+image of the year 2014 was processed for soil mapping and database generation. Visual image processing technique incorporating selected field check/training and collateral data was applied for this task. However, some digital image processing techniques such as spectral enhancement, band combination, NDVI, and image classification (maximum likelihood method) were also used as add-on data set to supplement the existing onscreen visual image processing. As per USDA soil classification system, soils of the study area were grouped into two orders, four suborders, five great groups, six subgroups, five families, and seven series. The study clearly demonstrates the usefulness of RS and GIS techniques for soil resource inventory, mapping, and database generation at micro level. During the post-field interpretation field check, it was observed that the salt-affected soil, waterlogging, soil erosion, deficiency of soil nutrients, etc. are the major human-induced soil related constraints in the study area that deteriorate the currying capacity of soil to support human population. The village level reliable soil database on spatial extent, types, and magnitude may prove a better input in micro level planning for proper management of degraded lands in the study area.

Acknowledgements The Author thankfully acknowledged to Indian Council of Social Science Research (ICSSR), New Delhi which provided financial assistance to conduct this study. The author is also grateful to Scientist In-charge, NRDMS centre Sultanpur and teaching staff of the Agriculture Science Faculty, Kamla Nehru Institute of Physical & Social Sciences, Sultanpur (UP) for constant suggestions during the course of the study.

References

Bishop YMM, Fienberg SE, Holland PW (1975) Discrete multivariate analysis theory and practice. MIT Press, Carnbridge, MA, p 557

Biswas RR (1987) A soil map through landsat satellite imagery in part of the Auranga catchment in Ranchi and Palamon district of Bihar, India. Int J Remote Sens 4:541–543

Biswas et al (1999) Prioritization of sub watershed based on morphometric analysis of drainage basin. Photonirvachak 27(3):155

DAC (1994) Draft report on status of land degradation in India. Department of Agriculture and Co-operation

Epema GF (1986) Processing thematic mapper data for mapping in Tunisia. ITC J 30–34
Karale RL, Venugopal KR, Hilwig FW(1970) Reconnaissance soil survey in the Ganges alluvial plain in Meerut District, UP. Report submitted to IPI, Dehradun
Kudrat M, Tiwari AK, Saha SK (1997) Modelling sediment yield for prioritization of sub-watersheds using remote sensing and GIS techniques. Geocarto Int 12:31–38
Karale RL (1992) Remote sensing with IRS-1A in soil studies: development, status and prospects. In: Karale RL (ed) Natural resources management—a new perspective. NNRMS, Bangalore, pp 28–143
Kudrat M, Prabhakaran B, Sastry TRSVS, Tiwari AK, Sharma KP, Manchanda ML (1995) Quantative estimation of soil loss through remote sensing: a case study of part of Chotanagpur Plateau, India. In: Varma CVJ, Rao ARG (eds) Management of sediments: philosophy, aims and techniques. Central Board of Irrigation and Power, New Delhi, pp 37–44
Kudrat M, Saha SK, Tiwari AK (1990) Potential use of IRS LISS II digital data in soil land use mapping and productivity assessment. Asian Pacific Remote Sens J 2:73–78
Kudrat M, Tiwari AK, Saha SK, Bhan SK (1992) Soil resource mapping sing IRS-1A LISS II digital data—a case study of Kandi area adjacent to Chandigarh (India). Int J Remote Sens 13:3287–3302
Karale RL, Bali YP, Rao KV (1981) Soil mapping using remote sensing techniques. In: Proceedings Indian academy of science and engineering sciences 3:197–208
Korolyuk TV, Shcherbenko HV (1994) Compiling soil maps on the basis of remotely sensed data digital processing: soil interpretation. Int J Remote Sens 15:1379–1400
Lal R (1995) Global soil erosion by water and carban dynamics. In: Lal R, Kimble J, Levine E, Stewart BA (eds) Soil management and greenhouse effect. Soil Sci Soc Am 41:39–51 (Special Publication)
Lee KS, Lee GB, Tyler J (1988) Determination of soil characteristics from thematic mapper data of a cropped organic–inorganic soil landscape. Soil Sci Soc Am J 52:1100–1104
Landsat 7 Science Data Users Handbook (2004) NASA. http://landsat.gsfc.nasa.gov/wp-content/uploads/2016/08/Landsat7_Handbook.pdf
Mirajkar MA, Srinivasan TR (1975) Landsat photo interpretation for preparation of small scale maps through a multistage approach. Photonirvachak 3:87–98
Milind R, Wadodkar Ravisankar T (2011) Soil resource database at village level for development planning. J Indian Soc Remote Sens 39(4):529–536
Mandal C, Mandal DK (1996) Qualitative assessment of soil erosion from soil survey data: a case study of Nagpur district. Geogr Rev India 58(1):29–40
Myneni RB, Hall FG, Sellers PJ, Marshak AL (1995) The interpretation of spectral vegetation indexes. IEEE Trans Geosci Remote Sens 33:481–486
Mishra A, Sharma SD (2003) Leguminous trees for the restoration of degraded sodic wasteland in eastern Uttar Pradesh, India. Land Degrad Dev 14:245–261
NRSA(1976) An application of satellite remote sensing techniques for integrated pilot survey of natural resources in parts of Punjab and Haryana—an abstract report. National Remote Sensing Agency, Secunderabad, India
NRSA (1978) Satellite remote sensing survey of natural resources of Andhra Pradesh. Project report. National Remote Sensing Agency, Secunderabad, India
NRSA(1979) Satellite remote sensing survey of natural resources of Haryana. Project report. National Remote Sensing Agency, Secunderabad, India
NRSA (1981) Satellite remote sensing survey for soil and land use in part of Uttar Pradesh. Project report. National Remote Sensing Agency, Hyderabad, India
Narayan V, Dhurua V, Babu R (1983) Estimation of soil erosion in India. J Irrig Drain Eng 109(4):419–433
Prakash A, Gupta RP (1998) Land-use mapping and change detection in a coal mining area—a case study in the Jharia Coalfield, India. Int J Remote Sens 19:391–410
Patil J (2003) Sustainable agricultural development: issues and programs. Financ Agric 35(1):35–38

Ravisankar T, Thamappa SS (2004) Satellite data interpretation and analysis for soil mapping. In: Venkatratnam L, Ravisankar T, Sudarshana R (eds) Soils and crops. NRSA Publication, Hyderabad

Rao BR, Fyzee M, Wadodkar MR (2004) Utility of remote sensing data for mapping soils at various scales and levels. In: Venkatratnam L, Ravisankar T, Sudarshana R (eds) Soils and crops. NRSA Publication, Hyderabad

Rawashdeh SA, Saleh B (2006) Satellite monitoring of urban spatial growth in the Amman Area, Jordan. J Urban Plan Dev 132.4(211):0733–9488

Sharma SD, Khan GH, Prasad KG (2001) Selection of suitable provenances of Dalbergia sissoo for sodic lands. Indian J For 24:58–64

Singer MJ, Ewing S (2000) Soil quality. In: Sumner ME (ed) Handbook of soil science. CRC Press, Boca Raton, FL

Saha SK, Kudrat M, Bhan SK(1991) Erosional soil loss prediction using digital satellite data and universal soil loss prediction-soil loss mapping in Siwalik Hills in India. In: Murai S (ed) Applications of remote sensing in Asia and Oceania. Asian association on remote sensing. Tokyo, pp 369–372

Soil Survey Staff (1975) Soil taxonomy. A basic system of soil classification for making and interpreting soil surveys. USDA/SCS. Agricultural Handbook, 436. U.S. Government Printing Office, Washington, DC, p 396

Soil Survey Staff (1994) Key to soil texonomy. USDA Soil conservation service, Washington DC, USA

Soil Survey Staff (1999) Soil taxonomy. A basic system of soil classification for making and interpreting soil surveys. In: Agricultural handbook 436. Natural resources conservation service, 2nd ed. USDA, Washington DC, USA, p 869

Soil Survey Staff (2004) R Burt (ed.). Soil survey laboratory methods manual. Ver. 4.0. USDA/NRCS. Soil survey investigations report no. 42. U.S. Government Printing Office, Washington, DC. http://soils.usda.gov/technical/lmm/. Accessed 24 Jan 2011

Soil Survey Staff (2009) R Burt (ed.) Soil survey field and laboratory methods manual. Ver. 1.0. USDA/NRCS. Soil survey investigations report no. 51. http://www.soils.usda.gov/technical/. Accessed 24 Jan 2011

Soil Survey Staff (2010) Keys to soil taxonomy, 11th ed. USDA/NRCS. U.S. Government Printing Office, Washington, DC. http://ftpfc.sc.egov.usda.gov/NSSC/Soil_Taxonomy/keys/2010_Keys_to_Soil_Taxonomy.pdf. Accessed 24 Jan 2011

Sellers PJ (1985) Canopy reflectance, photosynthesis, and transpiration. Int J Remote Sens 6:1335–1372

Sehgal JL, Abrol IP (1994) Soil degradation in India: status and impact. Oxford and IBH Publishing, New Delhi, India

Venkataratnam L, Rao KR (1977) Computer aided classification and mapping soils and soil limitations using landsat multispectral data. In: Proceedings of symposium on Remote sensing for hydrology, agriculture and mineral resources. Space Applications Centre, Ahmedabad, India, pp 101–104

Venkataratnam L (1980) Use of remotely sensed data for soil mapping. Photonirvachak 8:19–26

Chapter 8
Temporal Variation in Glacier's Area and Identification of Glacial Lakes in Sikkim

Mahfooz Alam and Sandeep Bhardwaj

Abstract Glacier's ice cover is changing over time and subsequently topography of the region and other factors result in the formation of glacial lakes. This paper aims to identify the temporal changes in glacier's area and glacial lakes with the help of remote sensing satellite images of Landsat. For the measurement of area covering glaciers, band rationing and filtering technique have been used with manual thresholding and manual post-processing by visual interpretation, and also conversion of raster to vector. For the identification of glacial lakes, NDWI technique is used with manual thresholding with the help of ERDAS IMAGINE 9.3 and QGIS 2.12.3 software. Results showing Glacier shrinkage about 22% ± 6% and the number of glacial lakes increased from 169 in the year 1992 to 261 in the year 2015. Glacial lakes can be seen as an indicator of glacial changes. The cause of shrinkage in these two decades can be said as increase in the temperature over the region and anthropogenic factors. Impact because of these changes could be a potential threat and may lead to disaster in the region.

Keywords Glacial lakes · Glacial shrinkage · Remote sensing techniques

8.1 Introduction

The report of the Intergovernmental Panel on Climate Change (IPCC) stated that 'glaciers are the best natural indicators of climate'. Many studies done by researchers, suggested that glacier changes may vary widely within a region based on climatic and topographical characteristics such as slope, aspects and elevation. Glacier changes are observed worldwide within the framework of the Global Terrestrial Network on Glaciers (GTN-G) of the Global Climate Observing Systems (Andreassen and Paul 2008).

M. Alam (✉) · S. Bhardwaj
Department of Geography, Kumaun University, Uttarakhand, India
e-mail: mahfoozalam@yahoo.com

S. Bhardwaj
e-mail: sandy891995@gmail.com

S. Sahdev et al. (eds.), *Geoecology of Landscape Dynamics*,
Advances in Geographical and Environmental Sciences,
https://doi.org/10.1007/978-981-15-2097-6_8

Increase in the number of lakes by melting of glaciers may leads to glacial lakes outburst floods that subsequently impact man as well as his environment. Such factors make it an important issue to be studied which can be possible using remote sensing and GIS techniques, for a larger region that may take more time and money if studied on field survey and other measurements.

Availability of glacier-related data was very meagre before the introduction of glacier study and analysis based on remotely sensed image. Remote sensing techniques have considerably improved the estimation of glacier's area change and also for the identification of water bodies such as lakes.

The objective of this paper is twofold:

- To present the variation or changes in Glacier's area from the year 1992 to 2015 of Sikkim (India), using satellite images of Landsat series.
- To identify the glacial lakes of Sikkim in the year 1992 and 2015 based on remote sensing and GIS techniques and tools.

8.2 Study Area

Our study focuses on glaciers and glacial lakes of Sikkim, which is located at the north-eastern part of India (Fig. 8.1). Sikkim covers a total geographical area of about 7096 km^2 and is centered at about 27.5330° N, 88.5122°.

Sikkim state borders china in the north, Bhutan in its east, Nepal in its west and Indian state west Bengal in its south. The terrain of the state is hilly and mountainous

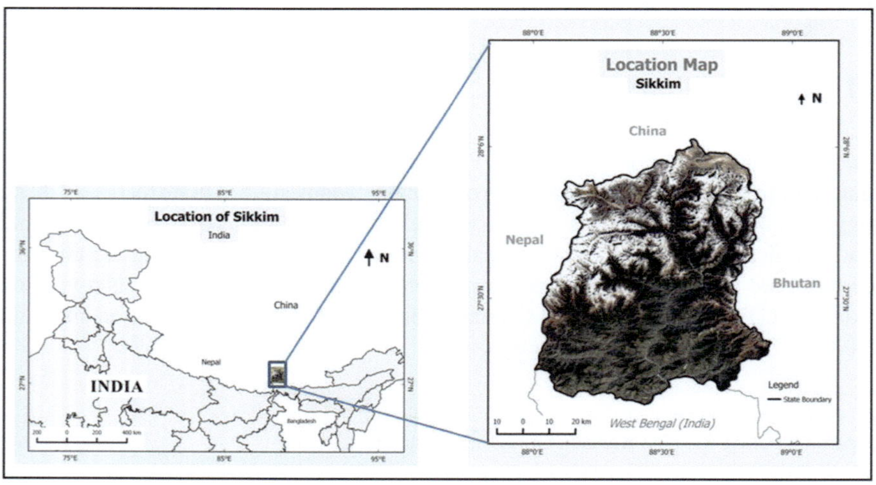

Fig. 8.1 Location map of study area. The background is Google earth imagery. *Source* Image ©2018 DigitalGlobe

with an elevation ranging from 280 to 8586 m. It is extended about 114 km from north to south and 64 km from east to west.

The average annual temperature of Sikkim is around 18 °C, minimum temperature goes down to less than 0 °C and the maximum temperature rarely exceed to 28 °C. The highest peak of the state is Kangchenjunga, which is the world's third highest peak situated on the border of Sikkim and Nepal. Mainly three major glaciers are there namely Zemu glacier, Lohnak glacier and Rathong glacier. Total glacier area in Sikkim cover an area of about 484 km^2, which is about 7% of the total geographical area of Sikkim.

8.3 Methodology

8.3.1 Data Source

Satellite imagery that has been used in study is downloaded from USGS-Earth Explorer that is of Landsat-TM5 scene from 23 September 1992 and Landsat-8OLI/TIRS scene from 7 September 2015 with the spatial resolution of 30 m, all these data sets used in the study are orthorectified; we selected the data of September month because there were no seasonal snow covers in this month that looked appropriate for the study of Glacier's area change. ASTER Global DEM Version2 Imagery of 17 October 2011 with spatial resolution of 1 arc second, for the full coverage of study area, two ASTER Global DEM Version2 Imagery were needed to download and acquisition date of both imagery were same. Other than that Google Earth Imageries of study area were also taken and Geometric rectification of these imageries were done in QGIS 2.12.3 with appropriate polynomial model and projection system for the study of both glacier cover as well as glacial lakes. All these remote sensing data sets that are used in this study are summarized in Table 8.1.

Table 8.1 Data used in this study

Sensor	Scene ID	Date	Spatial resolution
Landsat TM-5	LT51390411992267BKT00	23rd Sept. 1992	30 m
Landsat-8 OLI/TIRS	LC81390412015250LGN00	07th Sept. 2015	30 m
ASTER	ASTGTM2_N27E088 (DEM)	17th Oct. 2011	1 arc sec
	ASTGTM2_N28E088 (DEM)	17th Oct. 2011	1 arc sec

8.3.2 Glaciers Delineation

For the delineation of glaciers of Sikkim for the year 1992 based on landsat-TM5 Imagery, clean glacier outlines were extracted; the outlines were extracted by applying some techniques that has been already suggested in other studies that are as follows: first area of interest, that is, Sikkim clipped for further use then band rationing technique was applied TM3/TM5, that is, Red band/SWIR, these bands were used because maximum reflectance of clean ice in Red band and minimum in SWIR band (Dozier 1989), offers the possibility to discriminate clean ice from its surrounding terrain with a threshold in a reproducible and consistent manner (Albert 2002; Paul 2002). Other than this some other band calculation are also there such as TM4/TM5 and NDSI (Normalized difference Snow Index) but TM3/TM5 looked appropriate as compared to TM4/TM5 and NDSI. In resulting ratio image appropriate threshold were applied based on pixel value information and observation of clean ice and surrounded pixels.

After applying, threshold resulted image were showing some isolated pixel of clean ice, to remove these isolated pixel a 3 $\times$ 3 median filter were applied. Then we converted raster resulted image to vector layer. For the delineation of debris-covered glaciers, manual post-processing has been done by visual interpretation, this post-processing has been done using TM543 band composite as RGB, and this combination clearly discriminated the glacier area from cloud cover (Fig. 8.2) as compared to Natural colour composite, that is, TM321 as RGB (Fig. 8.3). In this manual correction process, debris-covered glaciers were digitized and edited manually. Finally, total glacier areas were extracted as vector layer that is used for the calculation of total area covering glaciers and debris-covered glaciers.

For the delineation of glaciers of Sikkim for the year 2015, based on landsat-8OLI/TIRS Imagery, clean glaciers were extracted by the same process as explained for the year 1992 but in this Landsat-8OLI/TIRS satellite imagery band 1 is for coastal aerosols, band 2 for blue, band-3 for green, band-4 for red and band-5 for NIR (near infrared), and band-6 for SWIR (Shortwave infrared), because of this characteristics of bands, here the formula for band rationing has changed to Band4/Band6 as (Red/SWIR) that were TM3/TM5 for Landsat-TM5. For the false colour composite of Band654 as RGB were used as it has showed the same composition that were there in Landsat-TM5 in the form of TM543, and for natural colour composite Band432 combination were used. In this scene there were cloud cover near glacier area, so for accurate extraction of glacier area Google Earth Imagery were used that were cloud-free as compared to Landsat-8 imagery then finally area of clean ice and debris-covered glacier were calculated.

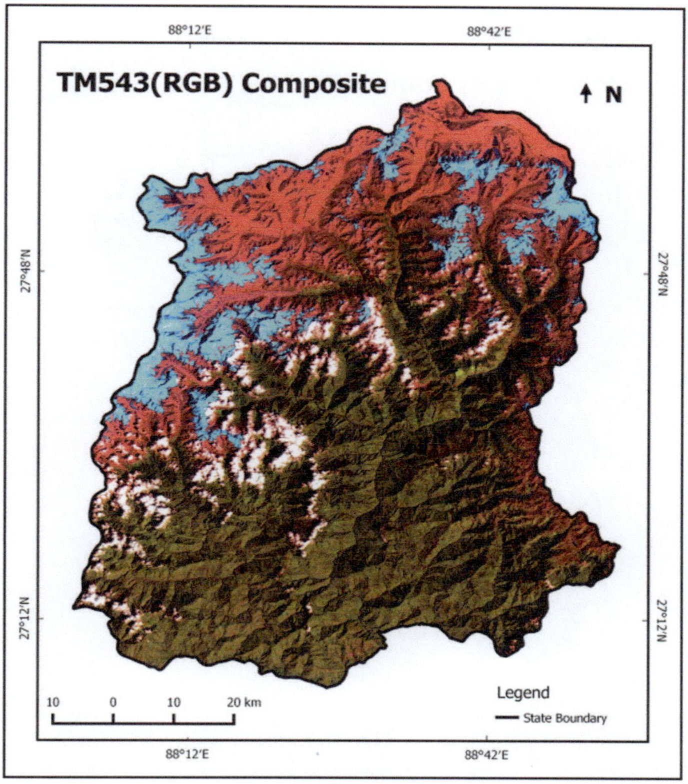

Fig. 8.2 TM543 composite image. *Source* Landsat TM

8.3.3 Identification of Lakes

For both the year 1992 as well as 2015, glacial lakes were automatically mapped using normalised difference water index (NDWI) that is NDWI = NIR − Blue Band/NIR + Blue Band, in the resulted image, water bodies appeared black with negative value, based on the observation of pixels value on image of water bodies and surrounded surface pixel, an appropriate threshold was decided for the separation of lakes from other surface features, after applying threshold on the resulted ratio image, it showed only water bodies on the image then conversion from raster to vector was executed and the water bodies were converted as vector layer; when we examined the vector layer with the base map it was observed that some of the shaded area were also classified as water bodies; to remove such discrepancy from the image again a manual correction process were done using visual interpretation method by putting Landsat image and Image from Google Earth as a base map so that it could map correctly with manual digitizing. Finally, a new vector layer of lakes of Sikkim was

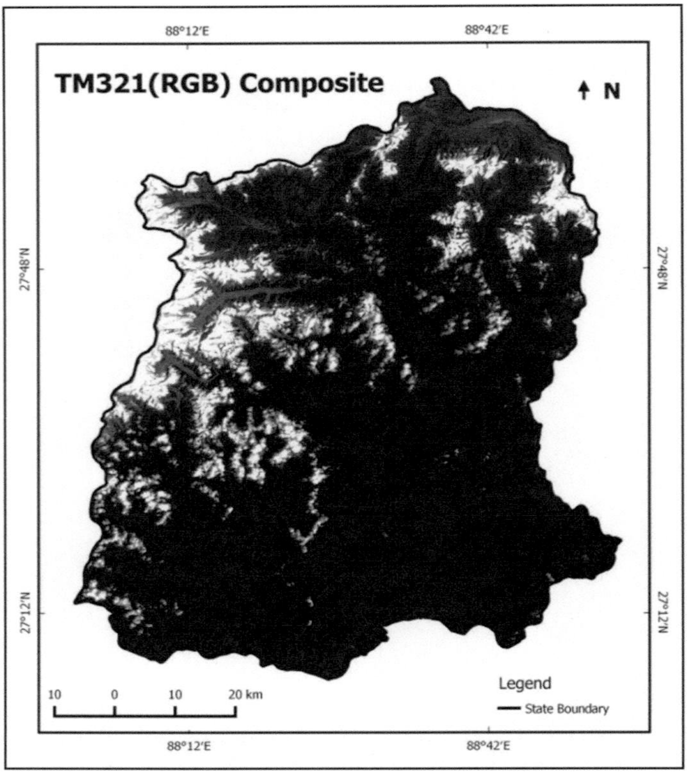

Fig. 8.3 TM321 Composite image. *Source* Landsat TM

identified. These processes were done for both years whether it was Landsat-TM-5, 1992 Image or Landsat-8OLI/TIRS from 2015.

8.3.4 Topographical Parameters

To extract the topographical parameters of the glacier-covered area there was a need to extract it with the DEM data. For the analysis of such parameters the advanced spaceborne thermal emission and reflection radiometer (ASTER) Global Digital Elevation Model (GDEM) Version2 data were used. It was also used for showing elevation range over the region (Fig. 8.4). For the creation of slope map, digital elevation model was used as input and same process was used for the creation of aspect map. The slope and aspect map were derived to analyze how it affects the changes in glacier's area.

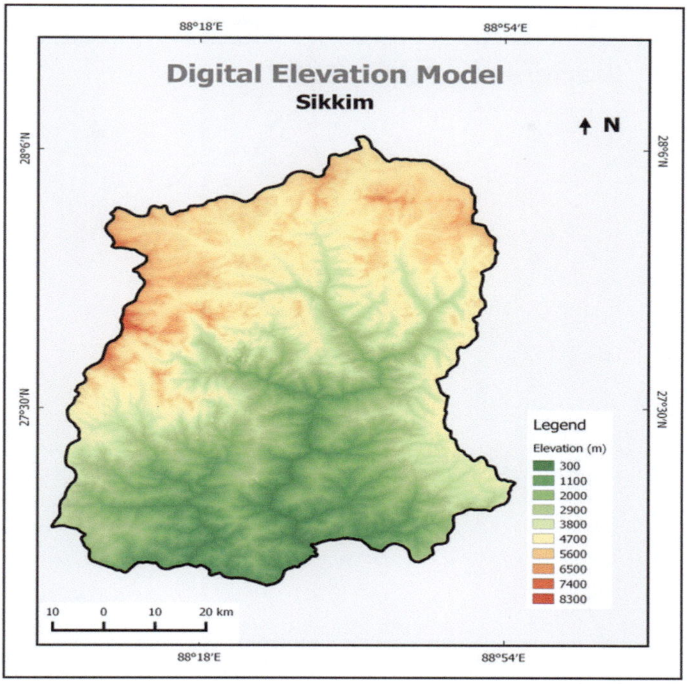

Fig. 8.4 DEM (Digital Elevation Model) of Sikkim. *Source* ASTER DEM

After the generation of these maps, the results of elevation, aspects and slope were compiled with the vector layers of glaciers and lakes using zonal statistics; this work is executed to analyze how such parameters affect the changes in glacier's area.

8.4 Results

According to the current world scenario, glacier's area is changing everywhere because of many reasons, global warming is a major one of them. Temporal data were used to detect changes in glacier's area that is shown in Fig. 8.5 for the year 1992, and for the year 2015 (Fig. 8.6). According to our study, Sikkim glacier area changes 22.18% from 1992 to 2015 (Table 8.2). In 1992 glacier area was 621.124 km^2, in overall glacier area clean ice is 537.873 km^2 and debris-covered glacier is 83.252 km^2. In 2015 glacier area is 483.374 km^2 in overall glacier area clean ice is 414.214 km^2 and debris-covered area is 69.125 km^2. The overall change in glacier area from 1992 to 2015 is 137.75 km^2.

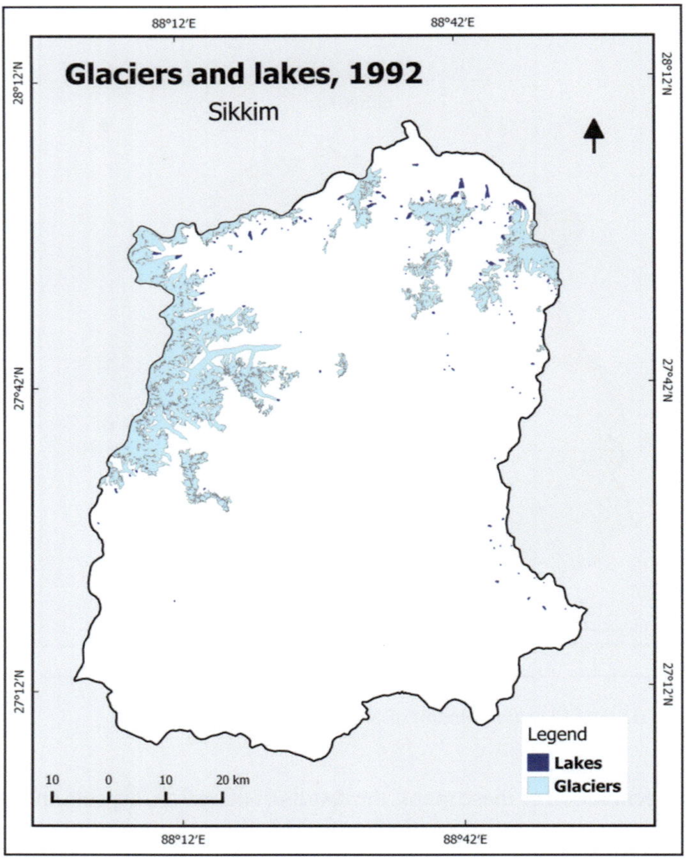

Fig. 8.5 Glacier and lakes in Sikkim (1992). *Source* Landsat TM

Glacier cover delineated by different analysts is showing a trend of decreasing glacier's area on different years (Fig. 8.9). Topographic characteristics in the glaciated area showing that mean slope on glaciated area is about 27° and mean aspect is about 158° with the elevation ranges from 3516 to 8237 m (Table 8.3).

In the region of changes in glacier's cover or loss of glaciers, majority of slope in that area is about 30° and aspect is southward with the elevation ranging from 3516 to 4500 m. These parameters are analyzed based on slope and aspect map showing in Fig. 8.7 and Fig. 8.8, respectively, (Fig. 8.9).

In our study, glacial lakes also changed from 1992 to 2015 and shows the increase in number of lakes. In 1992, numbers of glacial lakes were 169 and in 2015 number of glacial lakes has increased to 261 (Table 8.4). The changes of glacier's area and formation of lakes can be seen from images of Lohnak glacier (Figs. 8.10 and 8.11) in northwestern part of Sikkim.

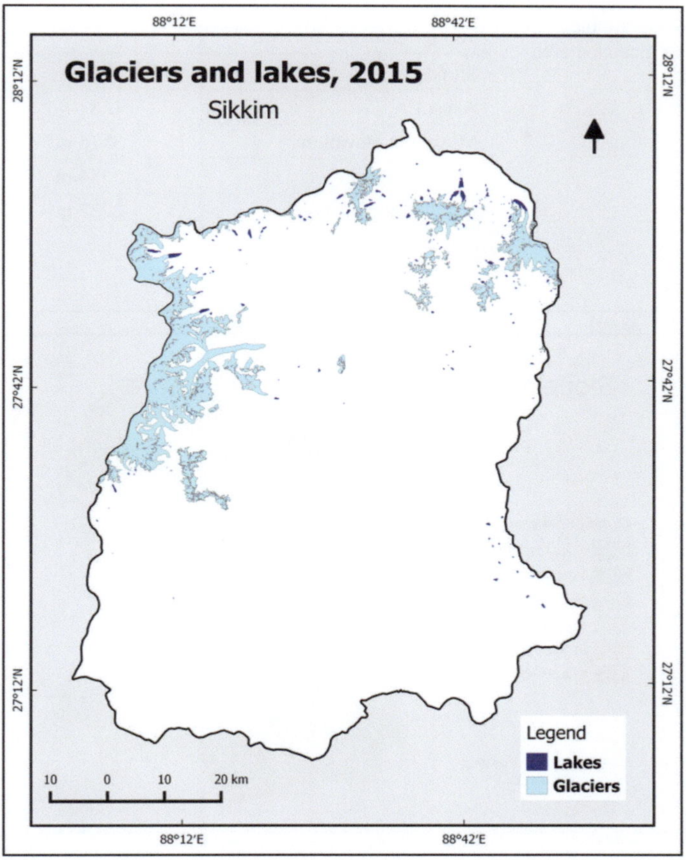

Fig. 8.6 Glacier and lakes in Sikkim (2015). *Source* Landsat-8

Table 8.2 Glacier's area and its changes from 1992 to 2015

Glaciers type	Area (km^2)		Area loss from 1992 to 2015	
	1992	2015	Change (km^2)	Change (%)
Glacier (clean ice)	537.872	414.249	−123.623	22.98
Debris-covered glacier	83.252	69.125	−14.127	16.97
Total glacier	621.124	483.374	−137.75	22.18

Source Landsat TM and Landsat 8

According to Indian Meteorological department, minimum temperature in Sikkim is rising at the rate of about 0.02 °C every year from last three decades. Because of the temperature increase at this rate, it could be the possible reason for melting of glacier that also has resulted in the increase of the area of glacial lakes.

Table 8.3 Topographic parameters of glaciated area

Topographic parameters	
Slope	27°
Aspect	158°
Minimum elevation	3516 m
Mean elevation	5384 m
Maximum elevation	8237 m

Source ASTER DEM

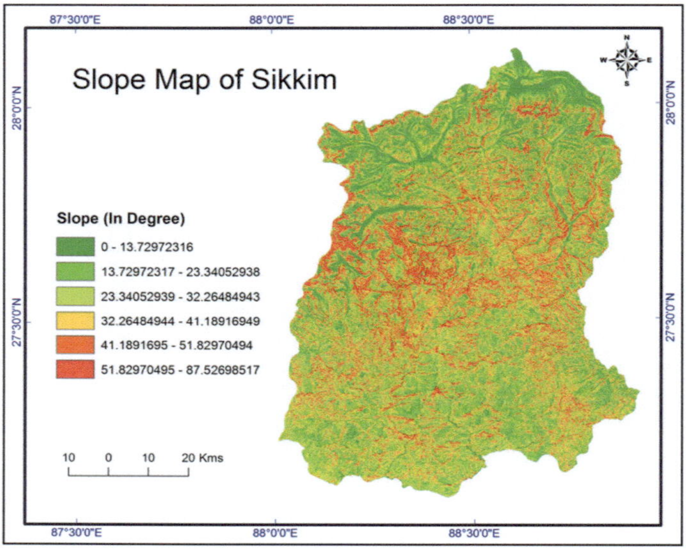

Fig. 8.7 Slope map of Sikkim. *Source* ASTER DEM

8.5 Conclusion

Remote sensing techniques have considerably improved the estimation of glacier's area change and also the identification of lakes in Sikkim. In this study semi-automated and manual methods have been used to extract the glaciers cover for the analysis of change. Glaciers in the Sikkim has been changed and decreased in its area by approximately 22%, and the number of glacial lakes has been increased up to 261 in 2015. Melting or shrinkage of glacier leads to unfolding of land and it also affects biodiversity. Minimum temperature increasing from last three decades may lead to disaster in the region by damaging property and life over a long distance on its flow area as a result of glacial lakes outburst floods.

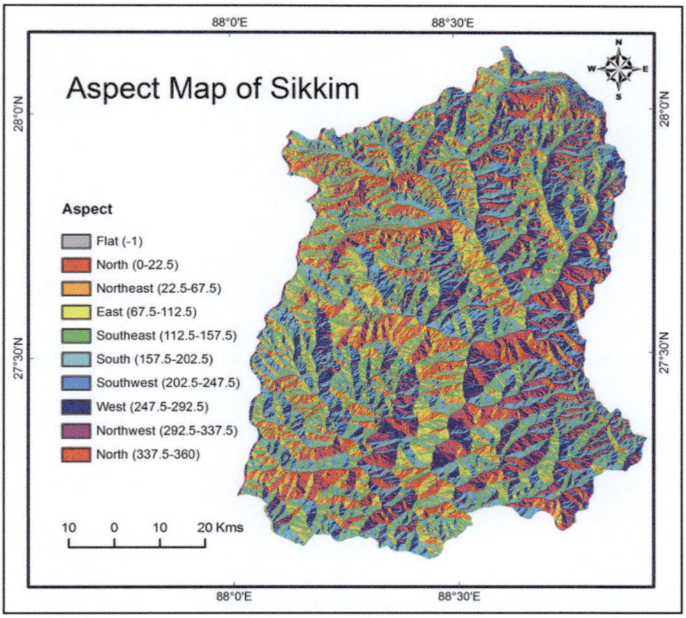

Fig. 8.8 Aspect map of Sikkim. *Source* ASTER DEM

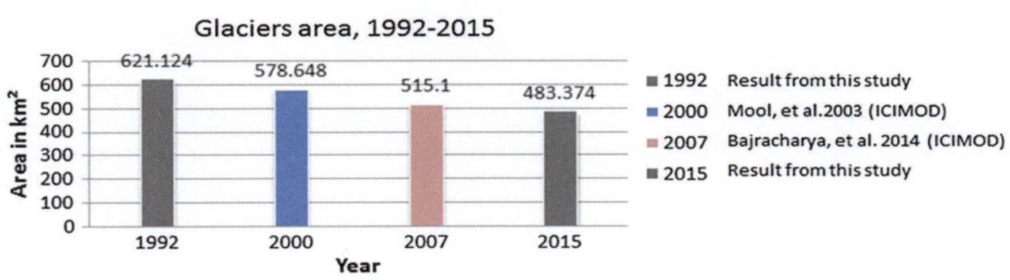

Fig. 8.9 Glacier's area calculated by different analysts. *Source* Landsat TM and Landsat 8

Table 8.4 Number of lakes in Sikkim

Year	Number of lakes
1992	169
2015	261

Source Landsat TM and Landsat 8

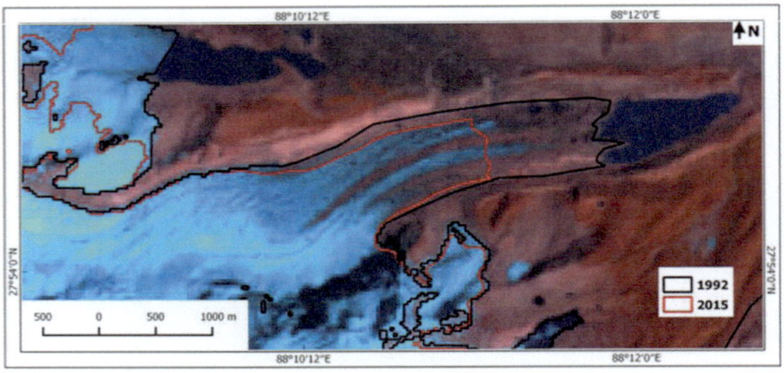

Fig. 8.10 Glacier outlines of Lohnak glacier for the year 1992 in black and 2015 in red with base image of Landsat-TM5 of 1992. *Source* Landsat TM

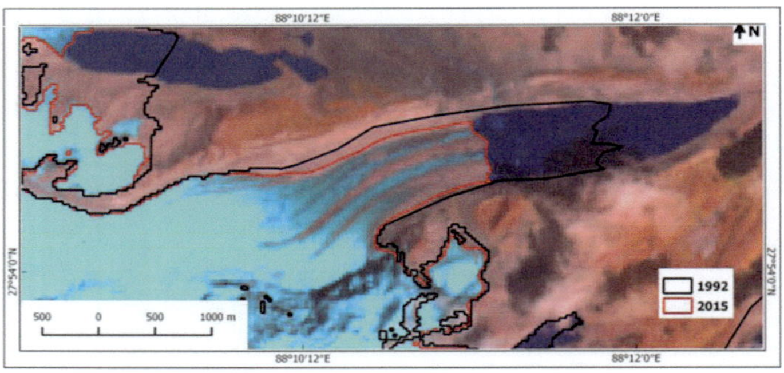

Fig. 8.11 Glacier outlines of Lohnak glacier for the year 1992 in black and 2015 in red with base image of Landsat-8 OLI/TIRS of 2015. *Source* Landsat 8

References

Albert TH (2002) Evaluation of Remote Sensing Techniques for Ice-Area Classification Applied to the Tropical Quelccaya Ice Cap, Peru. Polar Geogr 26(3):210–226

Andreassen, LM, Paul F, Kääb A, Hausberg JE (2008) Landsat-derived glacier inventory for Jotunheimen, Norway, and deduced glacier changes since the 1930s. The Cryosphere 2(2):131–145

Dozier J (1989) Spectral signature of alpine snow cover from the landsat thematic mapper. Remote Sens Environ 28:9–22

Paul F (2010) Changes in glacier area in Tyrol, Austria, between 1969 and 1992 derived from Landsat 5 Thematic Mapper and Austrian Glacier Inventory data. Int J Remote Sens 23(4):787–799

Raup B, Racoviteanu A, Khalsa SJS, Helm C, Armstrong R, Arnaud Y, (2007) The GLIMS geospatial glacier database: A new tool for studying glacier change. Glob Planet Chang 56(1–2):101–110

Chapter 9
Quantitative Morphometric Analysis of the Yerla River Basin, Deccan Trap Region, India

Namdev V. Telore

Abstract The evaluation of basin from the drainage morphometric parameters helps in understanding the physical behaviour of the catchments for various purposes. Remote Sensing and Geographic Information System (GIS) techniques were used to study quantitative morphometric analysis of the Yerla River basin of basaltic Deccan Trap region, India. Morphometric analysis is carried out using Survey of India topographical maps in ArcGIS software. Thematic maps such as drainage map, stream ordering map, contour map, elevation map, slope map, aspect map and digital elevation model (DEM) are prepared. SRTM data of 90 m spatial resolution is used to create DEM and ASTER data of 30 m spatial resolution is used to create longitudinal profile of the catchment area. Various linear, relief and areal morphometric parameters such as area, perimeter, stream order, stream length, stream number, bifurcation ratio, drainage density, stream frequency, drainage texture, length of basin, form factor, circulatory ratio and elongation ratio are computed. The dendritic type drainage network of the basin exhibits the homogeneity in texture and lack of structural control. The seventh order basin has low drainage density and poor stream frequency indicates coarse drainage. The mean bifurcation ratio indicates a hilly dissected basin. The form factor ratio and circularity ratio shows that the whole basin has an elongated shape. The compactness coefficient value shows less hazardous basin. Sinuosity indices value indicate transitional stage. Morphometric analysis helps to understand the geo-hydrological characteristics of the watershed. It is inevitable in development and management of drainage basin.

Keywords Drainage morphometry · Remote sensing · GIS · Yerla River Basin · Deccan traps · India

---

N. V. Telore (✉)
Department of Geography, Raja Shripatrao Bhagawantrao Mahavidyalaya, Aundh, Satara 415510, India
e-mail: nvtelore@gmail.com

S. Sahdev et al. (eds.), *Geoecology of Landscape Dynamics*,
Advances in Geographical and Environmental Sciences,
https://doi.org/10.1007/978-981-15-2097-6_9

9.1 Introduction

Morphometric analysis is widely used to assess the drainage characteristics of the river basins for watershed management plans. Morphometry is the measurement and mathematical analysis of the configuration of the Earth's surface, shape and dimensions of its landforms (Clarke 1966). In geomorphology, the development of a landscape can be known by measuring morphometry of drainage and relief properties. Horton (1945) is the pioneer worker who introduced the natural composition of drainages which was further modified and developed by Strahler (1950), Schumm (1956), Corley (1957), Scheidegger (1965), Woldenberg (1966), Morisawa (1968), Shreve (1969), Ebisemiju (1979), Chorley et al. (1984) and others. Influence of drainage morphometry is very significant in understanding the landform processes, soil physical properties and erosional characteristics.

Morphometric analysis of various basins has been studied by many scientists using conventional (Horton 1945; Smith 1950; Schumm 1956; Strahler 1956) and remote sensing techniques (Oguchi 1997; Nag 1998; Vittala et al. 2004; Chopra et al. (2005); Sreedevi et al. 2009; Saptarshi and Raghavendra 2009; Magesh et al. 2011; Agarwal et al. 2012; Sarmah et al. 2012; Chandrashekar et al. 2015; Pande and Moharir 2017; Sindhu et al. 2015; Rai et al. 2017; Malik and Shukla 2018; Suresh et al. 2018) stated that remote sensing and GIS techniques are essential for morphometric analysis of basins for watershed management. A comprehensive watershed management programme may have multiple objectives such as controlling runoff and utilizing the same for useful purposes, enhancing groundwater storage and appropriate use of the land resources in the watershed (Sebastian et al. 1995).

The knowledge of local geomorphology is used for avoiding environmental degradation, planning and management of watersheds. Bhatt and Ahmed (2014) used Cartosat DEM to evaluate the morphometric analysis to study flood-prone areas of upper Krishna basin. Morphometric knowledge is useful in watershed management to reduce accelerated erosion and sedimentation in the catchment areas (Chadha and Neupane 2011; Thomas et al. 2011). The basin characteristics are used in the successful implementation of watershed management programmes (Rao et al. 2012; Unde and Telore 2013; Singh et al. 2014). In the arid and semi-arid region, watershed management is extremely important to eradicate drinking and domestic water scarcity (Tideman, 2007). Panhalkar and Jarag (2015) analyzed that IDW is an appropriate technique for river bathymetry generation in comparison to Kriging and Topo to Raster techniques. Raju and Kumar (2001) classified micro watersheds based on morphological characteristics and carried out comparative analysis of various clustering techniques, which revealed that 13 micro watersheds out of 25 are commonly suggested by KCA, FCA and KNN. Doke et al. (2018) studied characteristics of morphogenetic regions of Maharashtra with reference to the climatic conditions prevailed over a long time and processes operated in the region.

Prioritization of sub-watersheds based on morphometric analysis of drainage basins using spatial information techniques was attempted by Nautiyal (1994), Biswas et al. (1999), Javed et al. (2009), Mishra and Nagarajan (2010), Sethupathi et al. (2011), Kanth and Hassan (2012), Panhalkar et al. (2012), Cunha and Bacani (2016) and Telore (2016). The use of free and open-source tools guarantees access to everyone and its increasing popularization opens new development perspectives in morphometric analysis and allied fields (Grohmann 2004; Manimozhi and Rengarajan 2007; Ahmed et al. 2010). Thus, detailed study of morphometric analysis of a catchment is of great help in understanding the influence of drainage morphometry of landforms and their characteristics.

9.2 Objective

To evaluate linear, areal and relief morphometric characteristics of the Yerla River Basin, Deccan Traps, India.

9.3 Methodology

Topographical maps prepared by Survey of India on 1:50,000 scale are used to generate the base map of the Yerla River basin. Morphometric analysis of a drainage system requires delineation of all existing streams. The study area is delineated using ArcGIS 10.5 software. All tributaries of different extents and pattern were digitized from topographical sheets of 1:50,000 scale. Digitization work is carried out for an entire analysis of drainage morphometry. Various thematic maps such as drainage map, stream ordering map, contour map, elevation map, slope map, aspect map and DEM are prepared. Shuttle Radar Topographic Mission (SRTM) DEM data of 90 m spatial resolution is used to create Digital Elevation Model (DEM) of the Yerla watershed in ArcGIS 10.5, ERDAS IMAGINE 2015 software. Longitudinal profile of the Yerla watershed is created using ASTER DEM data of 30 m spatial resolution in Global Mapper v15.1 software. Ground realities are checked with the help of handheld GPS during field visits. These parameters are measured from the toposheets. The various linear, relief and areal morphometric parameters such as area, perimeter, stream order, stream length, stream number, bifurcation ratio, drainage density, stream frequency, drainage texture, length of basin, form factor, circulatory ratio and elongation ratio are computed based on the formula suggested by (Horton 1945; Miller 1953; Schumm 1956; Strahler 1964) given in Table 9.1 and the results are summarized in Tables 9.2 and 9.3.

9.4 Study Area

The Yerla River is the largest left-hand seventh order stream of the Krishna River Basin in the Deccan Trap region (Fig. 9.1). The basin covers an area of 3067 km^2

Table 9.1 Formulae for computation of linear, relief and areal morphometric parameters

Morphometric parameters	Formula	References
Stream order (U)	Hierarchical order	Strahler (1964)
Stream length (Lu)	Length of the stream	Horton (1945)
Mean stream length (Lsm)	Lsm = Lu/Nu where Lu = Stream length of order 'U', Nu = Total number of stream segments of order 'U'	Horton (1945)
Stream length ratio (RL)	RL = Lu/Lu – 1 where Lu = Total stream length of order 'U' Lu – 1 = The total stream length of its next lower order	Horton (1945)
Bifurcation ratio (Rb)	Rb = Nu/Nu + 1 where Nu = Total number of stream segment of order 'U'; Nu + 1 = Number of segment of next higher order	Schumm (1956)
Drainage density (Dd)	Dd = L/A where L = Total length of streams of all orders A = Area of the basin (km^2)	Horton (1945)
Stream frequency (Fs)	Fs = N/A where N = Total number of stream A = Areas of the basin (km^2)	Horton (1945)
Drainage texture (Rt)	Rt = Nu/P where Nu = Total number of streams of all orders P = Perimeter of the basin (km)	Horton (1945)
Circulatory ratio (Rc)	$Rc = 4\pi A/Lp^2$ Where A = Area of the basin Lp = Perimeter of the basin	Miller (1953)
Elongation ratio (Re)	$Re = (2 \times (A/\pi)^{0.5})/Lb$ where A = Area of watershed, $\pi = 3.14$, Lb = Basin length	Schumm (1956)
Compactness coefficients (Cc)	$Cc = 0.2821\ P/A^{0.5}$ A = areas of basin (km^2), P = basin perimeter (km)	Horton (1945)
Basin relief (Bh)	Vertical distance between the lowest and highest points of watershed	Schumm (1956)
Relief ratio (Rh)	Rh = H/Lb where H = Total relief (relative relief) of the basin in km; Lb = Basin length	Schumm (1956)
Length of overland flow (Lg)	Lg = 1/Dd*2 where Dd = Drainage density	Horton (1945)
Chanel sinuosity	Channel sinuosity = O_L/E_L where O_L = observed path of a stream; E_L = expected straight path of a stream	Schumm (1963)

Table 9.2 Results of linear aspect of study area

Stream order	Number of streams	Stream length (km)	Bifurcation ratio	Mean stream length (km)	Stream length ratio
1	5987	4055.2262	4.2642	0.6773	0.3568
2	1404	1447.0295	4.6490	1.0306	0.5289
3	302	765.4509	4.6461	2.5346	0.5121
4	65	392.3704	3.8235	6.0364	0.3268
5	17	128.2608	3.4000	7.5447	0.7693
6	5	98.6800	5.0000	19.7360	0.6899
7	1	68.0795	–	68.0795	–
Total	7781	6955.0974			

Table 9.3 Results of areal aspect of study area

Basin area (km/km^2)	Basin length (km)	Basin perimeter (km)	Form factor	Circularity ratio	Elongation ratio	Drainage density	Stream frequency (streams/km^2)
3067.89	108.55	340.38	0.26	0.33	0.57	2.15	2.53

of Mahadev Hill range of the Western Ghat and represents semi-arid condition. The study area lies between latitudes of 16° 55′48″ and 17° 52′44″ N and longitudes of 74° 14′02″ and 74° 45′00″ E and receives 543.07 mm average annual orographic rainfall. The Yerla River flows to the south about 139.58 km through Khatav Taluka of Satara District and Kadepur, Khanapur, Tasgaon and part of Palus and Sangli Taluka's of Sangli District of Maharashtra State. The elevation of the watershed ranges from 540 to 1058 m that represents 516 m of high relative relief (Fig. 9.2). The drainage network is sub dendritic pattern. The study area is covered by compact, fine grained, massive basaltic lavas of Upper Cretaceous to Lower Eocene age and represents step like topography. The soils of the basin are residual, derived from the underlying basalts. Natural vegetation represents the southern tropical dry deciduous type.

9.5 Morphometric Analysis

9.5.1 Linear Aspects

Linear aspects that include stream order, stream length, mean stream length, stream length ratio and bifurcation ratio were determined and results are given in Table 9.2.

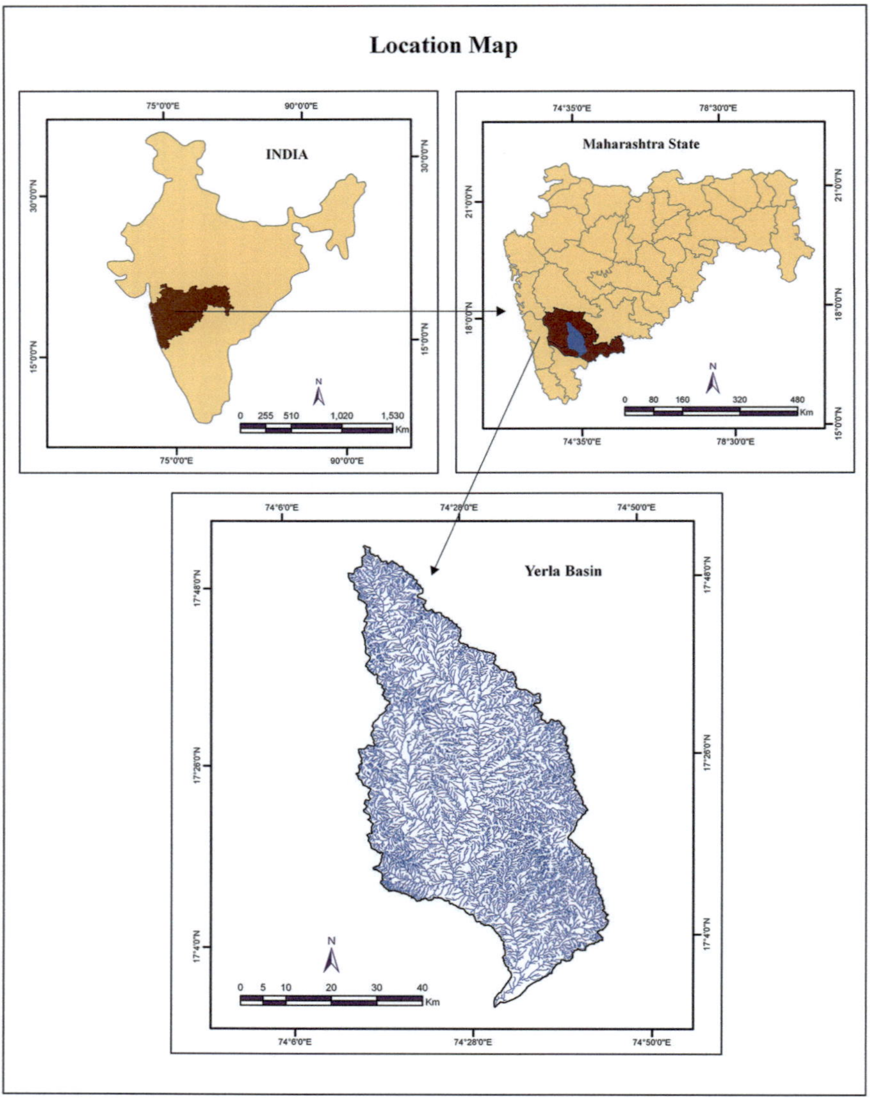

Fig. 9.1 Location map

9.5.1.1 Stream Order (Nu)

Stream ordering is the first step of quantitative analysis of the basin. Horton (1945) introduced a simple stream ordering system to describe the position of a stream in the drainage network in terms of the number of tributaries received. The Horton's stream ordering method was further modified by Strahler (1952). Under the Strahler scheme

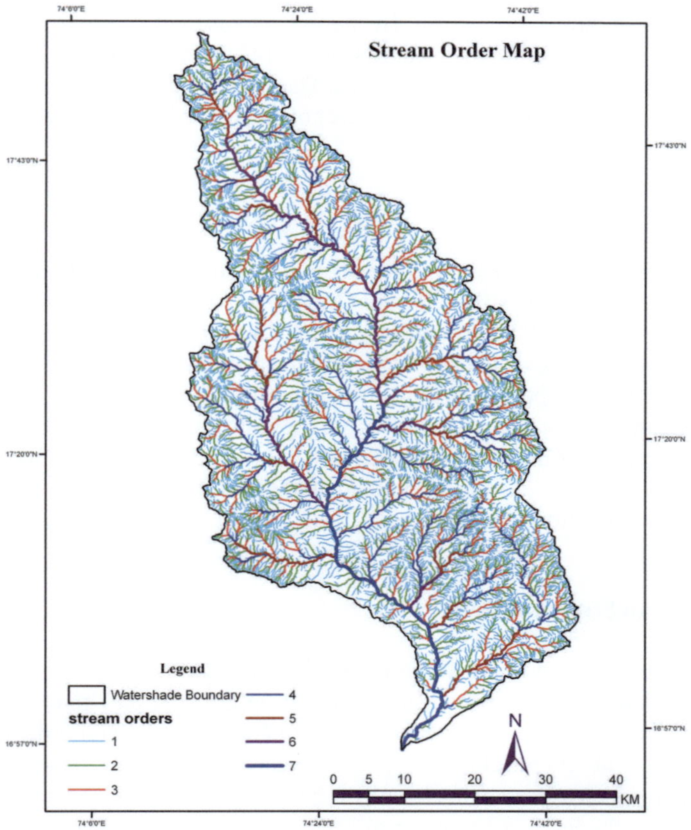

Fig. 9.2 Stream order map

all the fingertip tributaries are designated as first-order stream, two second-order stream produces a third-order stream and so on. To determine the stream ordering system of the Yerla watershed Strahler's (1952) stream ordering system has been used. The stream ordering map (Fig. 9.2) shows that the Yerla River basin is a seventh-order stream. The number of streams (N) of each order (U) of the Yerla basin is shown in Table 9.2. It has been observed that the maximum frequency indicates in the first-order streams. It has also noticed that there is a decrease in stream frequency as the stream order increases. The variation in order and size of the micro watersheds is largely due to physiographic and structural conditions of the region.

9.5.1.2 Stream Number

The total order wise stream segments are known as stream number. Horton (1945) stated that the number of stream segments of each order form an inverse geometric sequence with stream order number. The basin contains total 7,781 streams in which first-order streams are 5987 followed by second (1404), third (302), fourth (65), fifth (17), sixth (5) and seventh (1) order streams. The strong negative correlation (−0.73) is observed between stream order and number of streams in the study area (Table 9.2).

9.5.1.3 Stream Length (L)

Stream Length is the total length of streams of a particular order. The total stream length of the Yerla River is 6955.09 km (Table 9.2). It reveals that the total stream length 58.31% consist of the first-order streams. The percentage of the second, third, fourth, fifth, sixth and seventh order streams stand at 20.80%, 11.01%, 5.64% 1.84%, 1.42% and 0.98%, respectively. The strong negative correlation (i.e. $r = -0.82$) between stream order and total stream length suggests inverse geometric relationship between these two parameters. Longer lengths of streams are generally indicative by flatter gradient and coarse texture. Smaller lengths of streams are generally indicative of larger slopes and finer texture.

9.5.1.4 Mean Stream Length

The mean stream length of the different order streams of the study area is presented in Table 9.2. The orderwise mean stream length varies from 0.67 to 68.07 km. The strong positive correlation coefficient ($r = +0.77$) between stream order and mean stream length is observed. This strong positive correlation follows Horton's Law of stream length which expresses the length of stream of a given order in terms of stream order and takes the form of a direct geometric series (Horton 1932).

9.5.1.5 Stream Length Ratio (RL)

The stream length ratio of the Yerla watershed is obtained after dividing the mean stream length of the next higher order by the lower order stream. The stream length ratio in the study area ranges between 0.32 and 0.76. It is observed that stream length ratio increases as the order of the stream increases. The stream length ratio of different orders of the study area is presented in Table 9.2.

9.5.1.6 Bifurcation Ratio (Rb)

Horton (1945) defined the bifurcation ratio as the ratio between the number of streams of any given order to the number in the next lower order. Table 9.2 contains the bifurcation ratio of the different order streams in the watershed which shows that it has a constant nature and ranges between 3.40 and 5. Higher bifurcation ratio indicates some sort of geological control and lower bifurcation ratio suggests that structure does not exercise a dominant influence on the drainage pattern. The mean is presented in Table 9.2. Bifurcation ratio of the area is 4.29 which indicates it is a hilly dissected basin.

9.5.1.7 Sinuosity Indices

Sinuosity of a stream denotes the degree of deviation of its actual path from expected theoretical straight path. The shape of the open link in terms of geometric structure of drainage line involves the calculation of deviation of observed path (O_L) from the expected path—almost a straight line (E_L) of a river from the source to the mouth (Schumm 1963). Sinuosity indices of the study area are 1.30 which indicates transitional stage.

9.5.2 Relief Aspects

Strahler (1969) stated that relief measures are indicative of the potential energy of the drainage system because of its elevation above the mean sea level. Absolute relief of the Yerla watershed is 1058 m. Relative relief of the area is 516 m that indicates high relative relief, which is well illustrated in DEM in Fig. 9.3. The watershed is characterized by an undulating topography with an average slope of about 0.47%. Slope map (Fig. 9.4) illustrated that all the margins of plateau have greater slope between 29.04 and 54.86° and plateau surface remaining from 0 to 7.53°. Slope is decreasing from source to outlet region of the study area. Steep slope is observed in the source region of mainstream and tributaries—Nani River and Kapur Nala. The direction of the slope on the entire relief of the study area is illustrated in aspect map (Fig. 9.5). The Yerla River flows north to south direction. The orientation of streams is mostly east to west and west to east observed on the valley sections. In the upper part, eastern streams flow towards west and southwest direction and western streams flows towards northeast and east direction. Same pattern is found in the Nani river which is a biggest western stream of the River Yerla. In the middle portion, eastern streams are approaching towards north and northwest direction. In the lower eastern part in Kapur, Nala streams are approaching towards south and southwest direction and in the western part streams are approaching towards east direction (Fig. 9.5 and Table 9.3).

9.5.2.1 Relief Ratio (Rh)

The relief ratio of mainstream relief to horizontal distance along the largest dimension of the basin parallel to the principal drainage line is termed as relief ratio (Schumm 1956) (Table 9.1 and Fig. 9.3). He observed direct relationship between the relief and channel gradient. There is also a correlation between hydrological characteristics and

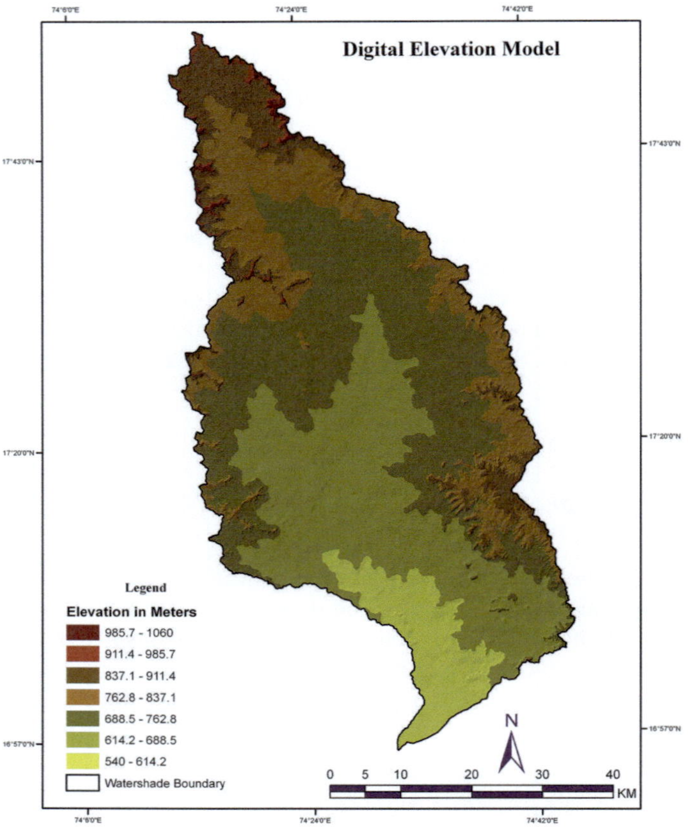

Fig. 9.3 Digital elevation model

the relief ratio of a drainage basin. The relief ratio normally increases with decreasing drainage area and size of basin of a given drainage basin. The value of the relief ratio of the study area is 520 m.

9.5.2.2 Profile Analysis

River profile is of great assistance to analyze relief and surface of terrain. The profile shows altitude against distance downstream. Longitudinal profile or valley thalweg gives a vivid picture of breaks in longitudinal course of the river. Longitudinal profile of the area is created using ASTER DEM data of 30 m spatial resolution in Global Mapper software and is shown in Fig. 9.6. Relative relief of the basin is 542 m and total length of river is 139.58 km. Longitudinal profile of the river is ideal. It is having concave shape in the upper portion near the source at Mol, then progressively

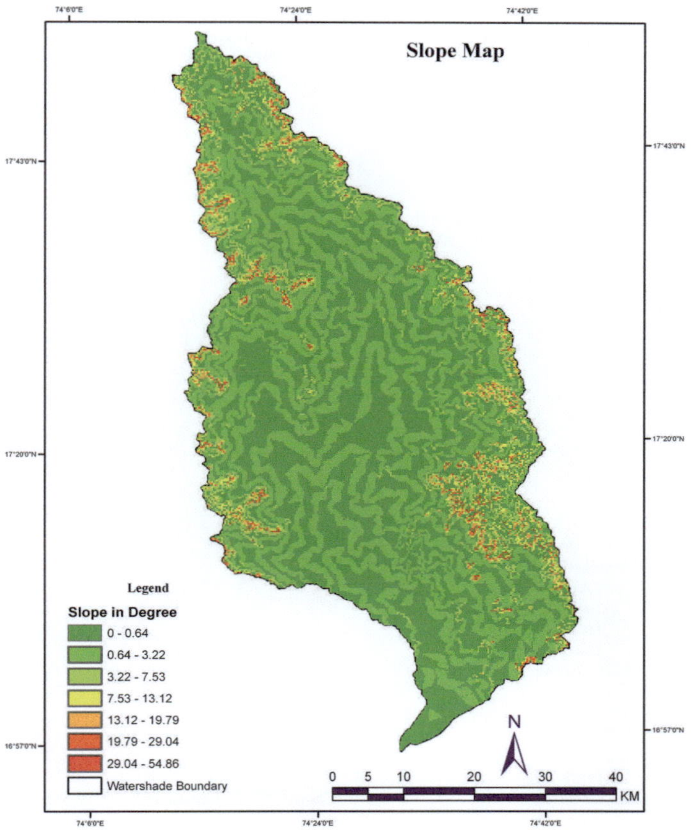

Fig. 9.4 Slope map

less gradient as the mouth is approached at Bramhanal i.e. Yerla—Krishna River confluence.

9.5.3 *Areal Aspects*

9.5.3.1 Drainage Area (A)

The drainage area of a watershed is the surface area located within the watershed basin boundary. The size of a drainage area has a significant impact on watershed management structures. Drainage area of the seventh order Yerla watershed is 3067.89 km^2 (Table 9.3). It is observed that mean areal extent of the drainage basin increases as the order of stream increases.

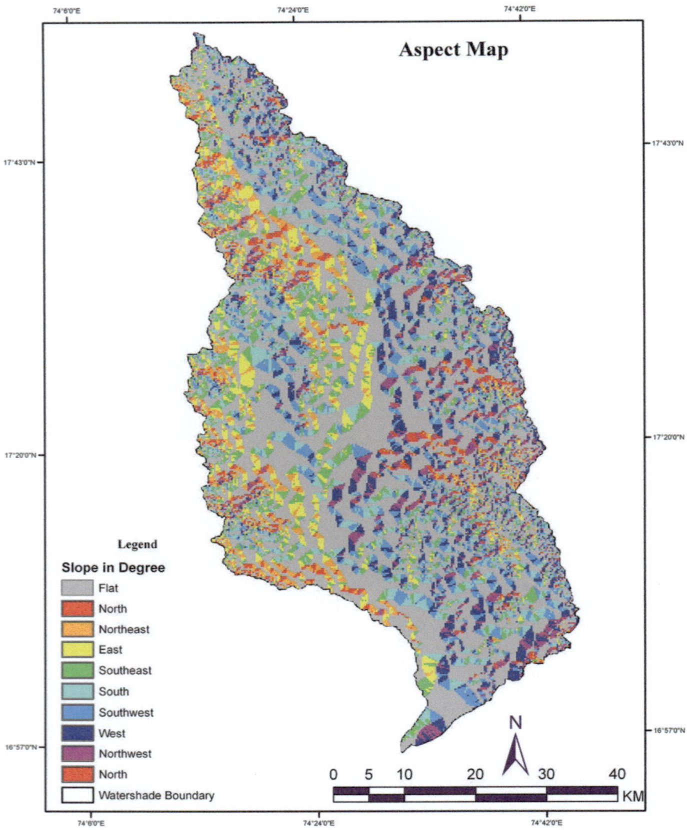

Fig. 9.5 Aspect map

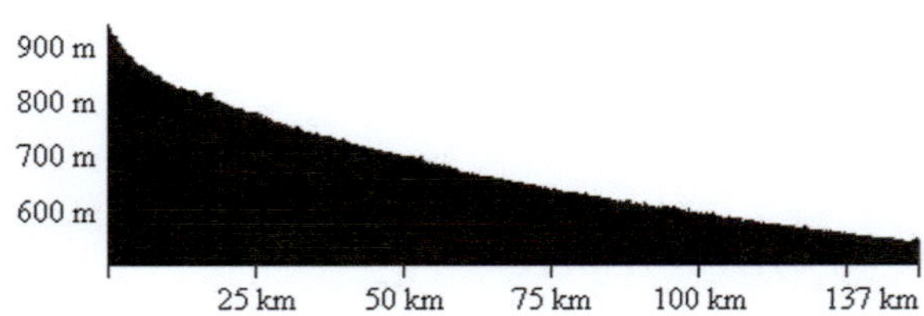

Fig. 9.6 Longitudinal profile

9.5.3.2 Basin Perimeter

The basin perimeter of the drainage basin is 340.38 km (Table 9.3). Generally, both drainage parameters, stream order and basin perimeter are strong positively correlated. Perimeter of the Yerla River increases as the order of the basin increases.

9.5.3.3 Basin Length

Maximum length between drainage basin reach and mouth has been considered as basin length. Table 9.3 shows that basin length of the study area is 108.55 km. It is observed that basin length increases as stream order increases.

9.5.3.4 Drainage Density (Dd)

Drainage density is the sum of stream lengths per unit area (Horton 1945). It expresses the closeness of spacing of stream channels. Density factor is related to climate, type of rocks, relief, infiltration capacity, vegetation cover, surface roughness and runoff intensity index. The amount and type of precipitation directly influences the quantity and characters of surface run-off. Kale and Gupta (2010) stated that semi-arid regions have a finer density texture than humid regions. The drainage density of this semi-arid region is calculated using formula given in Table 9.1 and found low i.e. 2.15 km/km^2 (Table 9.3) indicating coarse drainage texture. Low drainage density of study area indicates highly resistant or permeable subsoil material and low relief (Smith 1950).

9.5.3.5 Stream Frequency (Fs)

Stream or channel frequency is the total number of streams per unit area (Horton 1945). Scheidegger (1961) has suggested that stream frequency is related to drainage density. The number of streams in a drainage basin also depends on superficial material, amount of runoff, vegetation and topographic slope. All of which contribute in geomorphic character of the area. The drainage density of the area is 2.53 streams/km^2 which indicates poor stream frequency (Table 9.3).

9.5.3.6 Drainage Texture (T)

Thornbury (1969) defined drainage texture as the relative spacing of drainage lines. Drainage texture includes the drainage density and stream frequency (Horton 1945). Rock type is an important control on the drainage texture and density. In areas underlain by resistant and hard rock such as basalt the drainage is coarse textured. The drainage density values of the Yerla watershed are 2.15 km/km^2 indicating coarse drainage texture (Table 9.3 and Fig. 9.2). This is due to the fact that erosion in such rocks is difficult and all the streams require a large area to maintain a channel of given length. Therefore, the constant channel maintenance is high and density is low (Kale and Gupta 2010).

9.5.3.7 Circularity Ratio (Rc)

The circularity ratio has been used as an areal aspect and is expressed as the ratio of basin area of a circle having the same perimeter as the basin (Strahler 1964). Circularity ratio values approaching one indicate that the basin shapes are like circular and as a result, it gets scope for uniform infiltration and takes longer time to reach excess water at basin outlet. It is affected by geology, slope and land cover. The ratio is more influenced by length, stream frequency and gradient of various orders rather than slope conditions and drainage pattern of the basin. Basin shape determines how rapidly the runoff will reach the main river as well as the outlet. For elongated basins the runoff reaches slowly in the arrival of flow after heavy rains. Studies by Hack (1957) indicate that as the basins enlarge, the stream length increases and the basins become narrower and longer. Therefore, a majority of rivers have elongated basins. Miller (1953) has described the basin of the circularity ratios which have a range of 0.4–0.7, which indicates strongly elongated and highly permeable homogenous geologic materials. The circularity ratio of the whole Yerla watershed is 0.33 which indicates elongated shape and highly permeable homogenous geologic condition (Table 9.3).

9.5.3.8 Form Factor (Ff)

Form factor is defined as the ratio of basin area to the square of the basin length (Horton 1945). The values of form factor would always be less than 0.78 (perfectly for a circular basin). Smaller the value of Ff means more elongated will be the basin. The form factor ratio of the study area is 0.26 which indicates that the whole basin has an elongated shape (Table 9.3).

9.5.3.9 Compactness Coefficient (Cc)

Compactness coefficient is used to express the relationship of a hydrologic basin with that of a circular basin having the same area as the hydrological basin. A circular basin is the most hazardous from a drainage standpoint because it will yield the shortest time of concentration before peak flow occurs in the basin. In the study area, value of compactness coefficient is 1.34 which indicates less hazardous basin (Table 9.3).

9.5.3.10 Elongation Ratio (Re)

Schumm (1956) defined elongation ratio as the ratio between the diameter of the circle of the same area as the drainage basin and the maximum length of the basin. Analysis of elongation ratio indicates that the areas with higher elongation ratio values have high infiltration capacity and low runoff. Circular basins are more efficient in the discharge of runoff than elongated basins. The values of elongation ratio generally vary from 0.6 to 1.0 over a wide variety of climate and geologic types. Elongation ratio values close to 1.0 are typically of regions of very low relief, whereas elongation ratio values in the range 0.6–0.8 are usually associated with high relief and steep ground slope (Strahler 1964). Elongation ratio of the area is 0.57 which shows that the catchment falls in the less elongated category and showing high relief and steep ground slope (Table 9.3).

9.5.3.11 Length of Overland Flow (Lo)

It is the length of water over the ground before it gets concentrated into the define stream channels (Horton 1945). This factor basically relates inversely to the average slope of the channel and is quite synonymous with the length of sheet flow to a large degree. The length of overland flow approximately is equal to half of the reciprocal of drainage density (Horton 1945). The length of the overland flow of the study area is 2.23 km.

9.6 Conclusion

The dendritic type drainage network of the Yerla River basin, Deccan Trap region, exhibits the homogeneity in texture and lack of structural control. The seventh order basin has coarse drainage density and poor stream frequency. The mean bifurcation ratio of the area is 4.29 which indicates it is a hilly dissected basin. The form factor ratio and circularity ratio of the area is 0.26 and 0.33 which indicates that the whole basin has an elongated shape. The compactness coefficient value is 1.34 which shows less hazardous basin. Average slope of the area is 0.47%. Sinuosity indices are 1.30 which indicates transitional stage. The drainage density and channel frequency values of this semi-arid region are low i.e. 2.15 km/km^2 and 2.53 streams/km^2 which indicates coarse drainage texture and poor stream frequency. Longitudinal profile of the river is ideal. The results of this analysis would be useful in determining the effect of catchment characteristics such as size, shape, slope of the catchment and distribution of stream network within the catchment. The quantitative analysis of linear, relief and aerial parameters is found to be of immense utility in river basin evaluation for natural resource management.

Acknowledgements The author gratefully acknowledges the University Grants Commission, Government of India, New Delhi for providing Teacher Fellowship vide file no. F. No. 36-03/12 (WRO). Thanks to Professor Dr. Maya G. Unde, Department of Geography, Ahmednagar College, Ahmednagar for critical assessment of the work and anonymous reviewers and Editor-in-Chief for their many helpful suggestions.

References

Ahmed SA, Chandrashekarappa KV, Raj SK, Nishita V, Kavitha G (2010) Evaluation of morphometric parameters derived from ASTER and SRTM DEM-a study on Bandihole sub watershed basin in Karnataka. J Indian Soc Remote Sens 38(2):227–238

Agarwal KK, Prakash C, Ali SN, Jahan N (2012) Morphometric analysis of the Ladhiya and Lohawati River basin, Kumaun Lesser Himalaya, India. J Zeitschrift fur Geomorphologie 56(2):201–224

Bhatt S, Ahmed SA (2014) Morphometric analysis to determine floods in the Upper Krishna basin using Cartosat DEM. J Geocarto Int 29(8). https://doi.org/10.1080/10106049.2013.868042

Biswas S, Sudhakar S, Desai VR (1999) Prioritization of sub-watersheds based on Morphometric analysis of drainage basin, district Midnapore, West Bengal. J Indian Soc Remote Sens 27(3):55–166

Chadha DK, Neupane BR (eds) (2011) Significance of geomorphic analysis of watershed for optimization of recharge structures. UNESCO, New Delhi, pp 1–109. http://unesdoc.unesco.org/images/0021/002152/215277e.pdf

Chandrashekar H, Lokesh KV, Sameena M, Ranganna G (2015) GIS-based morphometric analysis of two reservoir catchments of Arkavati River, Ramanagaram District, Karnataka. Elsevier Aquat Procedia 4:1345–1353. https://doi.org/10.1016/j.aqpro.2015.02.175

Chopra R, Dhiman RD, Sharma PK (2005) Morphometric analysis of sub-watersheds in Gurdaspur district, Punjab using remote sensing and GIS techniques. J Indian Soc Remote Sens 33:531. https://doi.org/10.1007/BF02990738

Chorley RJ, Schumm SA, Sugden DE (1984) Geomorphology. Methuen, London

Clarke JI (1966) Morphometry from maps. Essays in geomorphology, Elsevier, New York, 235–274

Corley RJ (1957) Illustrating the laws of morphometry. Geol Mag 94(2):140–150

Cunha ERD, Bacani VM (2016) Morphometric characterization of a watershed through SRTM Data and geoprocessing technique. J Geogr Inf Syst 8(2):238–247. https://doi.org/10.4236/jgis.2016.82021

Doke A, Pardeshi SD, Pardeshi SS (2018) Identification of morphogenetic regions and respective geomorphic processes: a GIS approach. Arab J Geosci 11:20. https://doi.org/10.1007/s12517-017-3358-5

Ebisemiju FS (1979) A reduced rank model of drainage basin morphology. Geogr Ann Ser A Phys Geogr 61(1/2):103–112

Grohmann CH (2004) Morphometric analysis in geographic information systems: applications of free software GRASS and RS. Comput Geosci 30(2004):1055–1067. https://doi.org/10.1016/j.cageo.2004.08.002

Hack J (1957) Studies of longitudinal stream profiles in Virginia and Maryland. US geological survey professional paper, 294-B. https://pubs.usgs.gov/pp/0294b/report.pdfw

Horton RE (1932) Drainage basin characteristics. Trans Am Geophys Union 13:350–361

Horton RE (1945) Erosional development of streams and their drainage basins: hydrophysical approach to quantitative morphology. Bull Geol Soc Am 56:275–370. https://hydrology.agu.org/wp-content/uploads/sites/19/2016/06/Horton_GSA1945_ErosionalDevelopmentofStreams-andtheirDrainageBasins_-hydrophysical-approach-to-quantitative-Morphology.pdf

Javed A, Khanday MY, Ahmed R (2009) Prioritization of sub-watersheds based on morphometric and land use analysis using remote sensing and GIS techniques. J Indian Soc Remote Sens 37:261–274. https://doi.org/10.1007/s12524-009-0016-8

Kale VS, Gupta A (2010) Introduction to geomorphology. Orient Longman, Kolkata, pp 1–274

Kanth TA, Hassan Z (2012) Morphometric analysis and prioritization of watersheds for soil and water resource management in Wular catchment using Geo-spatial tools. Int J Geol Earth Environ Sci 2(1):30–41

Magesh NS, Chandrasekar N, Soundranayagam JP (2011) Morphometric evaluation of Papanasam and Manimuthar watersheds, parts of Western Ghats, Tirunelveli district, Tamil Nadu, India: a GIS approach. Environ Earth Sci 64(2):373–381

Malik MS, Shukla JP (2018) A GIS-based morphometric analysis of Kandaihimmat watershed, Hoshangabad district, M.P. India. J Geo-Marine Sci 47(10):1980–1985

Manimozhi C, Rengarajan R (2007) Use of remote sensing and geographical information system in drainage morphometric analysis of Thirumanimuttar upper subbasin in Salem and Namakkal District, Tamilnadu, India. Indian J Geomorphlogy 11 and 12(1/2):111–122

Miller VC (1953) A quantitative geomorphic study of drainage basin characteristics on the Clinch Mountain area, Virgina-Tennessee, Proj. NR 389–402, technical report 3, Columbia University. http://agris.fao.org/agris-search/search.do?recordID=US201400058936

Mishra SS, Nagarajan R (2010) Morphometric analysis and prioritization of sub-watersheds using GIS and remote sensing techniques: a case study of Odisha, India. Int J Geomat Geosci 1(3):501–510. http://www.ipublishing.co.in/jggsvol1no12010/EIJGGS2018.pdf

Morisawa ME (1968) Streams: their dynamics and morphology, New York, p 215

Nag SK (1998) Morphometric analysis using remote sensing techniques in the Chaka sub-basin, Purulia district, West Bengal. J Indian Soc Remote Sens 26(1/2):69–76

Nautiyal MD (1994) Morphometric analysis of drainage basin, district Dehradun, Uttar Pradesh. J Indian Soc Remote Sens 22(4):252–262

Oguchi T (1997) Drainage density and relative relief in humid steep mountains with frequent slope failure. Earth Surf Process Landform 22:107–120

Pande CB, Moharir K (2017) GIS based quantitative morphometric analysis and its consequences: a case study from Shanur River basin. Maharashtra India. Appl Water Sci 7(2):861–871. https://doi.org/10.1007/s13201-015-0298-7

Panhalkar SS, Mali SP, Pawar CT (2012) Morphometric analysis and watershed development prioritization of Hiranyakeshi Basin in Maharashtra, India. I J Envi Sci 3(1). https://doi.org/10.6088/ijes.2012030131052

Panhalkar SS, Jarag AP (2015) Assessment of spatial interpolation techniques for river bathymetry generation of Panchganga River basin using geoinformatic techniques. Asian J Geoinformatics 15(3)

Rai PK, Mohan K, Mishra S, Ahmad A, Mishra VN (2017) A GIS-based approach in drainage morphometric analysis of Kanhar River basin, India. Appl Water Sci 7:217–232. https://doi.org/10.1007/s13201-014-0238-y

Raju KS, Kumar DN (2001) Classification of microwatersheds based on morphological characteristics. J Hydro-Environ Res 5:101–109. https://doi.org/10.1016/j.jher.2010.09.002

Rao GT, Rao VVSG, Dakate R, Rao STM, Rao BMJ (2012) Remote sensing and GIS based comparative morphometric study of two sub-watershed of different physiographic condition, West Godavari District, A.P. J Geol Soc India 79:383–390

Sarmah K, Jha LK, Tiwari BK (2012) Morphometric analysis of a highland micro-watershed in East Khasi Hill District of Meghalaya, India: using remote sensing and GIS techniques. J Geog Regional Plan 5(5):142–150

Saptarshi PG, Raghavendra RK (2009) GIS-based evaluation of micro-watersheds to ascertain site suitability for water conservation structures. J Indian Soc Remote Sens 37:639–704. https://doi.org/10.1007/s12524-009-0057-z

Scheidegger AE (1961) General theory of dispersion in porous media. J Geophys Res https://doi.org/10.1029/JZ066i010p03273

Scheidegger AE (1965) The algebra of stream order number, US Geological Survey Professional Paper No 525 B

Schumm (1956) Evolution of drainage systems and slopes in badlands at Perth Amboy, New Jersey. Bull Geol Soc Am 67:597–646. https://pubs.geoscienceworld.org/gsa/gsabulletin/article-abstract/67/5/597/4811/evolution-of-drainage-systems-and-slopes-in?redirectedFrom=fulltext

Schumm SA (1963) Sinuosity of alluvial rivers on the Great Plains. Bull Geol Soc Am 74:1089–1100

Sebastian M, Jayaraman V, Chandrasekhar MG (1995) Space technology applications for sustainable development of watersheds. Technical report. ISRO, Bangalore

Sethupathi AS, Narsimhan L, Vasanthamohan V, Mohan P (2011) Prioritization of mini watersheds based on morphometric analysis using remote sensing and GIS techniques in a drought prone Bargr-Mathur sub watershed, Ponnaiyar River basin, India. J Geomat Geosci 2(2):403–414

Shreve RW (1969) Stream lengths and basin areas in topologically random channel networks. J Geol 77:397–414

Sindhu D, Ravikumar AS, Shivakumar BL (2015) Quantitative analysis of catchment using remote sensing and geographic information system. Elsevier Aquatic Procedia 4:1421–1428. https://doi.org/10.1016/j.aqpro.2015.02.184

Singh P, Gupta A, Singh M (2014) Hydrological inferences from watershed analysis for water resource management using remote sensing and GIS techniques. Egypt J Remote Sens Space Sci 17:111–121. https://doi.org/10.1016/j.ejrs.2014.09.003

Smith KG (1950) Standards for grading textures of erosional topography. Am J Sci 48:655–668

Sreedevi PD, Owais S, Khan HH, Ahmed S (2009) Morphometric analysis of a watershed of south India using SRTM data and GIS. J Geol Soc India 72:543–552

Strahler AN (1950) Equilibrium theory of Erosional slopes, approached by frequency distribution analysis. Am J Sci 248:673–696

Strahler AN (1952) Dynamic basis of Geomorphology. GSA Bull 63(9): 923–938 https://doi.org/10.1130/0016-7606(1952)63[923:DBOG]2.0.CO;2

Strahler AN (1956) Quantitative slope analysis. Bull Geol Soc Am 67:571–596

Strahler AN (1964) Quantitative geomorphology of drainage basins and channel networks. In: Chow VT (ed) Handbook of applied hydrology. McGraw Hill, New York, Section 4–11. http://geomorphometry.org/content/quantitative-geomorphology-drainage-basins-and-channel-networks

Strahler AN (1969) Physical geography, 3rd edn. Wiley, New York

Suresh D, Collins JC, Jayaprasad BK, Yarrakula K, Vishnavi V, Ghosh B (2018) Morphometric analysis of sub-watersheds in Gurdaspur district, Punjab using remote sensing and GIS techniques. Indian J Geo Marine Sci 47(10):1969–1979

Telore NV (2016) Geomorphology and watershed management: a study from Deccan trap region, India. Lap Lambert Academic Pub, Saarbrucken, pp 99–116

Tideman EM (2007) Watershed management: guidelines for Indian condition, 11th ed. Omega Scientific Pub, New Delhi, pp 7–17

Thomas J, Joseph S, Thrivikramji KP, Abe G (2011) Morphometric analysis of the drainage system and its hydrological implications in the rain shadow regions, Kerala, India. J Geogr Sci 21(6):1077–1088

Thornbury WD (1969) Principles of Geomorphology. 2nd edn. Wiley

Unde MG, Telore, NV (2013) Role of watershed development programme in sustainable development: Nidhal experience from Satara District, Maharashtra. Int J Shivaji University (Sc and Tech) 42:68–71 www.unishivaji.ac.in/uploads/journal/Journal_42/16.pdf

Vittala SS, Govindiah S, Honne Gowda H (2004) Morphometric analysis of sub-watersheds in the Pawagada area of Tumkur district, South India, using remote sensing and GIS techniques. J Indian Soc Remote Sens 32(4):351–362

Woldenberg MJ (1966) Horton's Law justified in terms of algometric growth and steady state in open systems. Bull Geol Soc Am 77:431–435

Chapter 10
Analysing the Capability of NCI Technique in Change Detection Using High- and Medium-Resolution Multispectral Data

Subhanil Guha, Himanshu Govil, Anindita Dey and Neetu Gill

Abstract In the present research work, neighbourhood correlation image (NCI) analysis technique was applied for identifying the changed and unchanged pixels in land use/land cover system in a small part of Dehradun city, India. Correlation image, slope image, and intercept image in the neighbourhood of a particular pixel for any two multi-temporal data sets provide significant information regarding changed pixels. The fundamental concept indicates a high correlation in the unchanged pixels and a low correlation in the changed pixels for same windows of two multi-date images. Various rectangular neighbourhood pixel windows (3×3, 5×5, 7×7, and 9×9) were used for NCI application. Results show that NCI analysis technique was almost equally effective for raw data or radiometric normalized data. Furthermore, the NCI technique becomes more effective in higher resolution with small pixel window size.

Keywords IKONOS · Multi-temporal · Neighbourhood correlation image (NCI) · Pixel window · Resolution

---

S. Guha (✉) · H. Govil
Department of Applied Geology, National Institute of Technology Raipur, Raipur, India
e-mail: subhanilguha@gmail.com

H. Govil
e-mail: himgeo@gmail.com

A. Dey
Department of Geography, Nazrul Balika Vidyalaya, Guma, West Bengal, India
e-mail: anindita.geo@gmail.com

N. Gill
Chhattisgarh Council of Science and Technology, Raipur, India
e-mail: gill.scorpio@gmail.com

S. Sahdev et al. (eds.), *Geoecology of Landscape Dynamics*,
Advances in Geographical and Environmental Sciences,
https://doi.org/10.1007/978-981-15-2097-6_10

10.1 Introduction

Change in land use and land cover of any urban area is a dynamic process because of rapid urbanization and immigration processes. Outskirts of city area can be considered as the most rapidly changed landscapes inside an urban area. Most of the observed land conversions are from other lands into built-up areas and agricultural land due to the necessity of residential space and occupational requirements. Land cover change is a dynamic phenomenon. A single type of land cover always tends to be changed. So, detection of the change in land use is a very important task in the present-day situation (Ridd and Liu 1998; Lunetta and Elvidge 2000). It is a process used to identify the differences in a natural or geographic feature within a definite time period (Singh 1989). The major goals of change detection are to detect the geographical area where the changes have occurred, to identify the changed land cover type and to calculate a number of changes occurred. Change enhancement methods and 'from-to' change information extraction methods can be considered as two major groups of change detection method (Jensen et al. 1993; Johnson and Kasischke 1998; Chan et al. 2001; Lu et al. 2004); Im and Jensen 2005). Generally, change enhancement methods (image ratioing, image differencing, principal component analysis, vegetation indices, etc.) give valuable information of spectral change and can estimate the amount of the change. Conversely, 'from-to' change information extraction methods (hybrid change detection, change vector analysis, post-classification comparison, etc.) impart pixel-based minute information about the categories of land cover change (Im and Jensen 2005).

Recently, some new approaches for change detection were discussed (Jensen 2005; Atkinson et al. 2012; Farooq and Govil 2013). A new change detection model was introduced for continuous land cover change detection on Landsat data (Zhu and Woodcock 2014). Traditional pixel-based change detection techniques and their comparison with the new object-based techniques were discussed by Hussaain et al. (2013). Tewkesbury et al. (2015) assessed the advantages and disadvantages of seven analysis units and six comparison methods within a change detection workflow. A new comprehensive change detection method (CCDM) was presented by Jin et al. (2013) for updating and monitoring the national land cover change. A generalized statistical model to solve binary change detection problems in multispectral satellite images with high radiometric resolution was proposed by Zanetti and Bruzzone (2016). Hao et al. (2015) proposed an unsupervised change detection approach using different trade-off parameters of an active contour model to reduce the effect of the parameters. A new super-resolution change detection method (SRCD) was explained for identifying changes in land cover at both high spatial and temporal resolutions (Li et al. 2016). Another new change detection method with higher accuracy in remote sensing technology based on a conditional random field was proposed (Cao et al. 2016). A completely new graph-based algorithm was evaluated by Vakalopoulou et al. (2016). Recently, robust change vector analysis (RCVA) technique with pixel neighbourhood effect was introduced, which is considered as an improvement over the conventional change vector analysis method (Thonfeld et al. 2016).

A number of studies were conducted on land cover change particularly in an urban area. Multi-temporal remote sensing images have been used to evaluate land cover changes in the Beijing metropolitan region (He et al. 2006; Zhang et al. 2002). Liu and Zhou (2005) worked on the future prediction of urban expansion. Jensen et al. (2005) evaluated CoastWatch change detection protocol in South Carolina. Various approaches were adopted by the researchers to carry out pixel-based change detection (Congalton 1991; Gong 1993; Jensen 2005; Ritter et al. 1990; Stauffer and McKinney 1978; Weismiller et al. 1977; Xiuwan 2002). Moreover, it was examined that in change detection analysis, pixel-based techniques are more suitable for low- or moderate-resolution image while object-oriented techniques are useful for high-resolution data (Niemeyer and Canty 2003). Four separate algorithms were analysed in bi-temporal space for detecting changes (Liu et al. 2004). Im et al. (2008) used image segmentation method and correlation image analysis method together to generate object-based change detection.

A sophisticated methodology was introduced based on neighbourhood image analysis and decision tree classification to identify the urban land use change using high-resolution data of Edisto Beach near Charleston with very good accuracy (Im and Jensen 2005). Here this new technique was generated on circular windows with 1, 2, 3, 4, and 5-pixel radius. Im et al. developed intelligent knowledge-based systems and discussed it with neighbourhood image analysis technique for detecting changes (Im et al. 2005). A threshold-based calibration approach was successfully evaluated in an automated binary change detection model using high-resolution QuickBird satellite image. Neighbourhood correlation images, object correlation images, multi-band difference images, and multi-band ratio images were used as single and multiple variables in this study (Im et al. 2007). Four optimum threshold search change detection algorithms were compared using neighbourhood correlation image, neighbourhood slope image, neighbourhood intercept image, and change vector magnitude image on IKONOS and QuickBird data (Im et al. 2008). The present study was concentrated on optimization and applicability of NCI technique on rectangular pixel windows with four different sizes (e.g. 3×3, 5×5, 7×7, and 9×9) using IKONOS data for detecting the change in urban area and validation of this technique in Indian context.

10.2 Methodology

10.2.1 Study Area and Pre-processing of Satellite Data

A small part of Dehradun city, Uttarakhand, India was selected for the present research work. For administrative, commercial, and residential purposes, the land use system of the city is continuously changing. It should be considered as a representative of a developing urban environment. For the purpose of the present research, multi-temporal data of near anniversary dates were required to keep the seasonal variation in atmospheric condition at the minimum level. IKONOS panchromatic

(PAN) data with 1 m resolution and multispectral (MSS) data with 4 m resolution for 4 November 2002 and 7 November 2005 (Source: Indian Institute of Remote sensing, Dehradun, India) were used in the present work (Fig. 10.1).

Several approaches were applied in the study to examine the suitability of NCI technique in detecting changes in land use. Raw image, calibrated image, and radiometric normalized image were used in both MSS and fused (PAN and MSS) mode. 2005 image was normalized with respect to 2002 image which was used as the reference image of the entire study. The capability of NCI technique was thoroughly examined through the following two different approaches (Fig. 10.2):

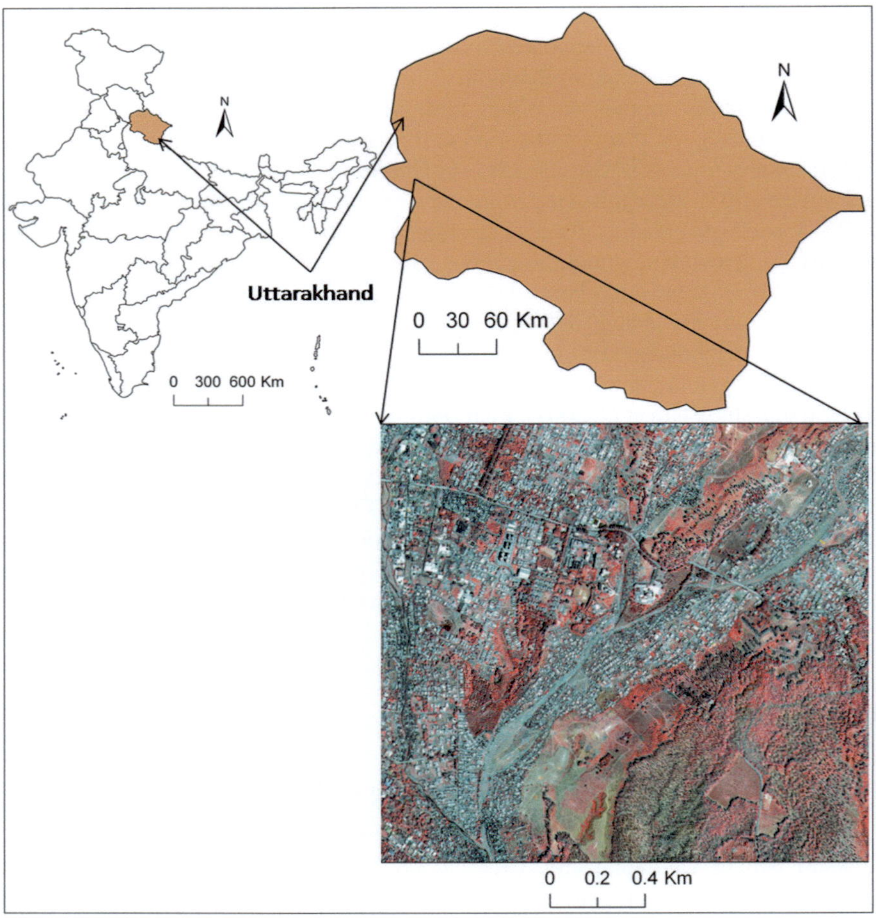

Fig. 10.1 Location of the study area (*Source* DigitalGlobe)

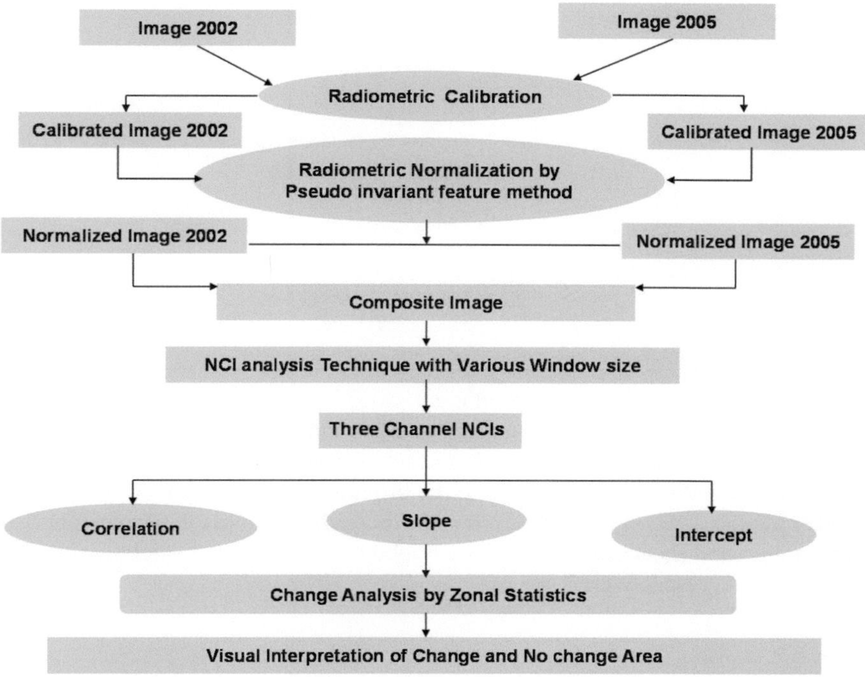

Fig. 10.2 Flow chart of methodology

1. NCI on radiometric calibrated and PIF-based radiometric normalized MSS image, and
2. NCI on radiometric calibrated and PIF-based radiometric normalized fused image.

10.2.2 Algorithm Description

In NCI technique, correlation, slope, and intercept images are generated to derive the information regarding land use/land cover change. The basic concept of this model is that the same pixel window should be highly correlated if the land cover remains unchanged and less correlated if the land cover is changed. The correlated multiple data sets provide information about the changed area. Slope image and intercept image also help to delineate the change information.

The following equations were used to determine correlation, slope, and intercept image:

$$r = \frac{\text{cov}_{12}}{s_1 s_2} \quad (10.1)$$

$$\text{cov}_{12} = \frac{\sum_{i=1}^{n} (BV_{i1} - \mu_1)(BV_{i2} - \mu_2)}{n - 1} \quad (10.2)$$

$$a = \frac{\text{cov}_{12}}{s_1^2} \quad (10.3)$$

$$b = \frac{\sum_{i=1}^{n} BV_{i2} - a \sum_{i=1}^{n} BV_{i1}}{n} \quad (10.4)$$

where r is the Pearson correlation coefficient; cov_{12} is the covariance between brightness values found in all bands of images 1 and 2 in the neighbourhood; s_1 and s_2 are the standard deviations of the brightness values found in all bands of images 1 and 2 in the neighbourhood; BV_{i1}, BV_{i2} are the brightness values of the ith pixel in all bands of images 1 and 2, respectively; n is the total number of pixel in the neighbourhood; and μ_1 and μ_2 are the mean of the brightness values in the neighbourhood in images 1 and 2, respectively.

10.2.3 Calibration and Radiometric Normalization of IKONOS Data

Radiometric calibration method was applied on both MSS and fused data by the conversion from DN to planetary reflectance is necessary (Table 10.1).

The formula of IKONOS spectral radiance is

$$L_\lambda = \frac{10^4 . DN_\lambda}{CalCoef_\lambda . Bandwidth_\lambda} \quad (10.5)$$

Table 10.1 Spectral radiance of IKONOS data

Band	CalCoefk DN*[mW/cm^{-2}–sr]$^{-1}$	Bandwidth (nm)
PAN	161	403
MS-1(Blue)	728	71.3
MS-2(Green)	727	88.6
MS-3(Red)	949	65.8
MS-4(NIR)	843	95.4

(*Data source* IKONOS Planetary Reflectance and Me an Solar Exoatmospheric Irradiance, by Martin Taylor)

For radiometric normalization of MSS and fused image of 2005 with respect to 2002 image, pseudo-invariant features method was used (Schott et al. 1988). It is an excellent and popular method to produce a normalized image with respect to the reference image.

10.2.4 Neighbourhood Configuration and NCI Generation

In general, the NCI technique was applied on the geometrically and radiometrically corrected IKONOS PAN and MSS image. Both the data sets were co-registered with <0.5 RMSE and radiometrically normalized using pseudo-invariant features (PIF) method in order to overcome the atmospheric effects that cause the apparent change in the imagery. Rectangular windows (3×3, 5×5, 7×7, and 9×9) were selected for neighbourhood configuration and generation of NCIs because they have more accurate information than a circular window.

10.3 Results and Discussion

10.3.1 NCI Value for Different Approaches

The correlation, slope, and intercept images were computed on 3×3, 5×5, 7×7, and 9×9 pixel window sizes for different sets of data. For both the approaches, a separate range of NCI value was obtained. In order to select the best approach, further comparative analyses were done on these NCI values.

In approach-1, NCI technique was applied on MSS IKONOS data. After radiometric calibration, the image was radiometrically normalized by pseudo-invariant features method (Fig. 10.3).

For 3×3 pixel window size, the range of correlation value lies between 0.99 and −0.99, which refers to almost perfect correlation. The range of correlation value (0.99 to −0.94) remains same for 5×5 and 7×7 windows. But 9×9 window has a lower correlation value (0.98 to −0.77). The correlation images for MSS data have been shown in Fig. 10.4.

After examining the MSS data, NCIs were also generated for fused IKONOS data. In approach-2, pseudo-invariant features-based normalization method was applied on calibrated fused IKONOS data (Fig. 10.5).

For the fused image, spatial resolution was increased from 4 m to 1 m by integrating the panchromatic image with the multispectral image. For 3×3 pixel window size, the range of correlation value presents exactly perfect correlation (+1 to −1). The correlation range was close to perfect (0.99 to −0.96) for 5×5 and 7×7 windows. Moreover, the 9×9 window also indicates strong (0.99 to −0.93) correlation. The correlation images for fused data have been shown in Fig. 10.6.

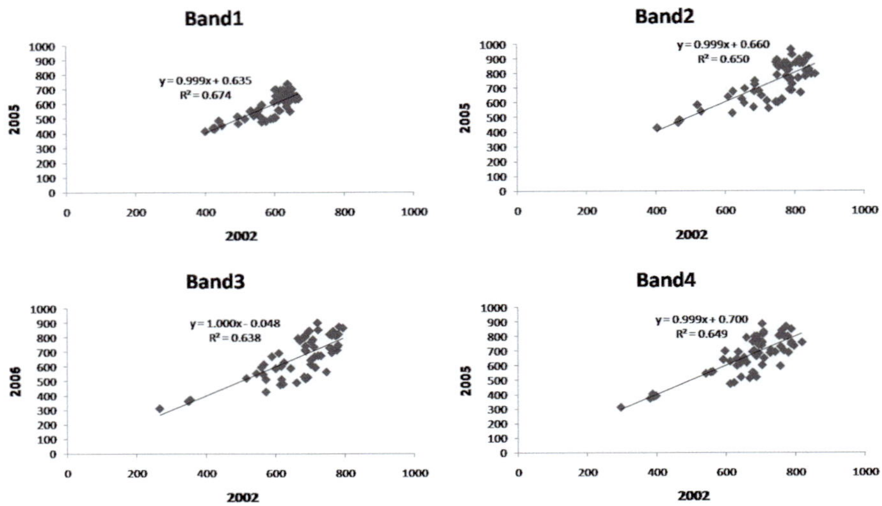

Fig. 10.3 Band-wise regression analysis of pseudo-invariant features of MSS data (*Source* DigitalGlobe)

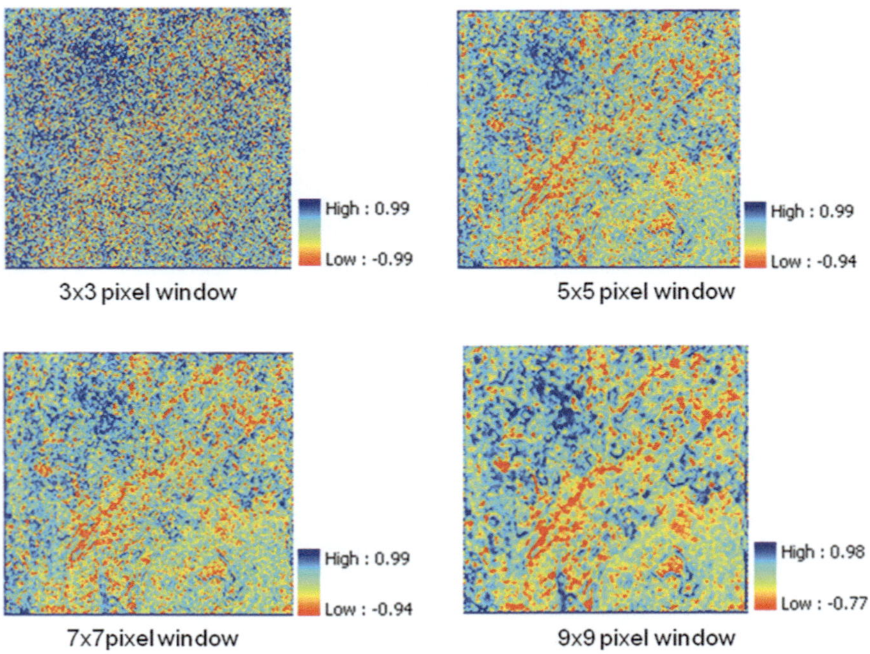

Fig. 10.4 NCIs on MSS data for various pixel windows (*Source* DigitalGlobe)

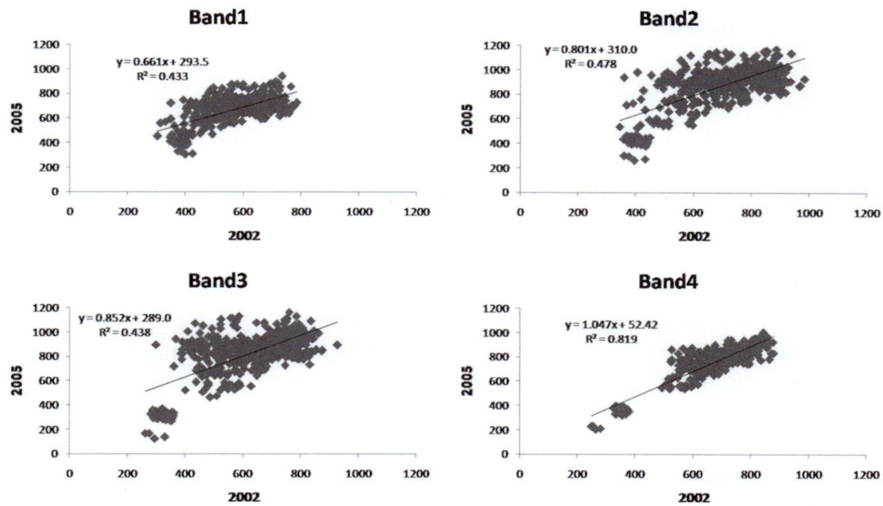

Fig. 10.5 Band-wise regression analysis of pseudo-invariant features of fused data (*Source* DigitalGlobe)

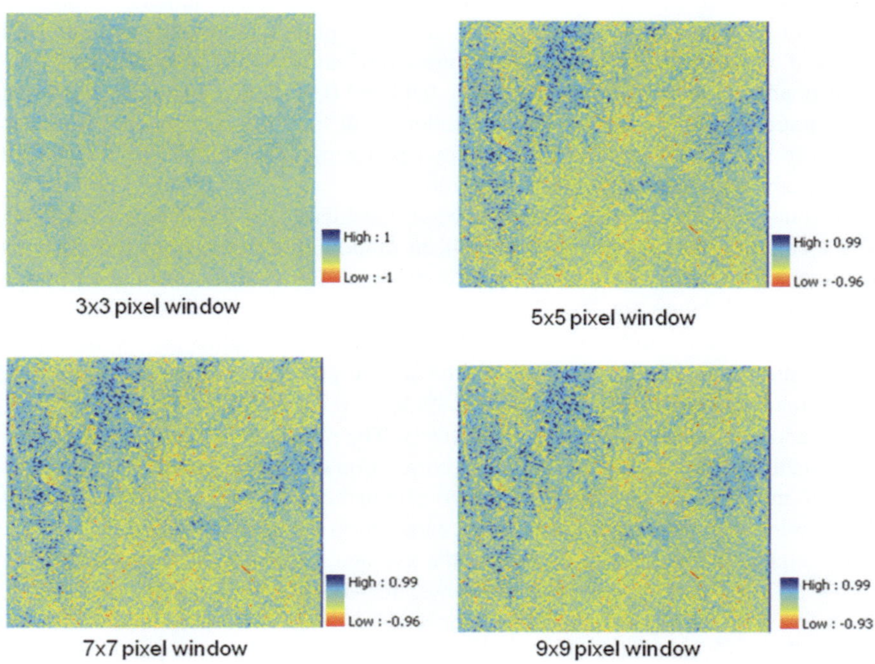

Fig. 10.6 NCIs on fused data for various pixel windows (*Source* DigitalGlobe)

10.3.2 Comparison Between Different Approaches for NCI Output

It was really a tough task to select the best approach for NCI analysis technique. It is undoubtedly encouraging that both the examined approaches have given almost same output. So, NCI technique should be considered as a robust technique specifically for urban change detection using high-resolution data.

NCI values are considerably changing due to the change in the pixel window size. The result is reflected in both the approaches. The fused image always provides better correlation than MSS image. In approach-1, MSS image has given a very high range of NCI value (0.99 to −0.99) for 3 × 3 pixel window size. This range is very close to the perfect correlation of two variables. This range has been gradually decreased with the increased window size. For 5 × 5 and 7 × 7 windows, the range (0.99 to −0.94) is exactly equal. But the range (0.98 to −0.77) is quite low for larger window size (9 × 9 pixel window size). Approach-2 also presents the same characteristics for the fused image. Fused image shows a range of exactly perfect correlation (+1 to −1) for 3 × 3 pixel window sizes. The range has changed for other window sizes but the difference is too insignificant. 5 × 5 and 7 × 7 window sizes depict same range of correlation value (0.99 to −0.96). Even 9 × 9 pixel window indicates a range of very strong correlation (0.99 to −0.93). In 3 × 3 pixel window, the average NCI values of changed and unchanged zones are −0.65 and 0.89 for MSS data while the corresponding values for fused data are −0.94 and 0.97. Thus, it is established that finer spatial resolution achieves much better result than the coarser one. It is also proven from the correlation image that any small geographical area like 3 × 3 pixel window always provides the best correlation.

A zonal statistics method has also been considered for the evaluation of this NCI technique. 40 sample zones have been randomly selected from the changed and unchanged areas of the MSS and fused image. For change zones, the average minimum NCI values for 3 × 3, 5 × 5, 7 × 7, and 9 × 9 pixel windows are −0.65, −0.28, −0.28, and −0.07, respectively. The 3 × 3 window shows the lowest average minimum value and establishes the best output. Again for unchanged zones, the average maximum NCI values are 0.89, 0.71, 0.71, and 0.66 for 3 × 3, 5 × 5, 7 × 7, and 9 × 9 pixel windows, respectively. The results indicate the proportional relationship of low window size with high correlation value. The smaller geographical area was more appropriate for NCI generation than the larger one. 3 × 3 pixel window size presents the most significant result compared to the others (Fig. 10.7).

For changed zones on fused image, the average minimum NCI values for 3 × 3, 5 × 5, 7 × 7, and 9 × 9 pixel windows are −0.94, −0.63, −0.63, and −0.51, respectively. In the case of unchanged zones, the average maximum NCI values are 0.97, 0.72, 0.72, and 0.60, for 3 × 3, 5 × 5, 7 × 7, and 9 × 9 pixel windows, respectively, indicating the strong inversely proportional relationship between pixel window size and correlation value. Being the smallest unit area, 3 × 3 pixel window size produced best NCI output (Fig. 10.7).

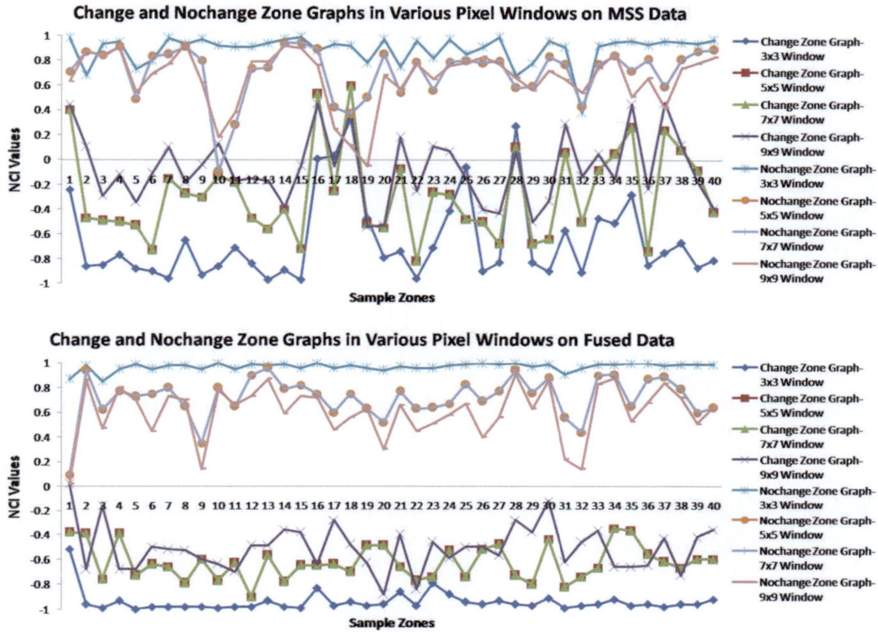

Fig. 10.7 Change and no-change zone graphs based on NCIs with various pixel windows for fused and MSS data (*Source* DigitalGlobe)

Figure 10.7 reflects a very close comparison between MSS data and fused data for all the used neighbourhood rectangular pixel windows based on the zonal statistics of change and no-change zones. The NCI values are inversely correlated to the pixel window size, i.e. small window size always has better correlation than large windows. 3×3 pixel window presents the best output.

Only 3×3 rectangular pixel window is considered for the evaluation of generated NCI values on MSS data and fused data (Fig. 10.8). From the zonal statistics graphs for change and no-change zones, it is clear that fused data always produce slightly better output though the difference is too negligible.

Lastly, it can be considered that NCI technique is more or less equally beneficial for raw MSS or fused image. A raw image can also provide almost same output like calibrated or normalized image. Here, NCI values were also examined with raw image and there were no differences between the raw image and the processed image. The average minimum and maximum NCI values were exactly same for the raw and the processed image. Apart from the correlation image, slope and intercept image individually may also act as an associated image in change detection analysis.

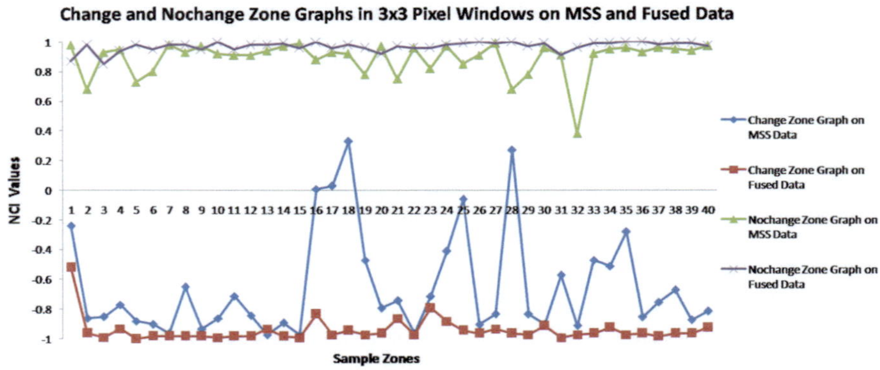

Fig. 10.8 Change and no-change zone graphs in 3 × 3 pixel window for comparing fused and MSS data (*Source* DigitalGlobe)

10.3.3 Visual Interpretation for Change and No-Change Zones on NCI

The correlation image is very helpful for the visual interpretation of change detection. In the correlation image, the bluish and the reddish pixels (i.e. higher correlation values) present the high positive and negative values, respectively. These high correlation values normally show the unchanged area. But the yellowish and the greenish pixels (i.e. lower correlation values) of the correlation image generally tend to be changed. If the slope and intercept images are considered, the picture will be just reverse. In that case, the high and low values of the intercept and slope images indicate the changed areas while unchanged areas are presented with moderate values. But the results may be different due to various reasons. Being an urban area, the land use features are heterogeneous in nature. Two different categories of land may have the same brightness value. So, it would be very difficult to identify and distinguish those features. For example, in the present study, some pixels of barren land and built-up area had almost same reflectance value. It really affects the change detection process. Here, mainly roads and water bodies remain almost unchanged and they are represented by high correlation value. Some forests and agricultural lands have also been changed to some extent and are represented by moderate correlation value (close to '0' value). The presence of shadow and moisture may also create some problems to distinguish the changed pixels from the unchanged, especially for small neighbourhood sizes. A large neighbourhood like 9 × 9 pixel window may identify the major land use conversions (e.g. deforestation). But a small neighbourhood like 3 × 3 pixel window size may detect the detailed minor changes in land use (e.g. an extension of a building) only based on the brightness value.

Visually it is very clear from Fig. 10.9 that NCI provides an excellent result with respect to raw or radiometrically corrected MSS or fused images to identify the change and no-change area. Apart from the correlation image, slope and intercept

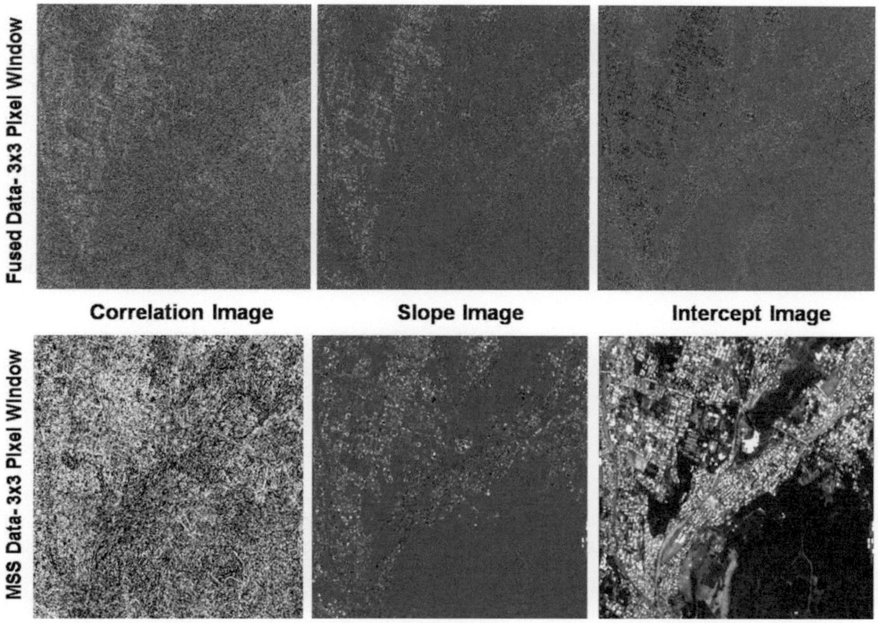

Fig. 10.9 Visual identification of changed/unchanged pixels on NCIs (*Source* DigitalGlobe)

images also play a very significant role in the identification of changed zone which is vividly identifiable from the greyscale image (Fig. 10.9). It is not necessary that high correlation value should always show the unchanged zone. If there is some confusion arises on the correlation image, slope and intercept images help a lot in change detection analysis. For the slope and intercept images, pixels with moderate brightness values remain unchanged or less changed while the brighter or darker pixels indicated the changed area.

10.4 Conclusion

From the final results and analysis of the research, it can be concluded that the NCI analysis technique is very much suitable and may be considered as one of the best techniques for change detection in an urban area using high-resolution remotely sensed data. Even, without radiometric calibration or normalization, NCI can show a very good result. Moreover, NCI technique provides a better output for a relatively small area and higher spatial resolution. From the research work, 3 × 3 mask size can be considered as the best mask size for evaluating NCI for detecting the change in the urban area. Fused data also provide a better result than the MSS data in generating

NCI. Finally, it may be concluded that NCI technique can play a very significant role through quick assessment of the suddenly damaged portion of the earth surface.

The study does not consider the NCI for different satellite sensors. So, one can also study the NCI to determine the upper limit of the optimal coarser resolution for NCI technique. The NCI can also be tested through different radiometric resolution (e.g. MODIS, AVHRR, Landsat 8) and different spectral resolution (e.g. MODIS, ASTER, Landsat 8).

Acknowledgements The authors are indebted to Indian Institute of Remote Sensing, Dehradun, India.

Disclosure statement No potential conflict of interest was reported by the authors.

References

Atkinson PM, Jeganathan C, Dash J, Atzberger C (2012) Inter-comparison of four models for smoothing satellite sensor time-series data to estimate vegetation phenology. Remote Sens Environ 123:400–417

Cao G, Zhou L, Li Y (2016) A new change-detection method in high-resolution remote sensing images based on a conditional random field model. Int J Remote Sens 37(5):1173–1189

Chan JC, Chan K, Yeh AG (2001) Detecting the nature of change in an urban environment: A comparison of machine learning algorithms. Photogramm Eng Remote Sens 67(2):213–225

Congalton RG (1991) A review of assessing the accuracy of classifications of remotely sensed data. Remote Sens Environ 37:35–46

Farooq S, Govil H (2013) Mapping regolith and gossan for mineral exploration in the Eastern Kumaon Himalaya, India using Hyperion data and Object Oriented Image Classification. Adv Space Res 53(12):1676–1685

Gong P (1993) Change detection using principal component analysis and fuzzy set theory. Canadian J Remote Sens 19(1):22–29

Hao M, Shi W, Deng K, Zhang H (2015) Fusion-based approach to change detection to reduce the effect of the trade-off parameter in the active contour model. Remote Sens Lett 6(1):39–48

He C, Okada N, Zhang Q, Shi P, Zhang J (2006) Modeling urban expansion scenarios by coupling cellular automata model and system dynamic model in Beijing, China. Appl Geogr 26:323–345

Hussain M, Chen D, Cheng A, Wei H, Stanley D (2013) Change detection from remotely sensed images: From pixel-based to object-based approaches. ISPRS J Photogramm Remote Sens 80:91–106

Im J, Jensen JR (2005) A change detection model based on neighborhood correlation image analysis and decision tree classification. Remote Sens Environ 99:326–340

Im J, Jensen JR, Hodgson ME (2008a) Optimizing the binary discriminant function in change detection applications. Remote Sens Environ 112:2761–2776

Im J, Jensen JR, Tulis JA (2005) Development of a remote sensing change detection system based on neighborhood correlation image analysis and intelligent knowledge-based systems. IEEE international geoscience and remote sensing symposium, Seoul. Korea, July 2005

Im J, Jensen JR, Tulis JA (2008b) Object-based change detection using correlation image analysis and image segmentation. Int J Remote Sens 29(2):399–423

Im J, Rhee J, Jensen JR, Hodgson ME (2007) An automated binary change detection model using a calibration approach. Remote Sens Environ 106:89–105

Jensen JR, Cowen DJ, Narumalani S, Althausen JD, Weatherbee O (1993) An evaluation of CoastWatch change detection protocol in South Carolina. Photogramm Eng Remote Sens 59(6):1039–1046

Jensen JR (2005) Introductory digital image processing: a remote sensing perspective, 3rd edn. Prentice Hall, Upper Saddle River, New York, p 525

Jin S, Yang L, Danielson P, Homer C, Fry J, Xian G (2013) A comprehensive change detection method for updating the National Land Cover Database to circa 2011. Remote Sens Environ 132:159–175

Johnson RD, Kasischke ES (1998) Change vector analysis: A technique for the multispectral monitoring of land cover and condition. Int J Remote Sens 19(3):411–426

Li X, Ling F, Foody GM, Du Y (2016) A superresolution land-cover change detection method using remotely sensed images with different spatial resolutions. IEEE Trans Geosci Remote Sens 54(7):3822–3841

Liu H, Zhou Q (2005) Developing urban growth predictions from spatial indicators based on multi-temporal images. Comput Environ Urban Syst 29(5):580–594

Liu YB, Nishiyama S, Yano T (2004) Analysis of four change detection algorithms in bi-temporal space with a case study. Int J Remote Sens 25(11):2121–2139

Lu D, Mausel P, Brondízio E, Moran E (2004) Change detection techniques. Int J Remote Sens 25(12):2365–2401

Lunetta RS, Elvidge C (2000) Remote sensing change detection: Environmental monitoring methods and applications. Taylor and Francis, New York, p 340

Niemeyer I, Canty MJ (2003) Pixel-based and object-oriented change detection analysis using high-resolution imagery. 25th ESARDA symposium on safeguards and nuclear material management, Stockholm

Ridd MK, Liu J (1998) A comparison of four algorithms for change detection in an urban environment. Remote Sens Environ 63:95–100

Ritter GX, Wilson JN, Davidson JL (1990) Image algebra: an overview. Comput Gr Image Process 49(3):297–331

Schott JR, Salvaggio C, Volchok WJ (1988) Radiometric scene normalization using pseudoinvariant features. Remote Sens Environ 26(1):1–14

Singh A (1989) Digital change detection Techniques using remotely sensed data. Int J Remote Sens 10:989–1003

Stauffer ML, McKinney RL (1978) Landsat image differencing as an automated land cover change detection technique. Report Contact, NASS-24356, General Electric Company, Beeltsvile, Maryland

Tewkesbury AP, Comber AJ, Tate NJ, Lamb A, Fisher PF (2015) A critical synthesis of remotely sensed optical image change detection techniques. Remote Sens Environ 160:1–14

Thonfeld F, Feilhauer H, Braun M, Menz G (2016) Robust Change Vector Analysis (RCVA) for multi-sensor very high resolution optical satellite data. Int J Appl Earth Obs Geoinf 50:131–140

Vakalopoulou M, Karantzalos K, Komodakis N, Paragios N (2016) Graph-based registration, change-detection, and classification in very high resolution multitemporal remote sensing data. IEEE J Sel Top Appl Earth Obs Geoinf 9:2940–2951

Weismiller RA, Kristof SJ, Scholz DK, Anuta PE, Momin SA (1977) Change detection in coastal zone environments. Photogramm Eng Remote Sens 43(12):1533–1539

Xiuwan C (2002) Using remote sensing and GIS to analyse land cover change and its impacts on regional sustainable development. Int J Remote Sens 23(1):107–124

Zanetti M, Bruzzone L (2016) A generalized statistical model for binary change detection in multispectral images. Geoscience and Remote Sensing Symposium. IEEE International

Zhang Q, Wang J, Peng X, Gong P, Shi P (2002) Urban built-up land change detection with road density and spectral information from multi-temporal Landsat TM data. Int J Remote Sens 23(15):3057–3078

Zhu Z, Woodcock CE (2014) Continuous change detection and classification of land cover using all available Landsat data. Remote Sens Environ 144:152–171

Chapter 11
Comparative Analysis of Multi-temporal Drought Indices Using Monthly Precipitation Data: A Study in the Southwestern Part of West Bengal, India

Pradip Patra

Abstract Drought is considered as one of the most important natural hazards of regional character. Although drought has different dimensions, its effects vary across different sectors. Its immediate impact is felt in agricultural sector, and the countries which have less irrigation facility usually face acute problems due to this phenomenon. The southwestern part of West Bengal is also considered as one of the drought-prone regions of India due to its complex climatic and physiographical characteristics. Thus, in this study, an attempt has been initiated to examine the drought condition of the region using multi-temporal drought indices including Standardized Precipitation Index (SPI), Rainfall Anomaly Index (RAI), and Effective Drought Index (EDI), and to find out the best drought index. 113 years (1901–2013) of mean monthly precipitation data recorded by India Meteorological Department (IMD) has been used for the purpose. Correlation matrix (Karl Pearson) shows that EDI is the best drought index with the highest correlation value ($R^2 \geq 0.72$) among the drought indices with different time steps, while SPI3 is the best suited meteorological drought index, having a higher correlation value ($R^2 \geq 0.95$) with 3 months of the time steps, and medium correlation value ($R^2 \geq 0.56$) among all the time steps. Therefore, SPI3 has been used to analyze an individual drought event quantitatively in each district of the region, and on the basis of the parameters like Drought Duration (DD), Drought Magnitude (DM), and Drought Intensity (DI), three equal climatic periods have been classified (1901–1937, 1938–1975, and 1976–2013) to find out whether the drought events have increased or not. The highest drought frequency has been recorded in Purulia and Paschim Medinipur districts, while DD and DM are the highest in Birbhum district.

Keywords Drought indices · SPI · RAI · EDI · Drought events

P. Patra (✉)
Department of Geography, University of Calcutta, Kolkata, India
e-mail: patrapradip1990@gmail.com

S. Sahdev et al. (eds.), *Geoecology of Landscape Dynamics*,
Advances in Geographical and Environmental Sciences,
https://doi.org/10.1007/978-981-15-2097-6_11

11.1 Introduction

Drought is a period of drying up of normal conditions and generally occurs when an area receives considerably less precipitation compared to the normal one. Drought is a slow onset, progress, and ending phenomena. It is the result of many factors, and accordingly, it can be classified into four categories like meteorological, hydrological, agricultural, and socio-economic (Palmer 1965; Wilhite and Glantz 1985; White and Walcott 2009). Defining and monitoring of drought are big challenges. So, preparedness for drought planning and its monitoring depend on the availability of spatio-temporal data from various sources, like meteorological, hydrological, vegetation information, etc.

Various drought indices have been propounded by different scholars, like Palmer's Drought Severity Index, PDSI; Palmer Hydrological Index, PDHI (Palmer 1965); Rainfall Anomaly Index, RAI (Van-Rooy 1965; Gibbs and Maher 1967); Crop Moisture Index, CMI (Palmer 1967); Surface Water Supply Index, SWSI (Shafer and Dezman 1982); Vegetation Condition Index, VCI (Kogan 1990); Standardized Precipitation Index, SPI (McKee et al. 1993); Temperature Condition Index, TCI; Vegetation Health Index, VHI (Kogan 1995); Normalized Difference Water Index, NDWI (Gao 1996); Effective Drought Index, EDI (Byun and Wilhite 1999); Reconnaissance Drought Index, RDI (Tsakiris et al. 2007); Temperature Dryness Vegetation Index, TVDI (Keshavarz et al. 2014; Wan et al. 2004); and Standardized Precipitation Evapotranspiration Index, SPEI (Vicente–Serrano et al. 2010).

Most of the drought indices have regional features and are applicable for a particular region only, e.g., RDDI (Gibbs and Maher 1967) used in Australia, PDI (Palmer 1967) widely used in USA, China Z-Indices (Wu et al. 2001) used by China Meteorological Society, Aridity Index by India Meteorological Organization (IMD), etc. Before SPI and EDI, PDSI was widely used by various researchers. But due to the lack of various climatic data, researchers often give importance to Drought Index (DI) which can be calculated using one climatic parameter, e.g., precipitation, and researchers have found that rainfall-based DIs are not only capable to detect meteorological drought, but it is also useful for detection of other kinds of droughts (Akbari et al. 2015; Rahiz and New 2014). After SPI (McKee et al. 1993), it has been widely used, especially since its recognition by WMO (Svoboda et al. 2012). Nowadays, EDI (Byun and Wilhite 1999) is also used by researchers as it gives emphasis on precipitation effectiveness. Most of the researchers are now use to calculate multiple Drought Indices (DIs), because it will help to find the characteristics of drought in different time periods, examine the sensitivity and accuracy of DIs, and make correlation among DIs so as to help us finding out the suitable DIs applicable for a particular area (Haied et al. 2017; Halwatura et al. 2016; Joetzjer, et al. 2013; Le et al. 2016; Khalili et al. 2011; Mahfouz et al. 2016; Pathak et al. 2016; Wang et al. 2015; Zarch et al. 2015). Keyantash and Dracup (2002) reviewed 14 well-known DIs which have been used for analyzing different types of drought and assessing their usefulness on measurement of drought severity and found that SPI is the most valuable estimator of drought severity. Morid et al. (2006) reviewed seven meteorological indices for drought monitoring in Iran; Kim et al. (2009) used daily precipitation and found how effectively EDI could be used to monitor drought. Vicente-Serrano et al. (2012) compared drought indicator ability to define drought at the global scale. Dogan et al. (2012) also compared multiple DIs using monthly precipitation data and found that EDI was the most suitable DI. In this context, it may be mentioned that the main aim of the Lincoln Declaration (2011) of DIs was to develop universal DIs and to prepare a guideline for an early warning system. In India, various works have been done to find out the best suitable DIs, e.g., district-wise multi-temporal DIs as computed by Jain et al. (2015) over the KBK districts of Odisha, and they have found that both SPI and EDI on an annual basis have responded in a similar way. Jain et al. (2015) also compared DIs and found that the EDI was the most suitable DI for the Ken river basin located in central India.

DIs are strong enough to define a drought, and the incorporated timescale is also helpful to analyze sensitive issues related to a drought. Keyantash and Dracup (2002) used only one month SPI, while Barua et al. (2011) used one month Percentage of Normal (PN) and SPI, but with the combination of multi-monthly DIs, which may result in multi-monthly precipitation, and play an important role to find out the cause and characteristics of drought. Thus, the present study attempts to compare multi-monthly DIs for determining drought and its severity over the western part of West Bengal, India.

11.2 Objectives

Main objectives of the study are: first, to analyze and compare the different kinds of multi-temporal Drought Indices (DIs) along with identifying the best suited DI for the region, and second, to assess the characteristics of historical drought events like intensity, frequency, duration, and magnitude besides finding out the most vulnerable drought-affected districts on the basis of such parameters.

11.3 The Study Area

West Bengal is the only state of India that connects both the Himalayas and the Bay of Bengal. Hence, climatic variability of West Bengal is very high. The state has experienced various climatic hazards ranging from high intensity and short duration phenomena like tropical cyclones, thunderstorms, and cloud bursts to low intensity long duration hazards, i.e., droughts. The study area (Fig. 11.1) consists of five districts situated in the southwestern part of West Bengal extending from 21°46′42″N to 24°36′04″N of latitudes and 85°47′21″E to 88°23′0″ E of longitudes covering an area of ~34,200 km^2, which is more than one-third area of the state. Precipitation characteristics of the region vary over place to place and time to time, although the mean precipitations range from 1328 mm in the west to 1550 mm in the southern

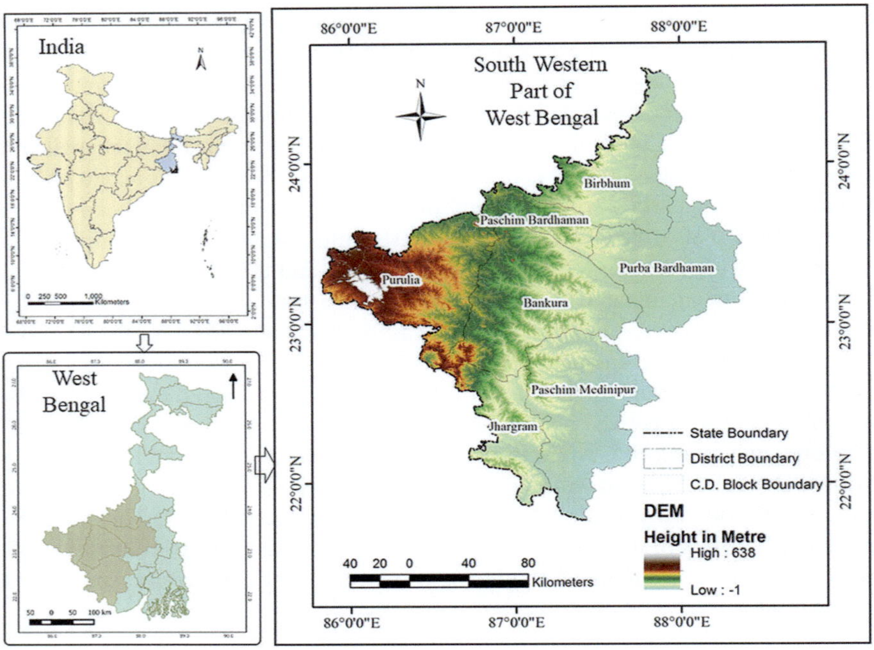

Fig. 11.1 Location map of the study area

part of the study area. However, the inter-annual variability of precipitation is very common, with the highest minimum annual precipitation of the region is 628 mm (in Bankura), while the maximum of 2347 mm (in Paschim Medinipur) recorded during 113 years (1901–2013). Summary of the precipitation characteristics of the study area is given in the Table 11.1. The maximum temperatures of the region also vary from place to place with the western side of the districts recording the highest maximum temperature of ~45 °C. High spatio-temporal variation of precipitation and intolerable temperature lead to high evapotranspiration and evaporation, which contribute to scarcity of water. Physiographically, the area is very significant, as it

Table 11.1 Precipitation characteristics of the study area (precipitations in mm)

Name of the districts	Minimum	Maximum	Range	Mean	SD	Skewness	Kurtosis
Birbhum	862.8	2163.8	1301.0	1379.4	275.94	0.37	−0.29
Barddhaman	973.0	1950.1	977.1	1329.3	217.04	0.41	−0.41
Purulia	652.4	2021.8	1369.4	1334.2	240.35	0.31	−0.03
Bankura	876.6	2140.5	1263.9	1363.6	266.43	0.51	0.00
Paschim Medinipur	907.1	2347.1	1440.0	1534.2	272.02	0.65	0.14

belongs to the transitional zone between the plateau fringe and the plain land; and the western portion of the area is thus drought-prone. Lateritic soil is the predominant soil type of the region, which has low water holding capacity, and the general slope is from the western part to the southeastern to increase the vulnerability of drought. The major economic activity of the region is agriculture with emphasis on cultivation of paddy, but much of it remains fallow due to scarcity of the water, especially in the dry seasons, although here the yield rates of paddy and other crops are found to be relatively high.

11.4 Data and Methodology

113 years of mean monthly precipitation data of five districts, namely, Birbhum, Barddhaman, Purulia, Bankura, and Paschim Medinipur have been collected from the India Meteorological Department (IMD) Pune. The data is almost complete as only 2–3% is missing data in the entire 113 years of period, and hence, it is quite reliable. The small amount of missing data have been estimated by simple arithmetic method and normal ration method, which is applicable in this case. Three meteorological drought indices (DIs) have been used to find out the regional drought scenario of the five districts. These three DIs are Standardized Precipitation Indices (SPI), Rainfall Anomaly Indices (RAI), and Effective Drought Indices (EDI), as discussed below.

11.4.1 Standardized Precipitation Indices (SPI)

SPI was developed by McKee et al. (1993) to estimate the intensity and duration of the drought event. In general, computation of SPI requires the fitting of a probability distribution of precipitation records for the timescale of interest in order to define probability of the precipitation. Then the fitted probability distribution is normalized to a standard normal distribution using the inverse normal function. In standard normal distribution, the mean of the variance of SPI for the location and desired time period are 0 and 1, respectively. Therefore, for any observed precipitation data, the SPI value is the deviation from the entire standard normal distribution. The multi-temporal SPI indices have been calculated as per McKee et al. (1993).

In 2009, the World Meteorological Organization (WMO) recommended all of the meteorological organizations situated in different parts of the world to use SPI along with their conventional index or indices to monitor drought. This is because SPI is flexible and it can be computed for multiple timescales. In shorter timescale, for example, 1, 2, or 3 months, SPI may play an effective role for early drought forecasting. Regional-scale drought can also be monitored using SPI.

However, the main demerit of SPI is that it totally depends on the availability of precipitation, but drought in a particular place is the result of various other factors like temperature and wind.

11.4.2 The Effective Drought Index (EDI)

White and Walcott (2009) propounded a new drought index called Effective Drought Index (EDI). EDI not only gives us the correct information about the onset, end, and stress of the drought, but it also emphasizes on the precipitation effectiveness. The original EDI was based on daily precipitation, but monthly precipitation also can be used (Dogan et al. 2012; Jain et al. 2015). Effective Precipitation (EP), as defined, is the outcome of the current month's precipitation and weightage of the previous month's precipitation, as in soil there remains some moisture from the previous month's precipitation. So, EDI is computed as the precipitation needed to return to the normal (PRN) condition. Similar to SPI, EDI is also used to classify drought events into four classes (Table 11.2).

PRN is calculated from monthly effective precipitation and its deviation from the mean for each month. Thus, the first step in the calculation of EDI is to calculate *EP*. If *Pi* is rainfall, "*m1*" is the month before the current month, and "*N*" is the duration of the preceding period, then effective precipitation for the current month (*EPj*) is given as

$$EP_j = \sum_{m=1}^{N}\left[\left(\sum_{i=1}^{m} P_i\right)/M\right]$$

For example, if N= 3, then EP= P1+ (P1+P2)/2+ (P1+P2+ P3)/3, where P1, P2, and P3 are rainfall amounts during the current month (j), previous month, and 2 months before, respectively. Then the average and standard deviation of *EP* values for each month are calculated, and the time series of *EP* values is converted to deviations from the mean (*DEP*)

$$DEP_j = EP_j - \overline{EP}_j$$

PRNj values are calculated using the formula:

$$PRN_j = \frac{DEP_j}{\sum_{i=1}^{N}\left(\frac{1}{i}\right)}$$

Table 11.2 Drought classification by DIs values

Drought category	SPI values	EDI value	RAI values
Mild drought	0 to −0.99	0 to −0.99	0 to −1.00
Moderate drought	−1.00 to −1.49	−1.00 to −1.49	−1.00 to −1.99
Severe drought	−1.50 to −1.99	−1.50 to −1.99	−2.00- to−2.99
Very severe drought	<-2.00	<-2.00	<-3.00

If N is 3 months, then the sum of the reciprocal is used to calculate, which is equal to 1/1+1/2+1/3. The final EDI value is obtained through dividing *PRN* by its standard deviation, which is given below:

$$DEP_j = PRN \big/ \sigma PRN$$

11.4.3 Rainfall Anomaly Indices (RAI)

The RAI was developed by Van-Rooy (1965). The positive and negative RAI values are computed by using the mean of 10 extremes values of precipitation. Let *M* be the mean of the 10% of the highest precipitation records for the period under study, "p" be the mean precipitation of all the records for the period, and the precipitation for the specific year, then the positive RAI (for positive anomalies) for that year is

$$\mathrm{RAI} = +\,3\frac{\mathrm{P} - \mathrm{p}}{M - \mathrm{p}}$$

Let *m* be the mean of the 10% of the lowest precipitation records for the period under study, and then the negative RAI (for negative anomalies) for the year is

$$\mathrm{RAI} = -3\frac{\mathrm{P} - \mathrm{p}}{m - \mathrm{p}}$$

11.5 Results and Discussion

From the monthly precipitation data, in total 3 DIs with 11 time steps (1 EDI, 5 SPI, and 5 RAI) have been computed (Table 11.3). As drought mainly occurs due to the deficiency of precipitation, multiple time steps is used here to establish the most suitable time step for drought delineation.

Table 11.3 shows that with the increasing time step of SPIs, drought month frequency of Mild Drought (MD) and Moderate Drought (MOD) has decreased, while the drought month frequency of Severe Drought (SD) and Very Severe Drought (VSD) have increased. In case of RAI, the result is reversed, with increase in frequency of MD and MOD, and decrease in frequency of SD and VSD. SPI-12 and EDI give us almost similar result so far as the different types of dry events are concerned, while RAI gives the most severe drought in comparison to the SPI or EDI. Hence, it can be concluded that RAI overemphasizes SD and VSD events. In case of inter-district comparison, Purulia recorded the highest number of VSD in all the DIs, and the frequency of MOD and SD also remained high in all other districts.

Table 11.3 Number of drought months in each DI of the selected five districts of West Bengal

DIs	Bankura				Birbhum				Barddhaman				Paschim Medinipur				Purulia			
	a	b	c	d	a	b	c	d	a	b	c	d	a	b	c	d	a	b	c	d
SPI3	612	149	21	2	605	154	26	4	597	140	33	2	603	165	23	0	595	145	27	3
SPI6	548	170	35	1	529	151	46	5	564	153	43	3	552	173	37	6	537	136	49	9
SPI12	495	138	69	4	461	168	57	5	512	158	58	2	572	128	45	12	526	122	51	17
SPI24	467	117	82	19	489	167	57	4	553	101	69	11	482	145	47	17	484	104	70	28
SPI48	489	120	54	25	474	115	64	16	442	105	103	14	537	95	55	19	510	114	66	23
EDI	531	157	47	1	522	155	56	0	522	163	48	4	548	147	41	5	552	143	31	13
RAI3	269	250	196	69	253	250	216	70	251	263	188	70	271	257	194	69	250	271	179	69
RAI6	270	243	173	68	238	264	162	68	300	232	162	69	267	254	179	68	247	273	144	68
RAI12	268	224	148	66	209	252	167	63	281	226	161	62	277	279	129	72	248	277	126	65
RAI24	306	200	105	74	283	205	165	64	334	231	95	74	268	214	150	59	316	203	95	72
RAI48	368	168	98	54	334	180	102	53	365	148	85	66	367	204	86	49	368	192	100	53

a. Mild Drought, b. Moderate Drought, c. Severe Drought, and d. Very Severe Drought

11.5.1 Comparisons of SPIs and EDI

There are 738 months of dry period (values less than 0) demarcated by EDI, of which 610, 627, 600, and 504 dry months belong to SPI3, SPI6, SPI12, and SPI24 at Purulia district. It is interesting that EDI with values of <-1.0 (169 months) is a very good match with SPI3 (187 months), SPI6 (133 months), and SPI12 (113 months) values while other indices have less dry period, especially SPI3 (102 months), SPI24 (79 months), and SPI48 (60 months), which means severe drought in this region may occur in the mid-temporal scale, while in shorter or long run time period, drought may not occur intensively. There are around 175 months of MOD period marked by SPI3 out of which SPI6 and SPI12, and EDI have also 93, 59, and 104 months of MOD period, respectively. It gives us good information that short-term precipitation loss is often recovered by high precipitation in later time period. In most of the cases, it is found that due to water scarcity in the pre-monsoon and the post-monsoon seasons, drought is marked by SPI3, but later on, huge precipitation in the monsoon period recovers short-term dry condition which is identified by EDI.

SPI with shorter temporal scale shows more MD event in comparison with the longer time step, while with increased time step Moderate Drought (MOD), Severe Drought (SD), and Very Severe Drought (VSD) events have increased. SPI with 12 months or higher time step shows lower MD, MOD, SD, and VSD compared to the EDI (Fig. 11.2a). SPI 9 and EDI have almost the same MD and SD events with 11 and 4% of drought events to the total.

11.5.2 Comparison of RAIs and EDI

From Fig. 11.2b, it is established that with the advancement of the temporal resolution MD event increases, while MOD to SD events decrease. In most of the cases, it is noticed that in very severe event, it did not change too much as it remains around 4–5% of the total events. Figure 11.2b shows that RAI3 has detected 5% of VCD events and 14% of SD events. In most of the cases, RAI has overperformed compared to the EDI. RAI3, RA6, and RAI12 show that around 30% of the drought months belong to moderate to very severe drought in comparison with the 15% of EDI. With the increased temporal resolution, moderate to very severe drought events have come down. As precipitation of the study area is mainly concentrated in during monsoon months, higher temporal scale gives us proper picture of water scarcity, which leads to the hydrological drought events.

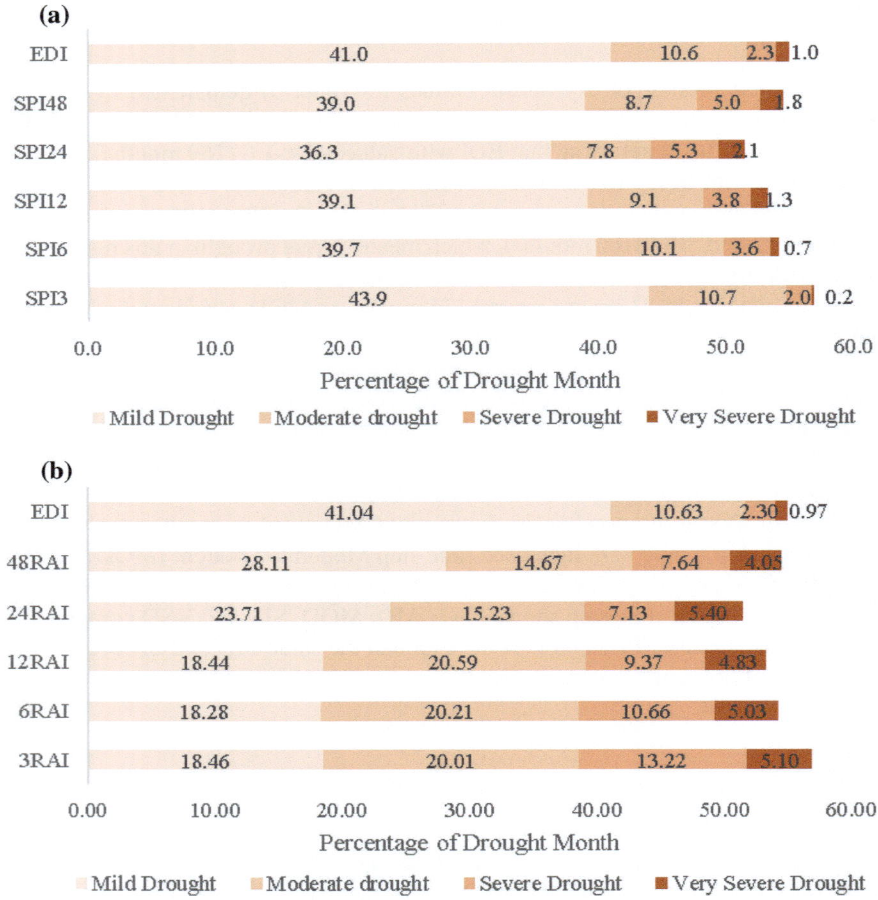

Fig. 11.2 **a** Comparison between EDI and SPIs. **b** Comparison between EDI and RAIs of Purulia district

11.5.3 Comparison Among the Different Drought Indices

To know the most suitable correlated drought index, 11 by 11 Pearson's correlation coefficient matrix (95% of confidence level and 5% of significance level) has been computed (Table 11.4). So, for each district, there are at about 121 correlation coefficient values. From the correlation matrix, it is found that with the increasing temporal scale, the correlation becomes weak as compared to the smallest temporal scale—although all of the correlations are good having the value above 0.55. From all the correlation matrices of five districts, two main characteristics are found: (1) Medium time step is highly correlated, while correlation with higher time step or lower time step is not good; and (2) DIs with the same time steps are highly correlated with

Table 11.4 Correlation matrix (Pearson) of multi-scale drought indices of Purulia

Variables	SPI3	SPI6	SPI12	SPI24	SPI48	EDI12	RAI3	RAI6	RAI12	RAI24	RAI48
SPI3	1.00	0.64	0.45	0.31	0.31	0.76	0.99	0.64	0.44	0.31	0.30
SPI6	0.64	1.00	0.66	0.45	0.42	0.86	0.63	1.00	0.66	0.45	0.40
SPI12	0.45	0.66	1.00	0.70	0.58	0.80	0.44	0.66	1.00	0.70	0.56
SPI24	0.31	0.45	0.70	1.00	0.74	0.56	0.30	0.45	0.70	1.00	0.72
SPI48	0.31	0.42	0.58	0.74	1.00	0.49	0.30	0.42	0.57	0.74	0.97
EDI12	0.76	0.86	0.80	0.56	0.49	1.00	0.75	0.86	0.80	0.56	0.48
RAI3	0.99	0.63	0.44	0.30	0.30	0.75	1.00	0.63	0.43	0.30	0.29
RAI6	0.64	1.00	0.66	0.45	0.42	0.86	0.63	1.00	0.66	0.45	0.40
RAI12	0.44	0.66	1.00	0.70	0.57	0.80	0.43	0.66	1.00	0.70	0.56
RAI24	0.31	0.45	0.70	1.00	0.74	0.56	0.30	0.45	0.70	1.00	0.72
RAI48	0.30	0.40	0.56	0.72	0.97	0.48	0.29	0.40	0.56	0.72	1.00
Average	0.56	0.65	0.69	0.63	0.60	**0.72**	0.55	0.65	0.68	0.63	0.58

Significance level alpha = 0.05

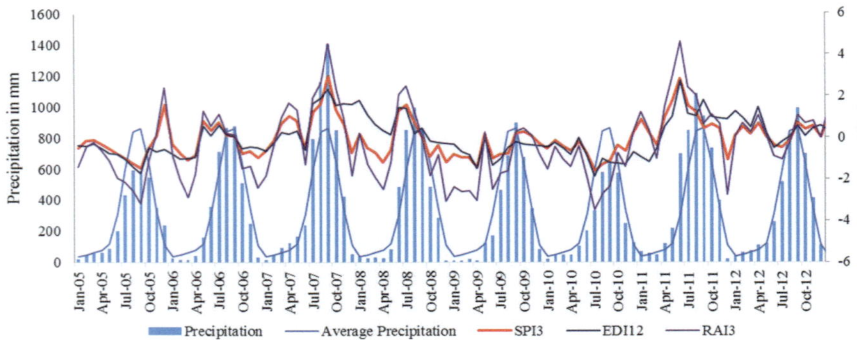

Fig. 11.3 Comparison of SPI3, EDI, and RAI3

each other, and correlation of one-time step with the nearby time step is also found to be good. It is also found that EDI is the best suited drought index having a good correlation of above 0.72 to the all of the multi-temporal DIs. The best correlation has been found between the SPI3 and RAI3 with 0.99, while the least correlation, that is, 0.29, is found between RAI3 and RAI48.

As EDI is computed on an annual basis (Table 11.4), it is highly correlated with SPI and RAI with the temporal step of 6 months to 24 months. Although the best correlated DI with EDI is SPI6 ($r^2 = 0.85$), for medium time step drought demarcation, i.e., from 9 months to 12 months, EDI is the best suited. But for hydrological drought delineation, SPI48 is the best of the DIs. While for shorter period drought detection, i.e., for 3 months' time step, SPI3 is the best suited drought index (Fig. 11.3) having the correlation value more than 0.56, while the correlation value of RAI3 is 0.55.

11.5.4 Discussion on Drought Events

11.5.4.1 Drought Events Classification

From the above discussion, it is clear that EDI is the best drought index, and SPI3 is the best meteorological drought index. Thus, SPI3 has been used here to identify internal characteristics of the drought events, like frequency, duration and magnitude.

(1) **Drought Initiation Time (DIT)**: Drought initiates whenever the precipitation shortfalls of the normal one and SPI3 having value less than zero can be considered as for identification of the drought event.

(2) **Drought Termination Time (DTT)**: The month when rainfall shortage becomes sufficiently small so that drought conditions no longer exists, and the value of SPI is found to be above zero.

(3) **Drought Duration (DD)**: The time period between DIT and DTT of a drought event (for example, DDb-c means the drought duration of b-c where DDb-c = DDb−DDc + 1 = 4 months).

(4) **Drought Magnitude (DM)**: It measures the cumulative deficiency of the precipitation within a drought period with respect to the defined threshold value (for example, DMb-c means the drought severity of b-c, where DMb-c = DMb−DMc).

(5) **Drought Intensity (DI)**: It is the average value of the precipitation deficiency throughout the drought event period. It can be measured by dividing DM by DD (for example, DIb-c = DMb-c/DDb-c).

For drought delineation, the period of 3 consecutive months having negative SPI value, and at least one month with SPI equal to or less than −1 (SPI<-1), is used as a threshold. Although in some cases prolonged dry period (negative SPI value) persists for 3 months or more with interruption of one-month rainfall (positive SPI), the total period is considered as one drought period.

11.5.4.2 Discussion on Historical Drought Events in the Study Area

(1) **Drought Frequency**

SPI3 is used to demarcate the frequency of drought events in the five districts. Purulia and Paschim Medinipur have recorded the highest frequency of drought events, with frequency of 67 in 113 years. Other districts like Birbhum and Bardhhaman have recorded 61 drought events each, while Bankura has recorded 60 drought events. It is also noticed that all of the districts have the highest number of drought events in the decades of 1920s, 1950s, and 1960s, although some differences have been found in regard to spatial characteristics of the droughts. The maximum number of drought events has been recorded in the period 1938–1975 in the districts of Paschim Medinipur and Barddhaman, while Purulia (Table 11.5) and Birbhum districts have the highest drought events in the period of 1901–1937, and Bankura district experienced it in latest period of 1976–2013.

(2) **Drought Duration**

The longest DD period of the study area is 38 months, recorded twice in Birbhum district from January 1964 to February 1967, and from November 1949 to December 1952. Consecutive 31 months of DD recorded in Paschim Medinipur district from May 1923 to February 1926. While in Bankura, the longest DD is 30 months recorded from March 1957 to August 1959. Barddhaman (27 months) and Purulia (26 months) districts recorded comparatively short duration of drought events. Every district of the study area shows a uniformity about the spatial extent of DD, as all of the districts show occurrence of the longest drought in the period of 1938–1975, while two other periods have a shorter duration of DD events.

(3) **Drought Magnitude**

The highest Drought Magnitude (DM) of 26.92 was recorded in Birbhum district during the period from January 64 to February 67, and DM of 26.77 from November

Table 11.5 Drought duration, drought magnitude and drought intensity in the district of Purulia

Onset month	End Month	DM	DD	DM	DI
June, 1965	July, 1967	21.72	26	21.72	−0.84
February, 1954	November, 1955	19.67	9	19.67	−2.19
August, 1914	September, 1916	19.22	26	19.22	−0.74
January, 1963	February, 1965	17.40	26	17.40	−0.67
October, 1971	April, 1973	15.11	19	15.11	−0.80
May, 1931	December, 1932	12.29	20	12.29	−0.61
December, 1909	September, 1911	12.02	22	12.02	−0.55
June, 1982	March, 1983	10.97	10	10.97	−1.10
September, 1988	April, 1989	9.94	8	9.94	−1.24
October, 2008	August, 2009	9.34	10	9.34	−0.93
November, 1933	February, 1935	9.31	16	9.31	−0.58
May, 1979	December, 1979	9.20	8	9.20	−1.15
October, 1950	April, 1952	8.81	17	8.81	−0.52
January, 1960	August, 1960	8.68	8	8.68	−1.09
June, 1969	September, 1970	8.60	16	8.60	−0.54
May, 1935	April, 1936	8.49	11	8.49	−0.77
December, 2009	November, 2010	8.17	12	8.17	−0.68
August, 1949	May, 1950	8.00	10	8.00	−0.80
December, 1902	September, 1903	7.61	10	7.61	−0.76
January, 1962	October, 1962	7.52	10	7.52	−0.75
November, 2004	October, 2005	7.41	12	7.41	−0.62
September, 1984	September, 1985	7.34	12	7.34	−0.61
March, 2000	March, 2001	6.91	13	6.91	−0.53
January, 1957	January, 1958	6.74	13	6.74	−0.52
May, 1906	February, 1907	6.46	10	6.46	−0.65

1949 to December 1952, while the highest DM of other districts was found to be in Bankura (DM of 23.69 from March 1957 to August 1959), Paschim Medinipur (DM of 23.32 from May 1923 to February 1926), Purulia (DM of 21.77 from June 1965 to July 1967), and Barddhaman (DM of 13.84 September 1953 to December 1955). Except Paschim Medinipur district, all other districts had high magnitude of drought occurrences in the period of 1938–1975, whereas during 1976–2013 drought occurrence became comparatively of medium magnitude as the DD within the period was less.

All of the districts had a very good relation between Drought Duration (DD) and Drought Magnitude (DM). The correlation between DD and DM indicate that DD was responsible for the higher DM with high correlation coefficient value ($R^2 >=$

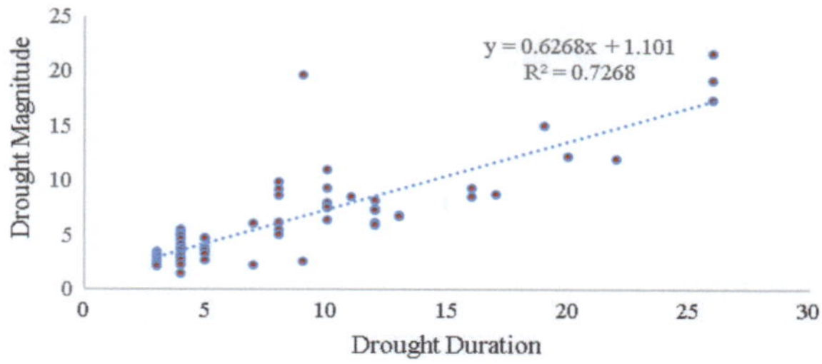

Fig. 11.4 Relationship between DD and DM of Purulia district

0.85). The relationship between DD and DM was not so good in case of Purulia with $R^2 = 0.73$ (Fig. 11.4) like other districts although as some high intensive but less duration drought occurred here.

11.6 Conclusions

The study reveals that how the DIs are useful for drought quantification in the southwestern part of West Bengal. The major findings of the study are:

(1) Comparison among multi-temporal SPI, EDI, and RAI indicates that all of the indices are highly correlated with each other for the same time periods, and the correlation increases at higher time periods.
(2) The correlation between SPI and RAI is high, compared to that between SPI and EDI or EDI and RAI for the same time period.
(3) RAI is not only the least correlated with EDI but it also overemphasized SSD and VSD events in all of the concerned districts.
(4) EDI is found to be the best suited drought index for long timescale, like 9–24 months of the time period, while SPI is the best suited for short time periods like 1 or 3 months' timescale.
(5) Overall, EDI is the best suited for agricultural drought, while SPI3 is the best for meteorological drought analysis. SPI3 also very well demarcates the drought scenario of recent years (i.e., 2000, 2010).
(6) A very high positive correlation has been found between DD and DM, with $R^2 > 0.85$ in all of the districts except Purulia, while $R^2 > 0.90$ has been found in the districts of Birbhum and Barddhaman.
(7) Out of 113 years of the time period, severe dry period has been noticed in the decades of the 1950s and 1960s. Moreover, in some cases, severe drought has occurred in two districts, while other districts remained comparatively less dry or wet. So, inter-district variation of drought within the southwestern part of West Bengal is one of the key findings of the present work.

Acknowledgements I would like to thank India Meteorological Department (IMD) for providing me the precipitation data, and the Centre for Scientific and Industrial Research (CSIR), Ministry of Earth Science (MoES) for funding this research work.

References

Akbari H, Rakhshandehroo GR, Sharifloo AH, Ostadzadeh E (2015) Drought analysis based on standardized precipitation index (SPI) and streamflow drought index (SDI) in Chenar Rahdar River Basin, Southern Iran. In: Moglen GE (Ed.) Watershed Management 2015: Power of the Watershed 11–22 American Society of Civil Engineers. https://doi.org/10.1061/9780784479322.002

Barua S, Ng AW, Perera BJ (2011) Comparative evaluation of drought indexes: case study on the Yarra River catchment in Australia. J Water Resour Plann Manag 137(2):215–226

Byun HR, Wilhite DA (1999) Objective quantification of drought severity and duration. J Climate 2747–2756. https://doi.org/10.1175/1520-0442(1999)012%3c2747:OQODSA%3e2.0.CO;2 retrived on 20.03.2017

Dogan S, Berktay A, Singh VP (2012) Comparison of multi-monthly rainfall-based drought severity indices, with application to semi-arid Koyna closed basin, Turkey. J Hydrol 470–471:255–268

Gao BC (1996) NDWI—A normalized difference water index for remote sensing of vegetation liquid water from space. Remote Sensing of Environment 257–266. https://doi.org/10.1016/S0034-4257(96)00067-3

Gibbs WJ, Maher JV (1967) rainfall deciles as drought indicators. Melbourne: Bureau of Meteorology (Australia)

Haied N, Foufou A, Chaab S, Azlaoui M, Khadri S, Benzahia K, Benzahia I (2017) Assessment and monitoring using meteorological indices in a semi-arid region. Energy Procedia 518–529. https://doi.org/10.1016/j.egypro.2017.07.064

Halwatura D, McIntyre N, Lechner AM, Arnold S (2017) Capability of meteorological drought indices for detecting soil moisture droughts. Journal of Hydrology: Regional Studies 396–412. https://doi.org/10.1016/j.ejrh.2017.06.001

Jain VK, Pandey RP, Jain MK, Byun HR (2015) Comparison of drought indices for appraisal of drought characteristics in the Ken River Basin. Weather Climate Extremes 8:1–11

Joetzjer E, Douville H, Delire C, Ciais P, Decharme B, Tyteca S (2013) Hydrologic benchmarking of meteorological drought indices at interannual to climate change timescales: a case study over the Amazon and Mississippi river basins. Hydrology and Earth System Sciences 4885-4895. https://doi.org/10.5194/hess-17-4885-2013

Keshavarz MR, Vazifedoust M, Alizadeh A (2014) Drought monitoring using a soil wetness deficit index (SWDI) derived from MODIS satellite data. Agric Water Manag 132:37–45

Keyantash J, Dracup JA (2002) The quantification of drought: An evaluation of drought indices. Bull Am Meteorol Soc 1167–1180. https://doi.org/10.1175/1520-0477-83.8.1167

Khalili D, Farnoud T, Jamshidi H, Kamgar-Haghighi AA, Zand-Parsa S (2011) Comparability analyses of the SPI and RDI meteorological drought indices in different climatic zones. Water Resour Manage 25(6):1737–1757

Kim DM, Byun HR, Choi KS (2009) Evaluation, modification and application of the effective Drought Index to 200-Year Drought climatology of Seoul, Korea. J Hydrol 378:1–12. https://doi.org/10.1016/j.jhydrol.2009.08.021 retrived on 15.03.2017

Kogan FN (1990) Remote sensing of weather impacts on vegetation in non-homogeneous areas. Int J Remote Sens 1405–1419. https://doi.org/10.1080/01431169008955102

Kogan FN (1995) Application of vegetation index and brightness temperature for drought detection. Adv Space Res 15(11):91–100

Le MH, Perez GC, Solomatine D, Nguyen LB (2016) Meteorological drought forecasting based on climate signals using artificial neural network—a case study in Khanhhoa province Vietnam. Procedia Engineering 154:1169–1175

Mahfouz P, Mitri G, Jazi M, Karam F (2016) Investigating the temporal variability of the standardized precipitation index in lebanon. Climate 4(27):1–14. https://doi.org/10.3390/cli4020027

McKee TB, Doesken NJ, Kleist J (1993) The relationship of drought frequency and duration to time scales. In: Eighth conference on applied climatology 179–184 Anaheim, California

Morid S, Smakhtin V, Moghaddasi M (2006) Comparison of seven meteorological indices for drought monitoring in Iran. Int J Climatol 26:971–985

Palmer WC (1965) Meteorological Drought. Washington D.C.: US Department of Commerce

Palmer WC (1967) Keeping track of crop moisture conditions, nationwide: the new Crop Moisture Index. Weatherwise 156–161. https://doi.org/10.1080/00431672.1968.9932814

Pathak AA, Channaveerappa, Dodamani BM (2016) Comparison of two hydrological drought indices. Perspect Sci 626–628. https://doi.org/10.1016/j.pisc.2016.06.039

Rahiz M, New M (2014) Does a rainfall-based drought index simulate hydrological droughts? Int J Climatol 34:2853–2871. https://doi.org/10.1002/joc.3879

Shafer BA, Dezman LE (1982) Development of a Surface Water Supply Index (SWSI) to assess the severity of drought conditions in snowpack runoff areas. In: Proceedings of the Western Snow conference 164–175 Fort Collins, CO: Colorado State University

Svoboda M, Hayes M, Wood D (2012) Precipitation index user guide. WMO-No. 1090. Geneva: World Meteorological Organization

Tsakiris G, Pangalou D, Vangelis H (2007) Regional drought assessment based on the reconnaissance drought index (RDI). Water Resour Manage 21(5):821–833

Van-Rooy MP (1965) A Rainfall anomaly index (RAI) independent of time and space. Notos 14:p43

Vicente-Serrano SM, Beguerı´a S, Lorenzo-Lacruz J, Camarero JJ, Lo´pez-Moreno JI, Azorin-Molina C, Revuelto J, Morán-Tejeda E, Sanchez-Lorenzo, A. (2012) Performance of drought indices for ecological agricultural and hydrological applications. Earth Interactions 16 p 10. https://doi.org/10.1175/2012EI000434.1

Vicente-Serrano SM, Begueria S, Lo´ Pez-Moreno JI (2010) A multiscalar drought index sensitive to global warming: the standardized. J Climate 1696–1718. https://doi.org/10.1175/2009JCLI2909.1

Wan Z, Wang P, Li X (2004) Using MODIS land surface temperature and normalized difference vegetation index products for monitoring drought in the southern great plains, USA. Int J Remote Sens 25:61–72

Wang KY, Li QF, Yang Y, Zeng M, Li PC, Zhang JX (2015) Analysis of spatio-temporal evolution of droughts in Luanhe River Basin using different drought indices. Water Science and Engineering 8(4):282–290. https://doi.org/10.1016/j.wse.2015.11.004

White DH, Walcott JJ (2009) The role of seasonal indices in monitoring and assessing agricultural and other droughts: a review. Crop and Pasture Science 599–616

Wilhite DA, Glantz MH (1985) Understanding the drought phenomenon: the role definitions. Water International 111–120

Wu H, Hayes MJ, Weiss A, Hu Q (2001) An evaluation of the standardized precipitation index, the China-Z Index and the statistical Z-Score. Int J Climatol 21:745–758

Zarch MA, Sivakumar B, Sharma A (2015) Droughts in a warming climate: a global assessment of Standardized precipitation index (SPI) and reconnaissance drought index (RDI). J Hydrol 526:183–195. https://doi.org/10.1016/j.jhydrol.2014.09.071

Chapter 12
Use of Remote Sensing and Geographic Information System on Agroforestry Ecosystem in Himalayan Region of Uttarakhand

Ganesh Datt Bhatt and Rajendra Singh Parihaar

Abstract The role of remote sensing and geographic information system in agroforestry ecosystem in the Himalayan region of Uttarakhand, India is presented in the research work. The satellite remote sensing data of Landsat MSS are procured from USGS and processed in ERDAS Imagine *ver.* 9.1 and Arc GIS *ver.* 10 software. The satellite data are classified as per supervised classification scheme. The satellite-based land use/cover thematic maps were prepared and validated in the field. The fieldwork was undertaken in the Himalayan region of Chamoli district of Uttarakhand. The maximum traditional agroforestry ecosystem was observed in the hill region as compared to Bhabhar region of Uttarakhand. The local people of Chamoli district are highly dependent on the natural resources for their daily livelihood. The economy of the local people depends on ecotourism, fruit trees, agricultural crops and dairy, etc. The local people are well educated having less employment opportunities. The youth are mostly engaged in labour work, shops, dairy production and ecotourism.

Keywords Agroforestry · Remote sensing · GIS · Himalayan region · Uttarakhand

12.1 Introduction

Acheamfour et al. (2015) stated that agroforestry is a unique land use system that intentionally blends perennial vegetation and herbaceous land cover types to enhance crop productivity, profitability and overall soil quality in an agroecosystem. Also, agroforestry combines trees with both annual crops and perennial pastures to increase sustainability of agricultural lands (Montagnini and Nair 2004; Nair et al.

G. D. Bhatt (✉)
School of Agriculture, Galgotias University, Plot No. 2, Sector 17 - a, Yamuna Expressway, Gautam Budh Nagar, Greater Noida 201310, Uttar Pradesh, India
e-mail: ganeshdattbhatt@gmail.com

R. S. Parihaar
Department of Botany, D.S.B. Campus, Kumaun University, Nainital 263001, Uttarakhand, India
e-mail: rajendraparihaar@gmail.com

S. Sahdev et al. (eds.), *Geoecology of Landscape Dynamics*,
Advances in Geographical and Environmental Sciences,
https://doi.org/10.1007/978-981-15-2097-6_12

2009; Kittur and Bargali 2013; Jhariya et al. 2015; Parihaar et al. 2015). Agroforestry is the intentional incorporation of woody vegetation with agricultural crops, grasses and farm animals on the same unit of land to produce ecological, economical and societal benefits and sustainable in nature (Rapid el et al. 2015; Armengot et al. 2016). Social factors like farmer economic and educational status (Anjichi et al. 2007; Matata et al. 2010; Yadav et al. 2016c), demography, social connections, culture and resource availability are important to understand why and how farmers select certain management practices (Seabrook et al. 2008). In Indian Himalaya, more than 90% people live in their villages, which are organized as independent socioecological systems (Agrawal and Chhatre 2006). Usually, the farmers have maintained close connections and equilibriums between agriculture, forestry and livestock, and these linkages are a basis of land use patterns. This indigenous agroforestry-based land use schemes not only support the livelihood through production of food, fodder and fuelwood, but also mitigate the impact of climate change through carbon sequestration (Singh et al. 2008; Bargali et al. 2009). Also, these systems add other ecosystem services like providing wildlife habitat (Jose 2009), maintaining biodiversity (Altieri 1999), reducing erosion (Lenka et al. 2012) and adding up microbial communities in soil (Banerjee et al. 2015). Fruit tree-based agroforestry land use systems constitute an important opportunity of organic matter addition to the soil ecosystem (Haile et al. 2008; Takimoto et al. 2008; Isaac et al. 2001; Rajper et al. 2016). Traditionally, forestry-based land use system in mountainous region is very near to natural ecosystems because it provides ecosystem services similar to the forest like biodiversity, availability of foodstuff and fibre, water resource and water cleansing, climate regulation and carbon sequestration, nutrient cycling, primary production of oxygen, soil formation, recreation and cultural services for the well-being of the people and society (Sharma 2007; Gosain et al. 2015; Singh and Singh 2016). Previous studies indicated that farmers make a modification to farming practices to take best advantage from the system. Therefore, land use options which increase livelihood security and reduce vulnerability to environmental changes are essential (Mendelsohn 2007).

Agroforestry ecosystem is related to tree species with agricultural crops and animals, and has been practiced since ancient times across the world in both the tropics and temperate regions. Traditionally, people are engaged in agroforestry practices for interdependent benefits of the three components; they are: trees, agricultural crops and livestock including food, fruit, fodder, fuel, fertilizer and fibre (6F) (Chavan et al. 2015). Most of the agroforestry ecosystems is part of indigenous traditional knowledge of local peoples (Parihaar et al. 2014; Kirtika et al. 2015). These agroforestry systems vary from one part of the Uttarakhand to other part of the world because of the diverse climatic, cultural and socio-economic conditions. The prominent traditional cultivation of local agricultural crops and fruits, and rearing of animals for dairy purposes have evolved long time. Management of agroforestry systems is practiced predominantly in smallholder farms and is influenced by a number of social factors besides economics. The objectives of the present study are—(i) to document the information about the existing traditional agroecosystems by the use of remote sensing and GIS techniques and (ii) to know about the changing pattern of the land use/land cover in Chamoli district of Uttarakhand, India.

12.2 Study Area

District Chamoli is the second largest district of Uttarakhand state of India (Fig. 12.1). Gopeshwar is the district headquarters. The study area lies between 79° to 80° E and 30° to 31° N, and is bounded by Tibet region in north and by districts of Uttarakhand,

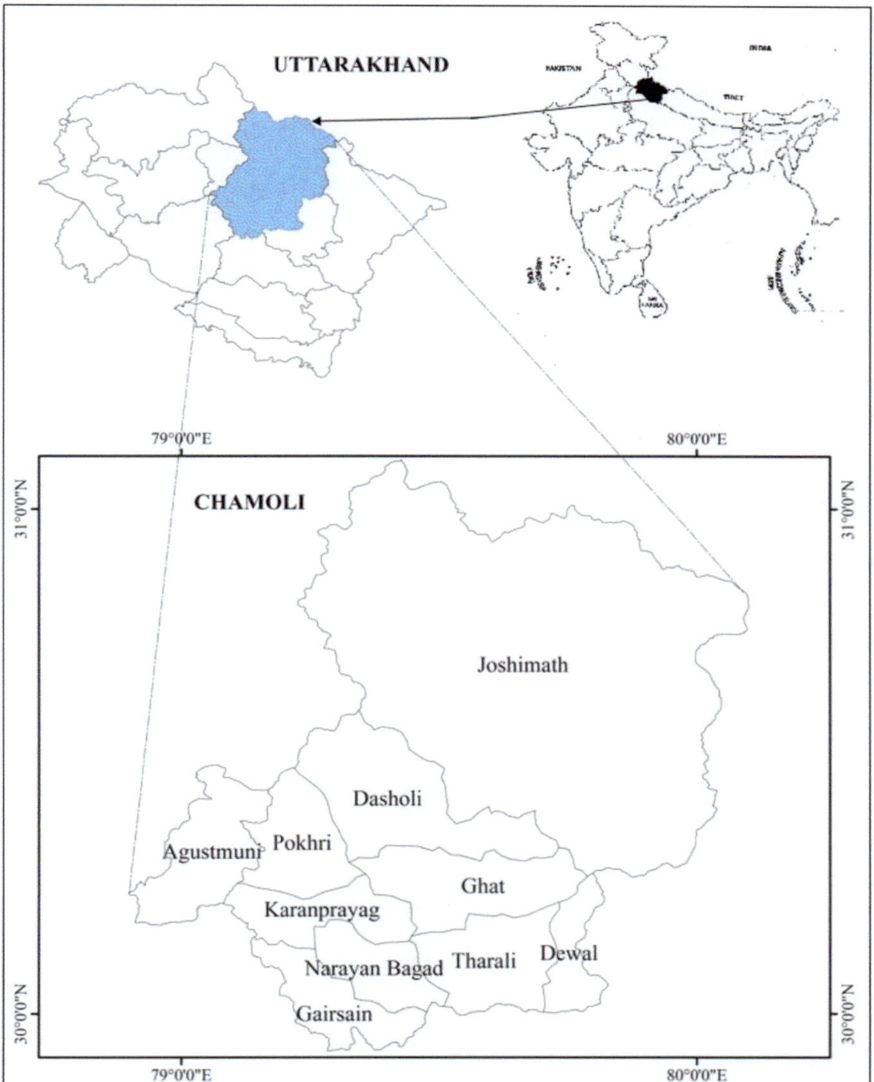

Fig. 12.1 Location of the study area

Table 12.1 Details of satellite data used in the present study

S. No.	Satellite	Sensor	Path	Row	Date	Spatial resolution (m)	Band
1.	-do-	MSS	156	39	1976-11-19	60	4
2.	-do-	MSS	156	39	2006-11-19	60	4

(*Source* United States Geological Survey)

namely, Pithoragarh and Bageshwar in east, Almora in south, Garhwal in southwest, Rudraprayag in west and Uttarkashi in northwest.

District Chamoli has six tehsils, viz., Joshimath, Chamoli, Karnaprayag, Pokhari, Gairsain and Tharali. It is divided into nine development blocks which are Joshimath, Dasoli, Pokhari, Ghat, Karnaprayag, Tharali, Narayanbagar, Dewal and Gairsain. Elevation of district Chamoli is 800–8000 m. Winter season is November to March. Rainfall is heaviest during monsoon season from June to September (Basistha et al. 2008). Major rivers in the study area are Alaknanda and Ramganga. Dhauli Ganga, Birhi Ganga, Nandakini and Pindar are the main tributaries of the study area.

12.3 Materials and Methods

12.3.1 Satellite Data Process

The Landsat MSS satellite data of 1976 and 2006 with spatial resolution 30 m was procured from Unites States Geological Survey (Source: www.usgs.gov.in). The satellite data information briefly explained in Table 12.1.

The satellite data were corrected radiometrically using ERDAS Imagine software ver. 9.1. The alterations such as histogram equalization, contrast enhancement, noise correction, dark pixel subtraction and haze were completed following Lillesand et al. (2008). The managed classification approach was incorporated to produce land use/cover classification scheme. Training sites, extraction of signatures from imagery and classification of the satellite imagery were done. Features of training sites were digitized and validated in the field. After training site digitization, statistical characterizations of the information were shaped. The maximum likelihood classification techniques were applied for the imagery. The study areas were extracted using the study area boundary.

12.3.2 Land Use/Cover in Chamoli

The land use/cover map of Chamoli district was made using satellite Landsat MSS data of 1976 (Fig. 12.2) and 2006 (Fig. 12.3). In 1976 and 2006, maximum area was

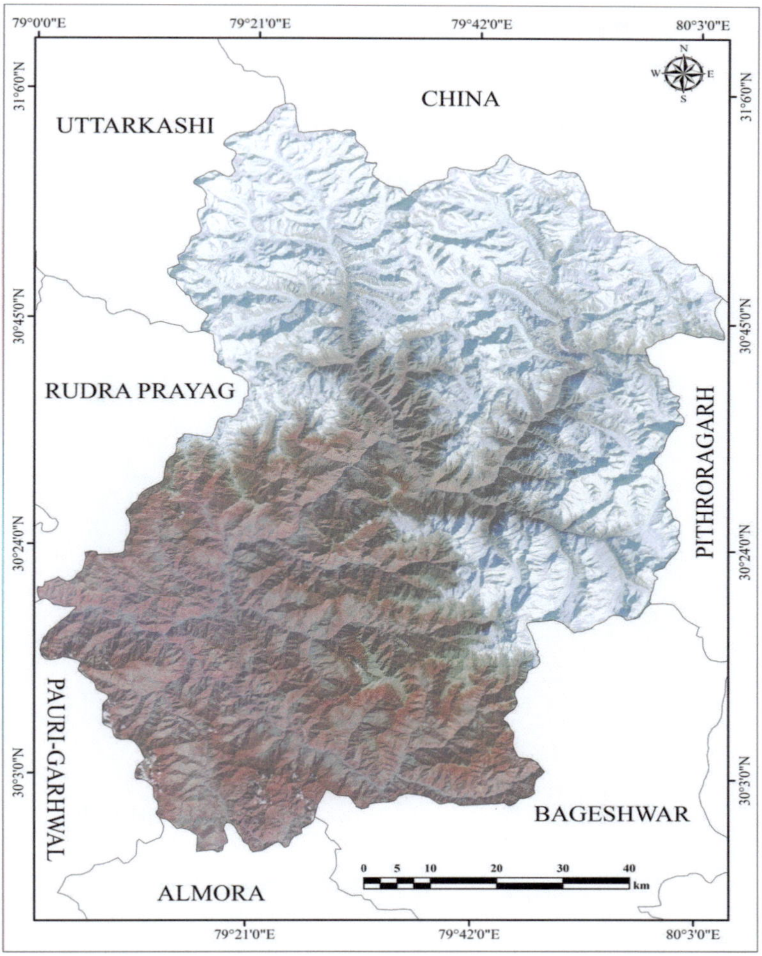

Fig. 12.2 Landsat MSS false colour composite map of 1976 (*Source* The United States Geological Survey)

covered mainly by forest and snow cover (Fig. 12.4), but the forest cover was observed less in 2006 (Fig. 12.5). It was observed that the forest cover in 1976 was 3966.40 km^2 (49.39%) followed by snow 2426.66 km^2 (30.22%) and agriculture 618.94 km^2 (7.70%). In the 2006, the forest cover was 3709.12 km^2 (46.19%) followed by snow 2500.82 km^2 (31.14%) and agriculture 643.78 km^2 (8.01%). The forest cover was reduced from 1976 to 2006 but snow and agriculture land was increased (Table 12.2).

There is an increase in the agricultural land in the study area because the wasteland and forest land are used for agriculture purposes. This can be simplified only through high spectral, spatial and temporal resolution qualities of remote sensing techniques.

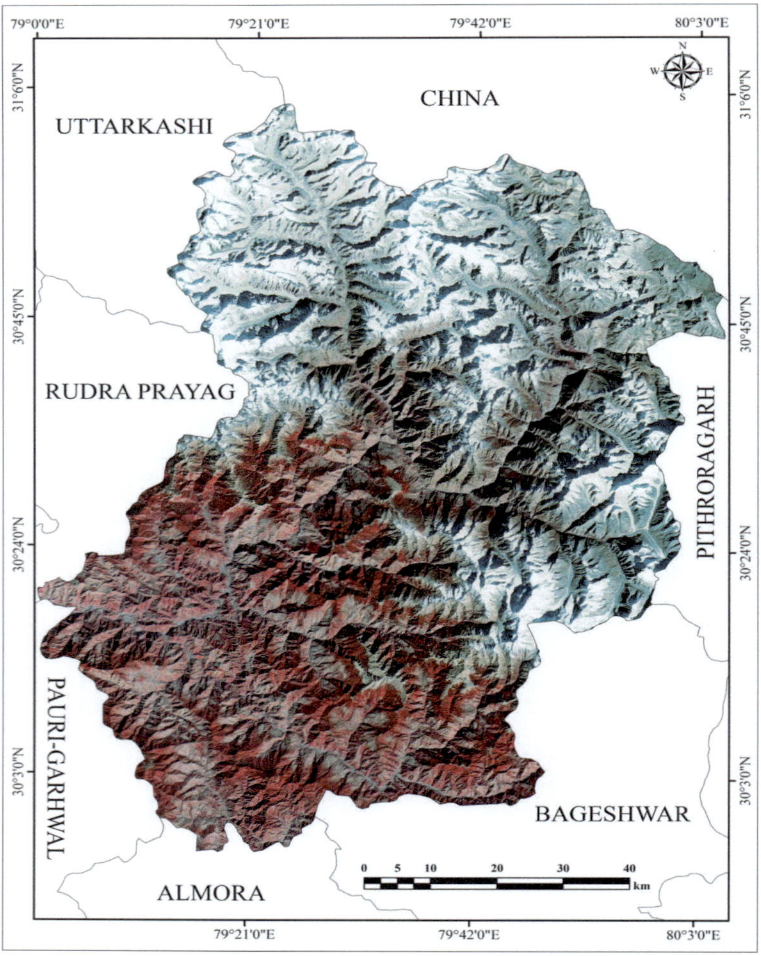

Fig. 12.3 Landsat MSS false colour composite map of 2006 (*Source* The United States Geological Survey)

Indeed, the precise database pertaining to forest cover information is an imperative input of formulating various management plans, and also remote sensing technology can be effectively utilized for change detection and monitoring activities (Jessica et al. 2001).

12.4 Results and Discussion

The following points are discussed in this section.

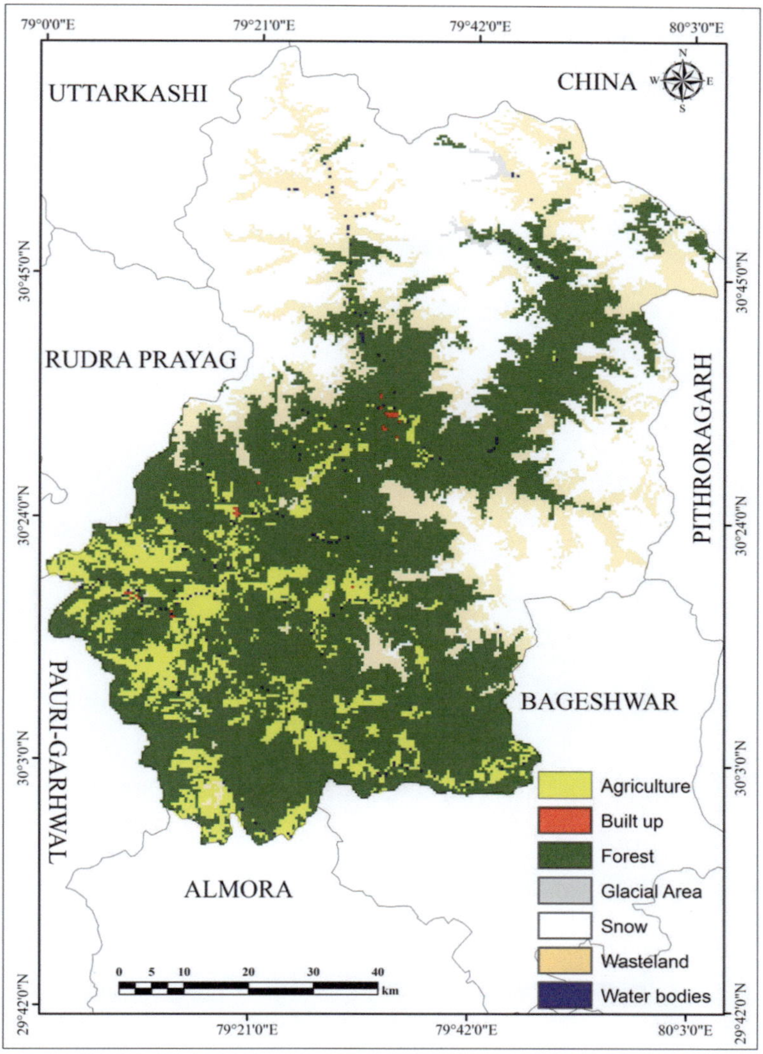

Fig. 13.4 Land use/cover map of 1976 (*Source* The United States Geological Survey)

12.4.1 Agri-Silviculture System

Agri-silviculture is a type of production technique which supplies woods, foodstuff and animal products from a single organizational unit where good agriculture practices are completed by the efficient use of trees. The system is the combination of the two components, i.e. agriculture and silviculture. This involves the establishment of an agriculture-cum-silviculture cycle so as to be a valuable symbiosis, and it is

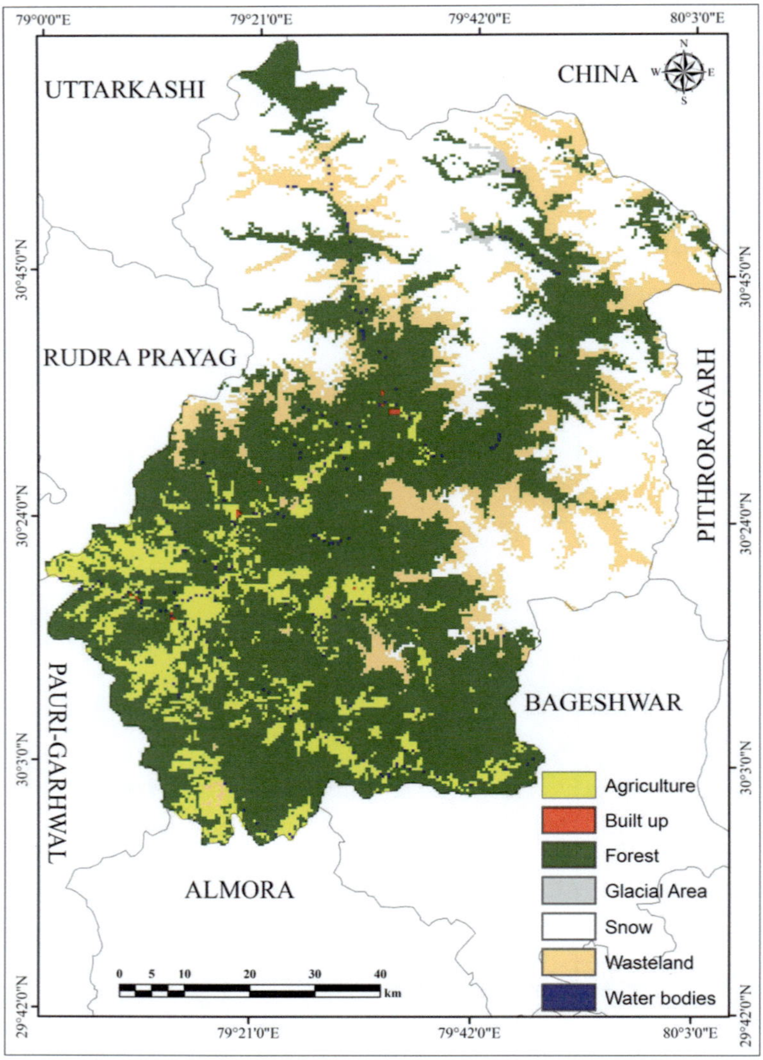

Fig. 13.5 Land use/cover map of 2006 (*Source* United States Geological Survey)

being accepted as a very popular system and simultaneously preventing the degradation of soil and vegetation. In this system, the agriculture crops such as *Triticum aestivum* L., *Solanum tuberosum* L., *Brassica nigra* (L.) K. Koch and *Pisum sativum* L. are grown during the winter season and *Oryza sativa* L., *Zea mays* L., *Solanum lycopersicum* L. and *Piper nigrum* L. in the summer season are cultivated on the permanent terraces prepared across the hill slopes, either as monoculture or polyculture, while fodder, fuel and timber trees like *Bauhinia variegate* L., *Celtisaus trails*

Table 12.2 Area covered under diverse vegetation types and land use categories in 2006

Vegetation types and land use	1976		2006	
	Area (km^2)	Area (%)	Area (km^2)	Area (%)
Agriculture	618.94	7.70	643.78	8.01
Built up	3.81	0.04	5.43	0.06
Forest	3966.40	49.39	3709.12	46.19
Glacial area	20.30	0.25	19.83	0.24
Snow	2426.66	30.22	2500.82	31.14
Wasteland	966.98	12.04	1124.25	14.00
Water bodies	26.88	0.33	26.74	0.33
Total	8030.00	100.00	8030.00	100.00

(*Source* The United States Geological Survey)

L., *Eucalyptus tereticornis* Sm., *Ficus hispida* L.f., *Ficus palmate* Forssk., *Grewia optiva* J.R. Drumm. Ex Burret, *Morus alba* L., *Populus deltoides* Marshall, *Tectona grandis* L.f., etc., were deliberately grown on the bunds of terraces (Table 12.3).

12.4.2 Agri-Horticulture System

Agri-horticulture land use system is made up by the combination of two systems. One is agriculture and another is horticulture. Those areas where fuel and fodder are easily available from other sources, land use practice by the system is very common. In the agri-horticulture system, the agriculture crops are grown in the interspaces of horticulture trees like *Malus domestica* Borkh., *Sorbus domestica* L., *Prunus armeniaca* L., *Prunus persica* (L.) Batsch, *Pyrus bourgaeana* Decne., etc., are grown in the hills of Garhwal region, *Psidium guajava* L., *Mangifera indica* L., *Litchi chinensis* Sonn., etc., in the Bhabhar belt. The production of fruit is the dominant component in the agro-horticulture system for commercial as well domestic use (Table 12.3).

12.4.3 Agri-Silvi-Pastoral System

The agri-silvi-pastoral system includes the trees and grasses that are grown together for production of fodder, fuel and timber, and soil conservation. Animal rearing is an important component of the agri-silvi-pastoral system where milk production along with bio-fertilizers for enhancing and maintaining the soil fertility and productivity is obtained. Along with grasses, trees fodder is also utilized for stall feeding. Traditionally, agri-silvi-pastoral system is considered as highly sustainable system in

Table 12.3 Characteristics of agroforestry-based ecosystems

Agroforestry system	Characteristics	Woody/fodder plant species			Herbaceous plant species		
		Common name	Botanical name	Uses	Common name	Botanical name	Uses
Agri-silviculture	In this system, principal crops like wheat, rice, maize, jaun (barley), etc., were grown with multipurpose trees like Bhimal, Kachnar, Padam, Timil and oaks. Generally, trees were present on field boundaries, terrace risers and in agricultural fields in scattered manner. Cultivation was done on varying degree of slopes on terraced fields. The repeated fragmentation and sub-division had reduced the size of the holding. Chilly and ginger were the main Kharif cash crops while potato, garlic and onion were the major Rabi cash crops	Bhimal	*Grewia optiva* J.R. Drumm. Ex Burret	Fu, Fo, Fi	Gehu	*Triticum aestivum* L.	Ed, Fo, Com
		Chir	*Pinus roxburghii* Sarg.	Ed, Ti, Com, Fu	Dhan	*Oryza sativa* L.	Ed, Fo, Com
		Padam	*Prunus cerecoides* D. Don	Fu, Fo, Ti, Reli	Makka	*Zea mays* L.	Ed, Fo, Com
		Tilonj	*Quercus floribunda* Lindl.	Fu, Fo, Ti	Maduwa	*Eleusine coracana* Gaertn.	Ed, Fo, Com
		Banj	*Q. leucotrichophora* A. Camus	Fu, Fo, Ti	Jhangora	*Echinochloa frumentacea* Link.	Ed, Com
		Gethi	*Boehmeria regulosa* Wedd.	Fo	Jaun	*Hordeum vulgare* L.	Ed, Fo
		Kharik	*Celtis australis* L.	Fu, Fo	Alu	*Solanum tuberosum* L.	Ed, Com

(continued)

Table 12.3 (continued)

Agroforestry system	Characteristics	Woody/fodder plant species			Herbaceous plant species		
		Kachnar	*Bauhinia variegate* (L.) Benth.	Fo, Ed, Med	Adarak	*Zingiber officinale* Roscoe	Sp, Com
		Timil	*Ficus roxburghii* Steud.	Fo. Ed	Sarson	*Brassica campestris* L.	Ed, Oil, Fo, Com
		Liptis	*Eucalyptus tereticornis* Sm.	Ti, Com	Urd	*Phaseolus mungo* L.	Fo
		Poplar	*Populus deltoids* Marshall	Com	Chilly	*Capsicum annum* L.	Ed, Com
		Sagwan	*Tectona grandis* L.F.	Ti, Com	Onion	*Allium cepa* L.	Ed, Com
		Sahatut	*Morus alba* L.	Fo, fu,			
Agri-horticulture	In this system, in hills fruit trees such as apple, apricot, pear, etc., were grown in systematic fashion in orchards with field crops. Turmeric and ginger were cash crops in peach gardens while potato was ideal cash crop for apple orchard. While in Tarai-Bhabhar area fruit trees like mango, litchi, guava, peach, amla, etc., were grown intermixed with sugarcane and other crops	Khumani	*Prunus armeniaca* (L.) Dumort.	Ed, Com	Rajma	*Dolichos lablab* L. Sweet	Ed, Fo.
		Plum	*Prunus domestica* L.	Ed, Com	Adarak	*Zingiber officinale* Roscoe	Sp, Com, Medicine
		Santra	*Citrus Sp.*	Ed, Com	Haldi	*Curcuma longa* L.	Sp, Com, Med
		Amrud	*Psidium guajava* L.	Ed, Com	Potato	*Solanum tuberosum* L.	Ed, Com

(continued)

Table 12.3 (continued)

Agroforestry system	Characteristics	Woody/fodder plant species			Herbaceous plant species		
		Seb	*Malus domestica* Borkh.	Ed, Com	Beans	*Vicia faba* L.	Ed
		Aru	*Prunus persica* (L.) Stoke	Ed, Com			
		Naspati	*Pyrus communis* L.	Ed, Com			
		Kafal	*Myrica esculenta* Buch.	Ed, Com			
		Darim	*Punica granatum* L.	Ed, Com			
		Aam	*Mangifera indica* L.	Ed, Com			
		Litchi	*Litchi chinensis* Sonn.	Ed, Com			
		Amla	*Phyllanthus Emblica* L.	Ed, Com			
Agri-horti-silviculture	This system was managed for the production of fruits, crops, fodder and fuel or wood for packaging materials. Fruit trees were planted at regular interval within the fields, fodder and timber trees were left on the field bunds while the annuals were grown in the interspaces of the trees	Woody plants and herbaceous species were almost similar as described for above agri-silvi and agri -horticulture systems.					

(continued)

Table 12.3 (continued)

Agroforestry system	Characteristics	Woody/fodder plant species			Herbaceous plant species		
Home gardens	This is a traditional component of the rural ecosystem that has been practiced for a long time by farmers. In this system, several plant species (mainly fruit, fodder, vegetable and spices) were grown and maintained near homestead by the household members and their products were primarily intended for the family consumption. Plant species were selected on the basis of climate and season. This system is an important source of food and nutrition and also provide secure livelihood	Banana	*Musa paradisiaca* L.	Ed, Com	Muli	*Raphanus sativus* L.	Ed, Com
		Papaya	*Carica papaya* L.	Ed, Med, Com	Arbi	*Colocasia esculenta* (L.) Schott	Veg, Com
		Kathal	*Artocarpus heterophyllus* Lam.	Ed, Com	Onion		Veg,Com.
		Darim	*Punica granatum* L.	Ed, Com	Garlic	*Allium sativum* L.	Sp, Com
		Aam	*Mangifera indica* L.	Ed, Com	Rai	*Brassica rugosa* L.	Green Veg, Sp, Oil
		Litchi	*Litchi chinensis* Sonn.	Ed, Com	Palak	*Spinacea oleracea* L.	Green Veg

(continued)

Table 12.3 (continued)

Agroforestry system	Characteristics	Woody/fodder plant species			Herbaceous plant species		
		Amrud	*Psidium guajava* L.	Ed, Com	Beans	*Vicia faba* L.	Ed
		Aaru	*Prunus persica* (L.) Stokes	Ed, Com	Tomato	*Lycopersicum esculentum* L.	Ed, Com
		Santra	*Citrus Sp.*	Ed, Com	Karela	*Momordica charantia* L.	Ed
		Malta	*Citrus sp.*	Ed, Com	Mater	*Pisum sativum* L.	Ed
		Lemon	*Citrus lemon* (L.) Burm. F.	Ed, Com	Kaddu	*Cucurbita maxima* Duchesne	Ed, Com.
					Bringal	*Solanum melongena* L.	Ed, Com
					Lauki	*Lagenaria siceraria* (Molina) Standl.	Ed, Com
					Dhania	*Coriandrum sativum* L.	Sp, Med, Com

(continued)

Table 12.3 (continued)

Agroforestry system	Characteristics	Woody/fodder plant species			Herbaceous plant species		
Agri-silvo-pastoral (S5)	In this system, agriculture, trees, shrubs and grasses were grown and maintained for fuel, fodder, timber and other benefits. This system was mostly managed by landowner on community land of the village	Apple	*Malus domestica* Borkh.	Ed, Com	Mater	*Pisum sativum* L.	Ed, com
		Kharsu	*Quercus leucotrichophora* A. Camus	Fo, Fu, Ti, Com	Alu	*Solanum tuberosum* L.	Ed, Com
		Baj	*Quercus floribunda* Lindl.	Fo, Fu, Ti, Com	Muli	*Raphanus sativus* L.	Com, Med, Ed
		Naspati	*Pyrus communis* L.	Ed, Com	Makka	Zea mays L.	Ed, Fo, Com
		Deodar	*Cedrus deodara* (Roxb.) G. Don	Com, Ti			

Abbreviation: Ed—Edible; Fo—Fodder; Com—Commercial; Ti—Timber; Sp—Spices; Med—Medicine; Veg—Vegetables

Garhwal Himalaya. Generally, open grazing has been replaced by stall feeding, and the size of animal heard has been reduced since 5–10 years. Adoption of girl education, small family size, small land holding, growing of commercial and cash crops are some of the reasons for stall feeding and small size animal heard (Table 12.3).

12.4.4 Agri-Horti-Silviculture System

Agri-horti-silviculture system is a common practice by the farmers of Garhwal region including the cultivation of agriculture crops along with forest and horticulture trees in the same land. The preparation of agri-horti-silviculture (agriculture crop + horticulture tree + forest) system provides stable and quality output to farmers and fodder, fuelwood for packaging materials. At regular intervals, fruit trees are planted in the field and fodder; timber trees are left on the field bunds although the annuals were grown in the interspaces of the trees. The majority of farmers has followed the agri-horti-silviculture system which is economically viable and ecologically sustainable (Table 12.3).

12.4.5 Home Gardens

Home gardens include organization of multipurpose trees and shrubs in association with annual and perennial agricultural vegetable crops and consistently livestock within the compound of distinct houses. Homestead gardens including trees are one of the known agroforestry practices which is ecologically workable and diversifies source of revenue of local people. Home garden is defined as land use system involving planned management of multipurpose trees and shrubs in intimate association with biannual and perennial agricultural crops and invariable livestock within the compound of individual houses; the whole tree crop and animal unit is intensively managed by family labour (Kumar and Nair 2006). They supply and supplement necessities and generate secondary income. They tend to be located close to dwellings for convenience and security (Ninez 1996). The whole crop tree-animal unit is being managed by the family members. In the home garden system, supplementary vegetables, fruits and spices are grown in mixture for own consumption as well as for marketing. Home gardens play a major role in getting better the health as the vegetables are a good source of vitamins, proteins, minerals and fibre, all of which are necessary for human activity and functioning. One of characteristic of home garden is high diversity of plant species that can be planted to reduce risks of crop failure, thereby increasing biological stability (Hames 1983; Bhatt et al. 2014). The species composition of home gardens can, however, differ from one garden to another, depending on climate and economic status of the owner (Kathleen and Mark 1990). A few species that are grown in the study area are: *Raphanus sativus* L., *Colocasia esculenta* L., *Allium cepa* L., *A. sativum* L., *Brassica campestris* L., *Spinacea oleracea* L., *Vicia faba* L., *Pisum sativum* L., *Solanum tuberosum* L. and *Zingiber officinale* Roscoe (Table 12.3).

12.4.6 Farming Systems

Agroforestry farming system in Chamoli district is mostly monsoon-based rains (July to September), which account for about the rain fed and traditional. The important agricultural crops in the study area are: *Triticum aestivum* L., *Leersia oryzoides* (L.) Sw., *Vigna mungo* (L.) Hepper, *Macrotylom auniflorum* (Lam.) Verdc., *Eleusine coracana* (L.) Gaertn., *Phaseolus lunatus* L., *Brassica rapa* L., *Glycine max* (L.) Merr., *Pisum sativum* L., *Hordeum vulgare* L., *Vigna radiata* (L.) R.Wilczek and *Macrotylom auniflorum* (Lam.) Verdc. The agroforestry tree species in the study area are: *Quercus oblongata* D.Don, *Grewia oppositifolia* Roxb. ex DC., *Celtisaus tralis* L., *Ficus auriculata* Lour., *Ficus racemosa* L., *Bauhinia variegata* L. and *Toona ciliate* M.Roem. The horticultural crops in the study area are: *Citrus aurantiifolia* (Christm.) Swingle, *Citrus pseudolimon* Tan., *Juglans regia* L., *Morus alba* L., *Prunus armeniaca* L., *Prunus persica* (L.) Batsch, *Punica granatum* L. and *Pyrus pyraster* (L.) Burgsd (Table 12.3).

12.5 Conclusion

It is suggested, if proper interaction were made between the local people, villagers and farmers, agriculture department for better understanding and development for sustainable development of ecosystem in the study area. The ground-based policy preparation and management should be prepared for better agroforestry productivity for livelihood. Agroforestry practices such as agri-silviculture and agri-horticulture system for wood, food and fruit, production, wind and soil protection by boundary and contour planting; fodder production as well as soil and water conservation through silvo-pastoral system; and home gardens, agri-silvi-horticulture for food, fruit and fodder should be promoted for the sustainable developments of the nature and natural resources.

Acknowledgements The authors gratefully acknowledge Prof. S.S. Bargali, Department of Botany, Kumaun University, Nainital for his valuable suggestions, support, encouragement and help for preparation of the manuscript.

References

Acheamfour MB, Chang SX, Carlyle CN, Bork EW (2015) Carbon pool size and stability are affected by trees and grassland cover types within agroforestry systems of western Canada. Agri Ecosyst Environ 213:105–113

Agrawal A, Chhatre (2006) Explaining success on the commons: community forest governance in the Indian Himalaya. World Dev 34(1):149–166

Altieri MA (1999) The ecological role of biodiversity in agro ecosystems. Agri Ecosyst Environ 74:19–31

Anjichi VE, Mauyo LW, Kipsat MJ (2007) The effect of socioeconomic factors on a farmer decision to adopt farm soil conservation measures: an application of multi variate logistic analysis in Butere/Mumias District Kenya. In: Bationo A, Waswa B, Kihara J, Kimetu J (eds) Advances in integrated soil fertility management in Sub-Saharan Africa: challenges and opportunities. Springer, Dordrecht, pp 915–920

Armengot L, Barbieri P, Andres C, Milz J, Schneider M (2016) Cacao agro forestry systems have higher return on labor compared to full-sun monocultures. Agron Sustain Dev 36:70

Banerjee S, Baah-Acheamfour M, Carlyle CN, Bissett A, Richardson AE, Siddique T, Bork E, Chang S (2015) Determinants of bacterial communities in Canadian agro forestry systems. Environ Microbiol 1462–2920

Bargali SS, Bargali K, Singh L, Ghosh L, Lakhera ML (2009) *Acacia nilotica* based traditional agro forestry system: effect on paddy crop and management. Curr Sci 96(4):581–587

Basistha A, Arya DS, Goel NK (2008) Spatial distribution of rainfall in Indian Himalayas—a case study of Uttarakhand region. Water Resour Manag 22(10):1325–1346

Bhatt GD, Sinha K, Deka PK, Kumar A (2014) Flood hazard and risk assessment in Chamoli district, Uttarakhand using satellite remote sensing and GIS techniques. Int J Innov Res Sci Eng Technol 3(8):15348–15356

Bhatt GD, Uniyal S, Yadav S, Deka PK (2015) Remote sensing and GIS tools used for changed detection techniques in Chamoli district, Uttarakhand, India. Int J Innov Res Sci Eng Technol 4(9):8109–8123

Chavan SB, Keerthika A, Dhyani SK, Handa AK, Ram N, Rajarajan K (2015) National Agro forestry Policy in India: a low hanging fruit. Curr Sci 10(25):1826–1834

Gosain BG, Negi GCS, Dhyani PP, Bargali SS, Saxena R (2015) Ecosystem services of forests: carbon stock in vegetation and soil components in a watershed of Kumaun Himalaya. India. Int J Ecol Environ Sci 4(3–4):177–188

Haile SG, Nair PKR, Nair VD (2008) Carbon storage of different soil-size fractions in Florida silvi pastoral systems. J Environ Qual 37:1789–1797

Hames R (1983) Monoculture, Poly culture and Poly variety in tropical forest Swidden cultivation. Hum Ecol 2(1):11

Isaac ME, Harmand JM, Lesueur D, Lelon J (2001) Tree age and soil phosphorus conditions influence N_2 fixation rates and soil N dynamics in natural populations of *Acacia senegal*. Forest Ecol Manag 261:582–588

Jhariya MK, Bargali SS, Abhishek R (2015) Possibilities and perspectives of agroforestry in Chhattisgarh. In: Zlatic M (ed) Precious Forests—Precious Earth. Publisher- INTECH 237–257

Jose S (2009) Agro forestry for ecosystem services and environmental benefits: an overview. Agro Forest Syst 76:1–10

Kathleen L, Mark B (1990) *Tropical home gardens*. United Nations University Press, Tokyo, pp 105–114

Kittur B, Bargali SS (2013) Perspectives of agroforestry: present and future facets. J Prog Agri 4(2):91–94

Kumar BM, Nair PKR (2006) Tropical Home gardens: a time-tested example of sustainable agro Forestry-*Advances in agro forestry*. Springer Science, Dordrecht, the Netherlands, p 377

Lenka NK, Dass A, Sudhishri S, Patnaik US (2012) Soil carbon sequestration and erosion control potential of hedge rows and grass filter strips in sloping agricultural lands of eastern Indian. Agric Ecosyst Environ 158:31–40

Lillesand TM, Kiefer RW, Chipman JW (2008) Remote sensing and image interpretation. 6th Eds. Wiley, New York 756

Matata PZ, Ajay OC, Oduol PA, Agumya A (2010) Socio-economic factors influencing adoption of improved fallow practices among smallholder farmers in western Tanzania. Afr J Agr Res 5(8):818–823

Mendelson R, Seo SN (2007) Changing farm types and irrigation as adaption to climate change in South America, *World Bank Policy Research Series Working Paper*, Washington DC, USA 4161

Montagnini F, Nair PKR (2004) Carbon sequestration: An underexploited environmental benefit of agro forestry systems. Agro Forest Syst 61:281–295

Nair PKR, Kumar BM, Nair VD (2009) Agro forestry as a strategy for carbon sequestration. J Plant Nutr Soil Sci 172:10–23

Ninez V (1996) Household level food production. Int Potato Center. Lima, Peru. 24:182–196

Padalia K, Kiran B, Bargali SS (2015) How does traditional home-gardens support ethnomedicinal values in Kumaun Himalayan bhabhar belt, India? Afr J Tradi Complement Alter med 12(6):100–112

Parihaar RS, Kiran B, Bargali SS (2014) Diversity and uses of Ethno-medicinal plants associated with traditional agroforestry systems in Kumaun Himalaya. Indian J Agr Sci 84(12):1470–1476

Parihaar RS, Kiran B, Bargali SS (2015) Status of an indigenous agroforestry system: a case study in Kumaun Himalaya India. Indian J Agr Sci 85(3):442–447

Rajper AM, Udawatta RP, Kremer RJ, Lin C, Jose S (2016) Effects of pro-biotics on soil microbial activity, biomass and enzymatic activity under cover crops in field and green house studies. Agro Forest Syst https://doi.org/10.1007/s10457-016-9895-1

Rapidel B, Ripoche A, Allinne C, Metay A, Deheuvels O, Lamanda N, Blazy JM, Valdés-Gómez H, Gary C (2015) Analysis of ecosystem services trade-offs to design agro ecosystems with perennial crops. Agron Sustain Dev 35:1373–1390

Seabrook L, McAlpine C, Fensham RJ (2008) What influences farmers to keep trees? a case study from the Brigalow Belt Queensland Australia. Landsc Urban Plan 84:266–281

Sharma R, Xu J, Sharma G (2007) Traditional agro forestry in the eastern Himalayan region: Land management system supporting ecosystem services. Trop Ecol 48(2):189–195

Singh L, Ghosh L, Bargali SS, Saxena RR (2008) Response of paddy planted under *Butea monosperma* trees in Chhattisgarh. Int J Ecol Environ Sci 34(1):63–66

Singh SP, Singh VP (2016) Addressing rural decline by valuing agricultural ecosystem services and treading food production as a social contribution. Trop Ecol 57(3):381–392

Yadav RP, Gupta B, Bhutia PL, Bisht JK (2016) Socioeconomics and sources of livelihood security in Central Himalaya, India: a case study. Int J Sust Dev World Ecol 23:1–9

Zhang JT, Ru W, Li B (2006) Relationships between vegetation and climate on the Loess Plateau in China. Folia Geobot 4:151–163

Chapter 13
UN SDGs and Context of Holy-Heritage Cities in India: A Study of Ayodhya and Varanasi

Rana P. B. Singh, Sarvesh Kumar and Pravin S. Rana

Abstract Following the UN SDGs (Sustainable Development Goals) Target 11.4, focussing on 'Transforming our World: The 2030 Agenda for Sustainable Development', cultural heritage and urban sustainability are now considered inseparable part of holistic development. Religion had played an important role for controlling power in Indian monarchy in the ancient past, and in contemporary India too it played a role in the formation of religious landscape and corporate identity of religious heritage, through commonly using processions, pilgrimage, religious assemblies, religious fairs (*melā*) and sacred places. Newly introduced concepts of pilgrimage tourism will help to revitalise these sacred cities. Situated in the Ganga river basin, Ayodhya and Banaras both have been primarily ancient *tīrthas* (riverfront sacredscapes) and salvific cities that record settlement continuity since at least ca 800BCE. Under the umbrella of holistic development, Government of India is promoting inclusive heritage sustainable development of heritage sites (cities), through active participation of stakeholders in the purview of SDGs and future development.

Keywords SDGs · Pilgrimage tourism · Ayodhya · Varanasi · Inclusive heritage development · Riverfront sacredscapes

R. P. B. Singh (✉) · P. S. Rana
Banaras Hindu University, Varanasi 221005, UP, India
e-mail: ranapbs@gmail.com
URL: https://banaras.academia.edu/RanaPBSINGH/Papers

P. S. Rana
e-mail: psranabhu@gmail.com

R. P. B. Singh
ACLA-Asian Cultural Landscape Association (SNU), Seoul, Korea

S. Kumar
Department of Geography, D.D.U. Gorakhpur University, Gorakhpur 273009, UP, India
e-mail: sarvesh1k@gmail.com

S. Sahdev et al. (eds.), *Geoecology of Landscape Dynamics*,
Advances in Geographical and Environmental Sciences,
https://doi.org/10.1007/978-981-15-2097-6_13

13.1 The Perspective: Culture, Cultural Heritage and Inclusive Heritage Development

The classical model of Sustainable Development (Gro Herlem Brundtland Report 1987; Fig. 13.1) has been critically examined after the passage of time and comprehended under the umbrella of Universal Forum of Cultures 2007, emphasising the base of Developing Civility—culture, language and religions for inter-faith dialogue. Three modes' model for culture and sustainable development has been structured engulfing sustainability (Fig. 13.2): **So**, Social; **En**, Environment; and **En**, Economic, where the three roles of Culture in Sustainable Development (the three circles representing the three pillars, **So**, **En** and **Ec**) are envisaged. In this model, Culture is *added* as the fourth pillar (a), Culture *mediating* between the three pillars (b) and Culture as the *foundation* for sustainable development (c). The arrows in the model indicate the ever-changing dynamics of culture and sustainable development (Dessein et al. 2015, p. 29).

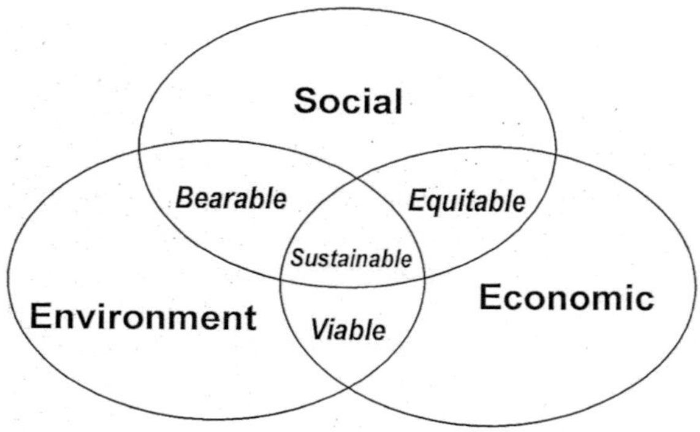

Fig. 13.1 Basic frame of sustainability (*Source* Kumar 2018, p. 215, after Brundtland Report 1987)

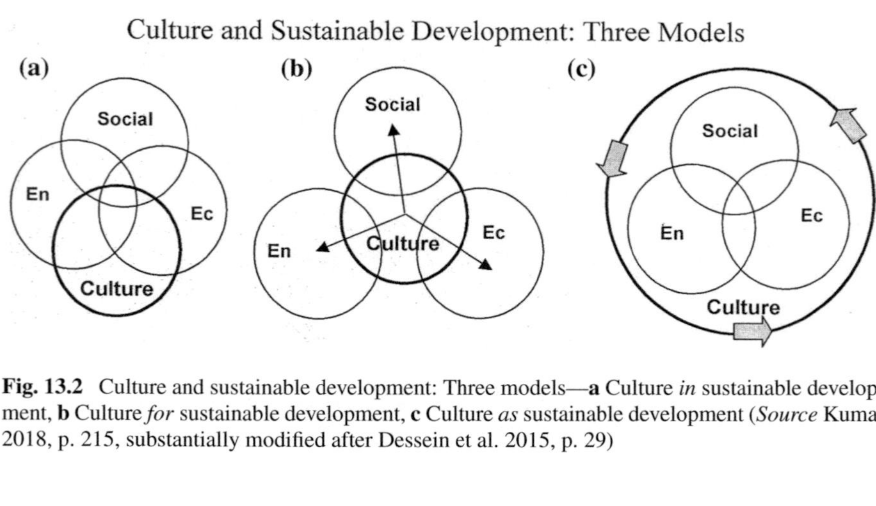

Fig. 13.2 Culture and sustainable development: Three models—**a** Culture *in* sustainable development, **b** Culture *for* sustainable development, **c** Culture *as* sustainable development (*Source* Kumar 2018, p. 215, substantially modified after Dessein et al. 2015, p. 29)

Within this perspective, the two heritage-holy cities of India, Ayodhya and Varanasi, are selected for illustrating their inclusive heritage development strategies and interlinking SDGs, especially Target 11.4, i.e. (i) vision for protecting and safeguarding cultural and natural heritages, and (ii) consideration of the local conditions, policies and orientation for preservation, protection and conservation of national cultural heritages.

13.2 Introducing the Two Heritage-Holy Cities: Ayodhya and Varanasi

It is important to note that not all of the geographical sites mountains, hills, rivers, caves, etc. are considered holy. They must possess some peculiarity together with sacral spirit and power of the place to qualify as a point where human beings can seek contact with the divine. Association of sacrality and the geographical setting enhances the sacred power of a place and therefore the awe or reverence to that. In fact, in the Indian case, one finds great attraction of sacred places that has initially not been due to a particular temple (alone), rather the geography of such holy places was significant (Singh 2015, p. 70, also Singh 2013, and Singh and Rana 2016).

The seven most sacred and salvific cities of the India include Ayodhya, Varanasi (Kashi), Mathura, Maya-Haridwar, Kanchi, Avantika-Ujjain, Puri, Dwarka (Fig. 13.3). The heritage-holy cities of Ayodhya and Varanasi are located in the Ganga river basin, which is served by the two main tributaries, the Sarayu (Ghaghara) and the Yamuna. The *Rig Veda* (RgV 1.23.20), one of the earliest texts dated ca 2000 BCE, narrates water as the possessor of life's infusing power, and the motherly qualities, thus considered as a spiritual endowment. The stories of the Ganga also associate the metaphysical, mystical and material worldviews of existence, continuity and maintenance of human life (Singh 2009a), thus the whole of its basin became sacred landscape (Singh 2015, p. 72).

Ayodhya (latitude of 26°47′16.67″ N and a longitude of 82°11′54.79″ E) is situated on the right bank of the holy river Sarayu (Ghaghara, a tributary of the Ganga) at a distance of 7 km east from Faizabad city (Fig. 13.3). Ayodhya was one of the famous cities and the first capital of the powerful Kosala Kingdom among the 16 Mahajanapadas of ancient India (Law 1944, p. 424). The city for a period of over two thousand years has borne witnessed to the presence of Jainism, Buddhism, Shaivism, Vaishnavism and Islam, and therefore Ayodhya consists of the sacred and religious places for Hindus together with Muslims, Jains, Buddhists and Sikhs (Shaw 2000, p. 698, also Kumar and Singh 2016).

Ayodhya is sacred and religious place even for Janis. There are five Jain temples located near birthplaces of the five Jain Tirthankaras (angels), viz., Adinatha or Rishabhadev temple in Muraitola Swargadvara, Ajeetnatha temple near Saptsagar, Abhinandananatha temple near to Saraya, Sumanthnatha temple near to the Ramkot and last one Anantnath temple near to Golaghat. Ayodhya also records more than

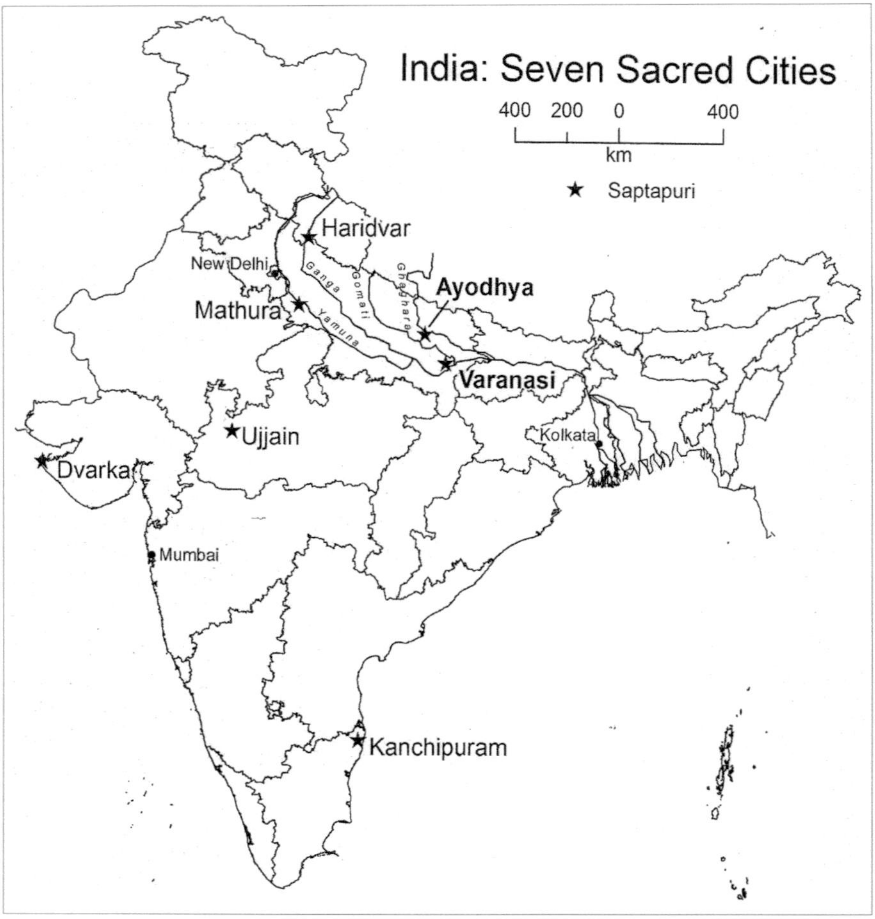

Fig. 13.3 India: Seven sacred holy-heritage cities (prepared by the authors)

hundred mosques, *mazars* (tombs), *idgahs* (prayer ground), Karbala (burial place), etc. related to Muslim sacred/ritual landscapes, that is how in metaphorically the city is called 'Chhoti Mecca' (Little Mecca). Muslims perform a variety of rituals in their sacred places.

Varanasi (latitude of 25° 19′3.52″ N and a longitude of 82° 58′26.09″ E) exists on the left side bank of the Ganga river (Fig. 13.3), and popularly called Kashi or Banaras. This city is popularly known as the Cultural Capital and Heritage city of India, and also as one of the oldest living cities in the world. While a number of cultures have risen high and fallen down, a number of cities disappeared in the abyss of time, Varanasi continued to grow and maintained its cultural and scholastic traditions of religious performances, learning, and arts and crafts.

Ashoka (272-242 BCE), who established the Buddhism as state religion, had paid a visit to Sarnath (northern part of the city), and under his patronage many monasteries, *stupas* and shrines were built. After the downfall of Mauryas, the prosperity of the city has gone into darkness till the rule of Kushana in the first century CE. However, again in the twelfth century, the city has recorded the glorious days, but due to several invasions the glories fell down.

In the early eighteenth century with the decline of the government in Delhi, Banaras first came under the rule of the Nawabs of Oudh in 1722, and later became the seat of Mansaram (1730–1738), the initiator of the present state of Kashi. Under the influence of the Marathas (1734–1785), a wave of cultural renaissance overtook Banaras who substantially rebuilt the city. For the first time in 1948, the Banaras Improvement Trust was formed for making 'Master Plan of Banaras', and thus in 1951 the first such plan was prepared. The ongoing Master Plan of Varanasi 2011–2031 incorporated the earlier strategies and structure for the future development on the line of SDGs and 'Smart City Development Plan'. The Comprehensive Development Plans (CDP) of both of the cities, Ayodhya and Varanasi, are in process of development under GOI missions of HRIDAY (*Heritage city Development and Augmentation Yojana*) and PRASAD (*Pilgrimage Rejuvenation And Spiritual Augmentation Drive*).

Both of these historic cities are sacred places, of course primarily for Hindus, but also they historically represent other religious groups, like Jainism, Buddhism, Sikhs and Islam/Sufis (Muslims). Banaras is predominated by the Shaiva imageries, while Ayodhya by Vaishnavite. These sacred-heritage cities record a number of rituals, festivities, pilgrimage journeys and ancient temples, river ghats (stairways and bathing places), holy water bodies and riverfront sites, and associated heritage values—those are the representative grandeur of art and tangible and intangible heritage repositories that should be used as resources for sustainable urban development. Presently, around 1.9 million pilgrims/tourists pay visit to each of these places every year. Of course, there appear many dilapidating religious heritage sites and monuments in lack of sustainable conservation and preservation strategy, bad administrative management and lack of people awareness and their active involvement.

13.3 Heritage Zones: Ayodhya and Varanasi

The idea of heritage zone is based on the importance of an area possessing representation of ancient and alive religious and sacred sites that maintained the continuity of age-old tradition and considered to be a basic resource for inclusive heritage development. These areas and zones are identified in the ongoing Master Plans of the heritage-holy cities of Ayodhya and Varanasi.

13.3.1 *Ayodhya*

Ayodhya is predominantly a sacred-heritage city that is how religious landscapes are the prevalent scene, and their concentration in specific area is taken to demarcate the specific zone. The ongoing framing of the Master Plan, which first shaped for 1983–2001 and now still in process for 2021, mentions scattered religious sites spotted in different parts, which can easily be bounded with four heritage zones (Fig. 13.4).

13.3.1.1 Ramkot Heritage Zone

In this zone, 55 properties are enumerated, including Hindu temples, *ghats* (riverfront bathing places), *kunds* (water pools), Jain temples, Muslim shrines and tombs, and passing on pilgrimage routes. The Ayodhya Act-1993 and Archaeological Survey of India's rules related to acquisition of land (buildings, shops, residential houses) near to Ramkot (Ramajanmabhumi) by provenance and rule under the Central Government of India prohibits any type of new constructions in the vicinity of Ramkot, to avoid interrupting security of the Ramajanmabhumi. However, there are incidences of illegal encroachments, taking benefit of religious sentiments.

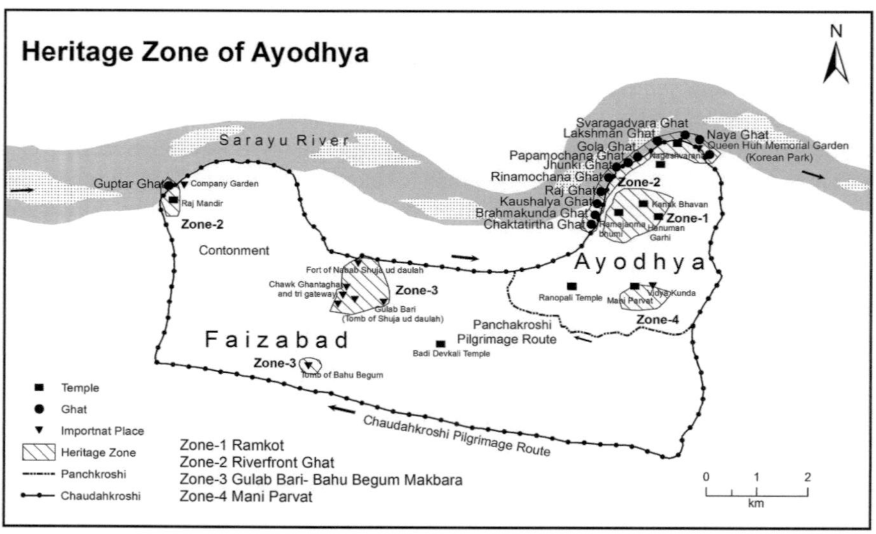

Fig. 13.4 Ayodhya: Heritage zone (compiled and prepared by the authors)

13.3.1.2 Riverfront Heritage Zone: Ghats and Temples

The bank of the Sarayu River at Ayodhya consists of 25 bathing places (*ghats*); those are counted as sacred and ritualistic sites for Hindu adherents. The riverfront buildings (monasteries and built heritage) are the distinct architectural grandeur, built mostly by kings and lords from different parts of India between eighteenth and twentieth centuries. This zone consists of two parts, viz., the eastern and the western. Most of the riverfront sacred-heritage sites are in the eastern part. Among these, Svargadvara Ghat ('door to the heaven'), spreads between the Sahastradhara and the temple of Treta Ke Thakur, is the most popular *ghat* for bathing and related rituals (Kumar and Singh 2015).

13.3.1.3 Gulab Bari–Bahu Begum Maqbara Zone

This zone contains about seven properties related to Nawab period, i.e. Gulab Bari or tomb of Suja-ud-daula, Tomb of Bahu Begum (Maqbara) and Tomb of Bane Khanam, and all are under the protection of Archaeological Survey of India. Tomb of third Nawab of Avadh Shuja-ud-daula is situated in the inner part of Gulab Bari. The construction of this tomb was initiated by Shuja-ud-daula in his Nawabi period and after death in CE 1775 he was buried in this tomb. The final construction of the tomb with dome and tower was accomplished by Bahu Begum, the wife of Nawab Suja-ud-daula in CE 1789.

13.3.1.4 Mani Parvat Zone

This zone contains more than 20 properties associated with Muslims, Buddhists and Hindus. Mani Parvat (mound) is the oldest Buddhist ruined site dated ca 400 CE that was seen by Faxian (Bakker 1986 II, p. 31), but presently it is known for Rama Sita temple and the ceremonial place for Shravan Jhula festival. This site is under the protection of Archaeological Survey of India. The Vidya Kunda lying 200 m east from the Mani Parvat and 1.75 km south-west from the Sita Kunda is situated in the inner courtyard of the Mahavidya temple complex.

Apart from the four heritage zones, pilgrimage and its associated sacred routes are the other common religious heritage zones in holy cities. Ayodhya has developed five pilgrimage routes in continuity of history, viz., Chaurasikroshi, Chaudahakroshi, Panchakroshi, Ramkot ki Parikrama and Antargrihi Parikrama; among these the two latter are minor and relatively less popular (Fig. 13.4). Three major pilgrimage routes define the three territorial limits of Ayodhya, viz., Chaurasikroshi, Chaudahakroshi and Panchakroshi. Like most of the pan-Indian holy centres, Ayodhya too displays a three-tier cosmology, respectively, as *macro* (i.e. *mandala*, the Outer one), *meso* (i.e. Kshetra, the Middle one) and *micro* (i.e. Puri/city, inner sanctum) cosmos, all demarcated by routes and linked archetypally by temples/shrines and are described in the ancient mythologies, which are frequently cited in various rituals and group-chanting and sacred walks.

13.3.2 Varanasi: Heritage Zones

In the Master Plans for Varanasi, for the first time some strategies of urban heritage and heritage zoning were proposed in the recent Master Plan (1991–2011; Singh 2009b, p. 327), focussing on to maintain and preserve the ancient glory of Varanasi, and to identify necessary facilities and infrastructure and various heritage complexes. The heritage zone as delimited in the Master Plan 2011 has been further incorporated in the ongoing Master Plan 2011–2031. Only a little over 2% of the total area is proposed under tourism and heritage zone. According to the zoning plan, five heritage zones have been identified (see Singh 2009b, p. 329–332; and Singh and Rana 2019, p. 155–158, Fig. 13.5).

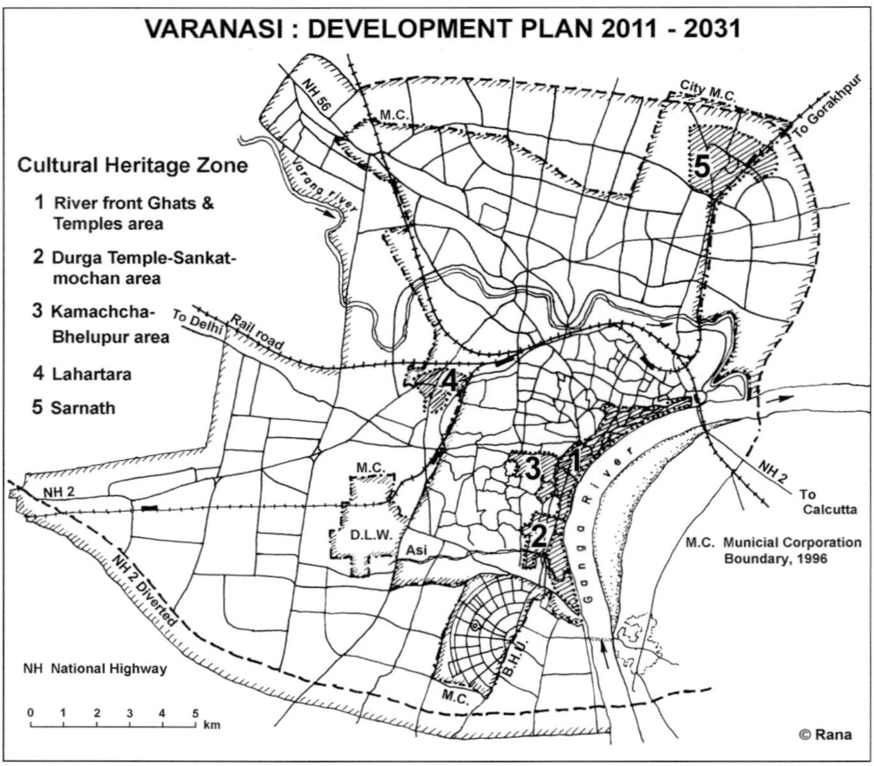

Fig. 13.5 Varanasi, Development Plan 2011–2031 (after Singh 2015, p. 102)

13.3.2.1 Riverfront Ghats (Stairways to the Riverbank)

The riverfront heritage covers the eastern portion of the city stretching within 200 m from the riverbank. Eighty-four riverfront *ghats* spread over a length of 6.8 km along the crescent-shaped bank of the River Ganga, from the south to the north. The riverfront is overlooked by lofty palatial buildings built between eighteenth and twentieth centuries, mostly by kings and lords from different parts of India; the area is primarily occupied by various shrines, temples and religious places.

13.3.2.2 Durgakund–Sankatmochan Area

This area contains about twenty temples and shrines and the water pools of Durgakund and Kurukshetra *kundas*, which are two historic sacred tanks dating from the late eighteenth century. Every Tuesday, and more frequently in the month of *Shravana* (July–August) and *Ashvina* (September–October), especially during the nine nights (*Navaratri*) in the light lunar fortnight (waxing), worshippers perform rituals honouring goddess in the Durga temple. This temple was built on the orthodox model of Hindu architecture. Towards the east near the Ganga river is the oldest sacred pond in Varanasi, called Lolark Kund, which was referred to in the epic *Mahabharata* (second century BCE) and which continue to attract a large mass of pilgrims, especially on its annual day of celebration falling on the *Bhadrapada* (August–September) 6th of the light-half lunar fortnight.

13.3.2.3 Kamachha–Bhelupura Area

This area records some of the old monasteries, ancient shrines and an ancient heritage site associated with the Jain Tirthankara Parshvanath, together with many monuments and buildings of the British period (eighteenth—nineteenth centuries). The historically notable temples and shrines of heritage values in this zone are Kamachha Devi, Krodhan Bhairava, Angareshi Chandi, Vatuka Bhairava and Vaidyanath Shiva.

13.3.2.4 Kabir Math (Lahartara) Area

This site was the birthplace of Kabir, a great saint-poet and social reformer of the sixteenth century. The Kabir Temple Complex is coming up as a great modern religious heritage and centre of solace and learning; of course it is turning to be a 'White Elephant'—less associated with the local society and culture. Under the heritage development project by the State Government, some works have already been going on.

13.3.2.5 Sarnath

This archaeological heritage site was famous for its sanctity, beauty and natural scenery, qualities that attracted the Buddha to give his first sermon here in 528 BCE. Following Muslim invasions and the downfall of the Gahadavalas Kings in the late twelfth century, the site was left in ruins and only came to light in CE 1793.

Varanasi is famous for its series and layers of sacred circuits (counting to 54), among which the Panchakroshi is the most popular. This pilgrimage circuit representing the cosmo-spatial *mandalic* territory (*kshetra*) of Kashi is a unique attribute of Varanasi, where exist 108 temples and shrines. Under the recently initiated heritage development project, partial works like improvement of roads, cleaning of the water pools and repairing of some of the roads are being completed. On the ground of pilgrimage tourism, this cosmic circuit should be given special emphasis, so also promote sustainable heritage tourism (for details see Singh and Rana 2018).

13.4 Ayodhya and Varanasi: Intangible Heritage on UNESCO Scale

The Intangible Cultural Heritage (ICH), conceived as the mainspring of humanity's cultural diversity, covers the practices, representations, expressions, knowledge, skills—as well as the instruments, objects, artefacts and cultural spaces associated therewith—that communities, groups and, in some cases, individuals recognise as part of their cultural heritage. The ICH has been transmitted from generation to generation, maintained and continued while also absorbing several aspects at different times as response and acceptance of the society.

The UNESCO has broadly classified ICH into five categories: (1) oral traditions and expressions, including language as a vehicle of the intangible cultural heritage; (2) performing arts; (3) social practices, rituals and festive events; (4) knowledge and practices concerning nature and the universe; and (5) traditional craftsmanship (see Singh 2015, pp. 125–126). While giving due consideration to the UNESCO Scale of ICH, in the Indian perspective of the ICH, specially referring to holy-heritage cities like Ayodhya and Banaras, various attributes of the ICH can be classified into ten broad categories (see Singh 2015, p. 127, also Singh et al. 2020, p. 252; see Table 13.1).

As discussed above, Ayodhya and Banaras, both are fully suited to be designated as sites of intangible cultural heritages (ICH) and their associated attributes should be taken care of in the inclusive heritage development in the frame of SDGs.

Table 13.1 Attributes of intangible cultural Heritage: Ayodhya and Varanasi

Se	Cultural Heritage attribute (selective)	Ayodhya (major/selected ones referred)	Varanasi/Banaras (major/selected ones referred)
1	Oral Traditions	Rama Katha (religious storytelling), Birha—folk songs, singing, dancing	Katha (religious storytelling), Birha—folk songs, singing, dancing
2	Performance Arts	Bhajan, Nautanki, Bharat Milap, Lok Nritya (folk dances), Ramajanma Badhai, Birha, Ramanami chanting, etc.	Classical like Dhrupad, Sanskrit Theatre, Folk dance and singings, Bharat Milap, instrumental and performances
3	Ramalila—Krishnalila (theatre form of the God's story; varieties, distinctive; historical)	at Bhagvatacharya Smaraksadan, Bade Sthan, Gola Ghat, Lakshmankila, Jankimahal trust, Ramajanaki Mandir—Sahabganj, Kothaparcha, Fatehganj, etc.	at Ramanagar, Chitrakut, Assi Ghat, Nati Emli, Chetganj, Khojwa, Shivpur, Dashavatar Lila; Krishnalila—(Assi Ghat), Nrisimhalila, etc.
4	Parikrama/Yatras, Pilgrimages	Chaurashikroshi Yatra, Chaudahkroshi Yatra, Panchakroshi Yatra, Ramkot ki Parikrama, Antergrihe Parikrama, Kalpvas	Chaurashikroshi; Panchakroshi; Avimukta; Nagar Pradakshina; Antargriha—Vishveshvara; Omkareshvara; Kedareshvara; Aditya (Sun) Yatra; etc.
5	Rituals and Festivals (selective) # For dates of selective Festivals, see Singh 2015, the Appendix: 1, p. 124	***HINDU Festivals***: Makara Samkranti, Matgajendra Mela, Chaitra Navaratri, Rama Navami, Sarayu Mahotsav, Savan Jhula Mela, Krishna Janmasthami, Hanuman Jayanti, Gudara ka Mela, Surya Kunda Mela, ***MUSLIM Festivals***: Muharraum, Idul-Fitr/Idul-Juha, Sabbe-rat, etc.	***HINDU Festivals***: Makara Samkranti, Maha Shivaratri, Ganga Dashahara, Krishna Janmasthami, Lolarka Chhata Mela, Naga Nathaiya, Surya Shashthi Chhatha (Karttika Purnima), Buddha Purnima; ***MUSLIM Festivals***: Muharraum, Idul-Fitr/Idul-Juha, Sabbe-rat, etc.
6	Traditional Art and Craftsmanship	Murti casting (sculpturing), wooden work, special sweets (e.g. Anarsa), special breakfasts—Kachori and Imarte, Samosa	Murti casting (sculpturing), Shringar (icon decorative art), Silver work, wooden toy making, Silk weaving (Banarasi Sari), etc.

(continued)

Table 13.1 (continued)

Se	Cultural Heritage attribute (selective)	Ayodhya (major/selected ones referred)	Varanasi/Banaras (major/selected ones referred)
7	Scholastic Traditions and schooling and discourses system	RML Avadh University, Narendra Dev University of Agriculture and Technology, Gurukul Sanskrit Mahavidyalaya, KPSL Saket Mahavidyalaya, Rama Nam International Bank	B.H.U., M.G.K. Vidhyapith, S. Sanskrit University, Central University of Tibetan Studies, Parshvanath Jain Institute, Institute of Textile and Weaving, Udai Pratap PG College, Darul Salfia Islamia
8	Indigenous Knowledge and Healing Tradition	Sanskrit study, meditation, Ramakatha Vyas Pitha (Rama Katha Vachak)	Nature therapy, Yoga centres and tradition, Ayurvedic medicine and centres
9	Memorials, icons and Saints' associated sites and related performances	Shri Rama, Gautam Buddha; Five Jain Tirthankara: Rishabhadev, Ajeetnatha, Abhinandananatha, Sumanthnatha, Anantnatha; Tulasi; Nanak/other Sikh saints; Muslim Mazars: Hazrat Noah, Shesh Paigambar, Hazrat Sayad shah (Adgada Mazar)	sites related to Shankarachaya, Tulasi, Kabir, Ravidas, Dadu, Nanak, etc., Rani Lakshmi Bai; Muslim Mazars: Sheikh Salim Chisti, Sheikh Ali Hazim, Shah Taaiyyab Banarasi, Ghazi Miyan Chandan Shahid
10	Birthplaces and memorials of Freedom Fighters and Literates, and related celebrations	Narendra Deva, Mangal Pandey, Chakbast and Mir Babar Ali Anis (Urdu Poet), Umrao Jaan, Begum Akhtar (singer of Ghazal, Dadra, and Thumri genres of Hindustani classical music)	Pt Gopinath Kaviraj, L.B. Shastri, Bismillah Khan, Pt. Ravi Shankar, Premchand, Bhartendu Harishchandra, Jaishankar Prasad, Ram Chandra Shukla, Birju Maharaj, Girja Devi, and others

Source Compiled by the authors, with additions; see also Singh 2015, p. 127

13.5 Shaping Sustainable Heritage City System

The structuring, shaping and operating sustainable heritage city (SHC) system may be arranged into six-tier operation, i.e. 1—Setting the vision for SHC of the venture, 2—Identifying the SDGs Target 11.4 and Inclusive Heritage Development, IHD, Strategies, 3—Achieving the Political Cohesion, 4—Building the SHC Frame, 5—Measuring the SHC's Potential and the Progress, and 6—Ensure Accountability and Responsibility (see Table 13.2). The structural shaping is rationally befitting to both of the cities, Ayodhya and Varanasi.

Table 13.2 Shaping sustainable heritage city (**SHC**) system

SHC attribute	Resultant function
1. Setting the vision for SHC of the venture ↓↑	Identifying—an inclusive heritage city vision is tune with the heritage city's identity and long-term inclusive heritage development, **IHD**, strategy; relevant multi-stakeholders and mechanisms; the existing governance and organisational mechanisms for historic heritage city solutions
2. Identifying the SDGs Target 11.4 and IHD Strategies ↓↑	Developing heritage city infrastructure (e.g. Internet of Things); Identifying and developing smart and sustainable heritage city services in the purview of SDGs Target 11.4 containing within the so-called 'Urban Sustainable Development Goals'
3. Achieving the Political Cohesion ↓↑	Local governments should obtain the necessary political approval and backing to ensure that the strategic programme is pursued as related to IHD. This includes the adoption of the programme/targets through consensus
4. Building the SHC Frame ↓↑	Improvement of existing traditional infrastructure and new infrastructure must be built under the IHD; developing an action plan for Public–Private–Partnership programmes; ensuring long-term services via good operation maintenance of Sustainable Heritage City, **SHC**
5. Measuring the SHC's Potential and the progress ↓↑	Consisting of monitoring and evaluating potential and work programme required to achieve the UN—SDGs Target 11.4, emphasising a set of sustainable development targets related to heritage cities and heritage sites and settlements
6. Ensure Accountability and Responsibility ↓↑	Involves evaluating, reporting and learning from SHC process and related experiences. The reflective process of evaluation will feed into a process of continuous learning, which in turn will influence and inform the IHD of the future vision and strategy for smart and SHC

Source Kumar (2018), p. 216, developed in Indian context taking in view the ITU-T FG-SSC (2015), and Habitat III Quito Report, October (2016)

In successful operation of Sustainable Heritage City System, like to other planning models, community participation is the key energy in functioning and maintenance of the system by making rational balance among social (religious and cultural heritage), physical (housing and community development) and economic (cultural heritage tourism) attributes of heritage-holy city; this can easily be schematized using set and superimposing sets (see Fig. 13.6).

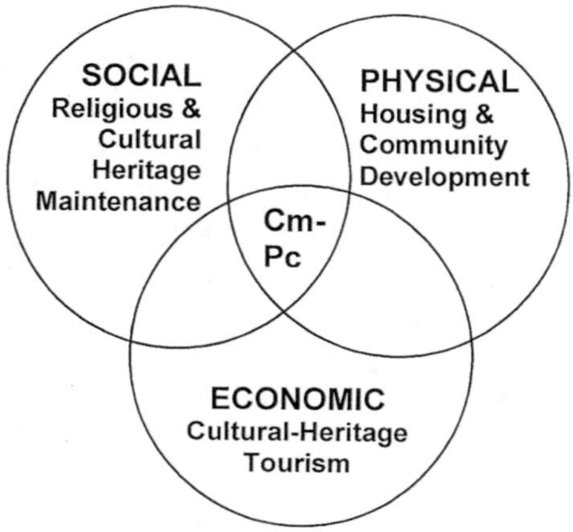

Fig. 13.6 Interacting Action Plan for Community Participation (prepared by the authors)

13.6 Heritage and Pilgrimage Inclusive Development: Framing HRIDAY and PRASAD

The Ministry of Tourism and Culture, and Ministry of Urban Development, Government of India, have recently initiated counter-depending missions of (i) *Heritage city Development and Augmentation Yojana* (HRIDAY), and (ii) *Pilgrimage Rejuvenation And Spiritual Augmentation Drive* (PRASAD), with an aim to strengthen and promote the heritage sites and centres of pilgrimage tourism in making the environment green and sustainable while befitting into the roots of culture and traditions. Through the newly formed government in Uttar Pradesh, an agreement between GOI Ministry of Culture and Tourism and government of U.P. has been made on 24 March 2017 to develop and transform the holy cities as special sites for heritage and religious tourism, which includes Varanasi, Ayodhya, Mathura, Gorakhpur and Agra. Special plans are also in process preparing conservation and rejuvenation of heritage and religious sites in these cities (see Singh and Rana 2019, p. 168).

13.6.1 National Mission of HRIDAY

The National mission on the '*Heritage city Development and Augmentation Yojana*' (HRIDAY), aims conserving and preserving the distinct and unique characters of the *heritage cities*, those continued their traditions of heritage (tangible, intangible, transitory and cultural landscapes. They would be used as a resource for sustainable development and ecological restoration on the line of achieving the SDGs (Sustainable Development Goals). The strategy of SDGs also includes cleanliness, planning, livelihood of the local people and economy (cf. Kumar and Singh 2019, p. 78). The protection, augmentation, management, authenticity and integrity of properties (both tangible, intangible, and transitory) are also important considerations, together with the marked specific characteristics.

13.6.2 National Mission of PRASAD

Aiming to beautify and improve the amenities and infrastructure at pilgrimage centres of all faiths, a National mission on '*Pilgrimage Rejuvenation And Spiritual Augmentation Drive*' (PRASAD) has been announced in the Union Budget 2014–2015 and an amount of Rs. 1,000 million (US$ 15 mill.) has been proposed for this initiative. Under PRASAD the old historical–cultural pilgrimage routes and associated sites would also be developed (see Kumar and Singh 2019, pp. 78–79).

These two schemes together planned to activate the following objectives (see Kumar and Singh 2019, p. 326):

- Enhancing heritage sensitive infrastructure in the purview of planning, development and implementation.
- Introducing service delivery and infrastructure provision that befit to core areas in historic cities (e.g. Ayodhya and Varanasi).
- Motivating tourist and pilgrims to have direct experience of uniqueness of city' heritage and improving sensitivity to preserve and revitalise heritage.
- Preparation and easy accessibility of Heritage Inventory Register, which may provide the historical accounts and a basis for urban planning, growth and service provision and delivery.
- Basic service delivery, including sanitation services like public conveniences, toilets, water taps, street lights with the use of latest technologies in improving tourist facilities/amenities to provided, having a system of monitoring and continuous appraisal.
- Local capacity building and scope for absorbing the contemporary problems related to inclusive heritage-based industry be implemented.
- An inter-changing and countering system of effective linkages between tourism and cultural facilities, including conservation of natural and built heritage, should be strengthened.
- Using appropriate and indigenous technologies for retrofitting historic buildings and urban heritage adaptive rehabilitation and maintenance, the structural plan to be chalked out (HRIDAY 2015).

The interfaces and reciprocity between pilgrimage and tourism are integral parts of human travel. That is how 'pilgrimage tourism' ('*Tīrthayātrā-Paryatan*') is considered as an alternative for the solution; of course, this is more inclined to metaphysical issue and life philosophy: meeting sacred and profane. Pilgrimage tourism is considered now as strategy for heritage awakening, deeper experiences and transferring the religiosity into global humanism and spirituality (see Rana 2014). The sustainable frame of pilgrimage tourism and heritage should be promoted in three ways: philosophical, organisational and managerial, which may fulfil the objectives of SDGs Target 4.7 and Target 11.4, focussing inclusive development together with protecting and safeguarding the cultural and natural heritages. In recent debate, the eco-healing approach to pilgrimage tourism is considered as a post-modernist way to consider pilgrimage as a bridge between recreation and spirituality; this way pilgrimage tourism will hope to provide a rational alternative for cultural awakening and strategy for poverty alleviation (cf. Singh, Rana and Kumar 2019, p. 80). As the 'caring for the place (the *Earth*)' is inherent in the pilgrimage tourism, it will also provide opportunity to intimately sense and deep feelings for the place and the people—their behaviour, their heritage and the present lifeworld in which they live, act and keep the glorious tradition alive, see Singh (2011a, 2011b).

The approach to study tourism so for has been the study of economic activity. However, it limits the scope and answer to many questions posed as a consequential result. On the line of 'commodification approach' proposed by Ashworth (1991, p. 111), the 'eco-healing package' may be considered that may extend the horizon of potential resources in pilgrimage tourism as an alternative tourism, expected that it will fulfil the objectives of PRASAD (see Singh 2004, p. 213).

13.7 Epilogue: Vision and Concluding Remarks

Unlike the Millennium Development Goals, as a result of a concerted effort from many sectors, the SDGs contain an explicit heritage target, Target 11.4, it calls for making cities and human settlements inclusive, and safeguard the world's cultural and natural heritage. The Target is contained within the so-called 'Urban Sustainable Development Goals', a set of sustainable development targets related to heritage cities and settlements. Taking in view the prevailing condition in holy-heritage cities in India (e.g. Ayodhya and Varanasi), the basic frame for holistic development will be chalked out. Conserving heritage renews a sense of identity can inspire new smart and sustainable system in pilgrimage city and sacred town development patterns, especially emphasising on the valorisation of the assets of the poor. In the other context of SDGs, the social sustainability of heritage city is based on social groups of local living peoples, pilgrims and *sadhus* (Hindu religious ascetic), resulting in social beliefs and cultural performances in the form of intangible cultural heritage. According to the ancient history, Ayodhya and Varanasi have been historically the holy-heritage cities where various social groups through the religious faiths of Hinduism, Buddhism, Jainism and Islam meet together in making multiple visitation sites of heritage importance and nexus of harmonious life.

On the line of the earlier proposal (that was not finalised yet), taking UNESCO Guidelines for inscription in the World Heritage List, cities of Ayodhya and Varanasi are also in the process of assessment as both fulfil the five criteria out of ten (see Kumar 2018, pp. 243–247, and Kumar and Singh 2017, pp. 58–65). The making of dossiers for getting Ayodhya and Varanasi in the UNESCO WHL is in process since last decades, while taking in view the strategies of Smart City Development and

SDGs (cf. Gidwani 2012); however, these are only the political agenda for mobilising people in their support, instead of real sense, prioritisation and action.

Cultural tourism plays a great role towards socio-economic changes and promotion of sustainability. According to Indian sentiment, the pilgrim's centres or *tirthasthāna* used to be visited by number of tourists to earn virtue (Dasgupta, et al. 2006, p. 11). As one of the largest industries, tourism is associated with many of the prime sectors of world's economy. Economic sustainability of Ayodhya and Varanasi is regulated by annual visit of tourists (of course, mostly pilgrims), recorded ca 1.9 million, in each of the places, in 2018. Taking this view government is trying to promote pilgrimage tourism in the purview of SDGs, taking in view 'heritage-making' ('*heritagization*' or '*patrimonialization*' in French). Pilgrimage defined as 'a journey resulting from religious causes, externally to a holy site, and internally for spiritual purposes and internal understanding' (Collins-Kreiner 2009, p. 153); that's how pilgrimage tourism will promote, sustain and maintain economic and cultural sustainability, together with maintaining belief systems. Both of the pilgrimage cities bear various interrelated phenomena to which people are affiliated with the belief system, faith and spiritual merits. This is also to be kept in mind that the sacred spaces vis-à-vis public spaces, in a way, will serve as peace plaza and places of spiritual awakening having 'the potential for healing communal strife and reviving urban art, (*cultural and*) folk practices. Heritage conservation can thus become an empowering tool for local communities and for the visitor an opportunity for spiritual growth' (Sinha 2014, p. 60), which is an ultimate aim of the urban areas.

A recent study remarks that 'If the urban SDG is to prove useful as a tool as intended for encouraging local and national authorities alike to make positive investments in the various components of urban sustainability transitions, then it must be widely relevant, acceptable and practicable' (Simon et al. 2016, p. 60). This is valid in the case of Indian heritage cities, where one always faces the problem of linking locality and universality. Additionally, central to this task has been the challenge of determining how to benchmark and measure performance according to the SMART criteria (i.e. specific, measurable, assignable, realistic and time-specific), based on specialist scholarship, the existing literature and practical experience of the site (see Birch 2015, p. 228), taking into account demand, pressure response and multifactor versus single factor, and also considering two metrics: the traditional dimensions of sustainability (equity, economics and environment) and later forming the Liveability Principles of Partnership for Sustainable Communities (PSC), while making bridge between the age-old traditions and high-tech smart city plans.

In a recent meeting ICOMOS (2016) having discourses on 'Heritage as Driver of Sustainability: Mission and Activities for 2017–18', it has been noticed that the SDGs focused on achieving representation of heritage in the major policy papers of Agenda 2030. Now that these policy papers are adopted at the highest, global level, and their goals and targets await being fulfilled, focus has shifted to 'implementation' through public participation and PPP (Public–Private–Partnership). Therefore, the mission that ICOMOS has given its Focal Point for the SDGs in the new term is to 'steer a coordinated process of advocacy to advance the implementation of UN Agenda 2030—SDGs and Habitat NUA (New Urban Agenda) from the perspective of cultural and natural heritage, within the framework of the ICOMOS mandate and inputs from strategic partners'. The NUA, a framework laying out how cities should be planned and managed to best promote sustainable urbanisation within the purview of culture and cultural heritage will keep the vitality and image of holy-heritage cities.

Activities towards accomplishing this mission include liaising with stakeholders at national, regional and local level (e.g. holy-heritage cities like Ayodhya and Varanasi) and across governance sectors (public, private, civil society, experts/academia), in particular for

1. *Localising implementation*, by providing guidance and direction to stakeholders to adopt tools of implementation,
2. *Monitoring implementation*, with a focus on Indicator Target 11.4.1 defined by the UN (expenditure and share of budgets allocated to heritage) to achieve consistent and comprehensive data collection,
3. *General advocacy*, for mainstreaming culture and heritage within sustainable development, by increasing visibility and outreach across the UN system and the public sphere.

From the perspective of historic preservation in the ancient culture like India, the goal looks great, the target thought to be good, but the indicator in view of the contemporary scenario, frankly, is terrible. In this context, the crucial and critical issues, mostly against viability and optimality, include inappropriate expenditure (public and private) used per capita on the preservation, protection and conservation of all types of heritage (cultural, natural, mixed, intangible and transitory), hierarchical gaps in the layers of government (national, regional and local/municipal), type of expenditure (operating expenditure/investment) and also type of private funding (donations in kind, private non-profit sector and sponsorship), and finally the resultant consequential issues that intensify the problems.

While making plans and strategies in this direction, the hard realities to be kept serious consideration, i.e. (1) while the SDGs set targets for nations, most of the implementation will need to take place at the city or regional levels; (2) the SDGs have much to say about '*what*' but much less about '*how*' or '*by whom*'; and (3) goals are great, but the real tests of success will be implementation, bottom-up experimentation, and localization! Considering all the diagnostic niches and consequences, let us March and make 'pilot models' at the level of holy-heritage city. The following six such procedures represent a direct commitment to heritage conservation in the spectrum of SDGs and NUA (Rypkema 2016):

- Leverage cultural heritage to strengthen social participation and the exercise of citizenship;
- Develop vibrant, sustainable and inclusive urban economies, building on cultural heritage;
- Support urban economies through promoting heritage conservation activities;
- Promote regeneration while preserving cultural heritage;
- Include culture as a priority component of urban plans and strategies that safeguard cultural heritage; and

- Support leveraging cultural heritage for sustainable urban development.

Following the path towards SDGs Target 11.4 in making holy-heritage cities vibrant and liveable centre of global harmony, spiritual awakening, peace and deeper understanding, public participation and education are pre-requisite (see Singh 2017, p. 26). This paper should be taken as a frame and appeal in this direction.

References

Ashworth GJ (1991) Heritage planning. Conservation as the management of urban change. GeoPers, Groningen, The Netherlands

Bakker HT (1986) Ayodhya: the history of Ayodhya from the 17th century BC to the middle of the 18th Century. Egbert Forsten, Groningen, the Netherlands

Birch EL (2015) Measuring U.S. sustainable development. Cityscape: J Policy Dev Res 17(1):223–232

Brundtland GH (1987) Our common future [a report of the World Commission on Environment and Development]. Oxford University Press, New York

Collins-Kreiner N (2010) The geography of Pilgrimage and tourism: transformations and implications for applied geography. Appl Geogr 30:153–164

Dasgupta S, Mondal K, Basu K (2006) Dissemination of cultural heritage and impact of pilgrim tourism at Gangasagar Island. Anthropologist 8(1):11–15

Dessein JK, Soini G, Fairclough, Horlings LG (eds) (2015, May 8) Culture in, for and as sustainable development: conclusions from the COST action IS1007 investigating cultural sustainability. Jyväskylä: University of Jyväskylä. http://www.culturalsustainability.eu/conclusions.pdf. Accessed 25 March 2017

Gidwani D (2012) Ayodhya on its way to becoming global heritage zone. *DNA* (a daily from New Delhi, edition Lucknow), 30 July. http://www.dnaindia.com/india/report-ayodhya-on-its-way-to-becoming-global-heritage-zone-1721508. Accessed 25 March 2017

HABITAT III (2016) Habitat III, Issue Papers 4- Urban Culture and Heritage, Quito Report. UNO, New York, October 2016. http://www.unesco.org/nw/fileadmin/MULTIMEDIA/HQ/CLT/pdf/ISSUE-Paper-En.pdf. Accessed 3 May 2017

HRIDAY (2015) Operational guideline for Heritage City Development and Augmentation Yojana. Ministry of Urban Development, Government of India, New Delhi

ICOMOS (2016, Nov. 24), Mission and Activities for 2017–18: Focal Point for the UN Sustainable Development Goals, Agenda, and 'Heritage as Driver of Sustainability'. https://www.icomos.org/images/DOCUMENTS/Secretariat/2016/ICOMOS_Focal_Point_for_SDGs_112016/ICOMOS_FP_SDG_Memo_201611241.pdf. Accessed 3 May 2017

ITU-T FG-SSC (2015) Smart Sustainable Cities: a Guide for City Leaders. Report (05/2015). ITU International Telecommunication Union, Geneva. https://www.itu.int/en/ITU-T/focusgroups/ssc/Documents/website/web-fg-ssc-0315-r6-city_leaders_guide.docx. Accessed 29 April 2017

Kumar, S (2018) Cultural landscape and heritage of Ayodhya-Faizabad: a geographical analysis. Unpublished Doctoral dissertation in Geography, Banaras Hindu University, Varanasi

Kumar S, Singh RPB (2015) Interfaces of Hindu Pilgrimage routes and agricultural landscape: a study of Ayodhya (India). In: Kohdrata N (ed) Proceedings, 4th ACLA international symposium on agricultural landscape of asia: learning, preserving, and redefining: 11–13 Sept. 2015. [Udayana University, Denpasar, Bali, Indonesia]: pp 38–49

Kumar S, Singh RPB (2016) Ritual Landscapes of Ayodhya (India) in the context of multicultural religious belongingness. In: Suparp T (ed) Proceedings, the 5th ACLA international symposium on 'Sacred Sites, Cultural Landscapes, and Harmonizing the World of Asia': 2–5 December

2016; Faculty of Humanities and Social Science; Lampang Rajabhat University, Muang District, Lampang Province, Thailand: pp 174–183

Kumar S, Singh RPB (2017) Rationales for inscribing "The Heritage sites, Riverfront and Pilgrimages Routes of Ayodhya" in the UNESCO World Heritage List. *Sthapatyam*. Indian J Architect Appl Sci (New Delhi) 3(9):51–78

Kumar S, Singh RPB (2019) Ayodhya: a study of Urban Governance and Heritage Inclusive Development. In: Sharma VR, Chandrakanta (eds) Making Cities resilient. Springer International Publisher AG, Dordrecht and New York, pp 317–338

Law BC (1944) Ayodhya in ancient India. J Ganganatha Jha Res Inst (Allahabad) 1(pt. 4):423–444

Rana PS (2014) Pilgrimage tourism: a study of Varanasi. SRME Publishers, Sagar MP (India)

Rypkema D (2016, 01 November) *Habitat III: Quito, SDGs, NUA, UN-Habitat, and Heritage*. http://forum.savingplaces.org/blogs/special-contributor/2016/11/01/quito-sdgs-nua-un-habitat-and-heritage. Accessed 29 April 2017

Shaw J (2000) Ayodhya's sacred landscape: ritual memory, politics and archaeological 'fact'. Antiquity 74(285):693–700

Simon D et al (2016) Developing and testing the Urban sustainable development Goal's targets and indicators—a five-city study. Environ Urban 28(1):49–63

Singh P (2004) Shiva-Kashi: puranic context and contemporary reference. Vishwavidyalaya Prakashan, Varanasi [in Hindi]

Singh RPB (2009a) The Ganga (Ganges) river: image and symbol of India. In: His Geographical thoughts in India: snapshots and vision for the 21st century. Planet Earth & Cultural Understanding Series, Pub. 2. Cambridge Scholars Publishing, Newcastle upon Tyne UK: pp 266–302

Singh RPB (2009b) Banaras: making of India's heritage City. Planet Earth & Cultural Understanding, Series Pub. 3. Cambridge Scholars Publishing, Newcastle upon Tyne U.K

Singh RPB (2011a) Varanasi, India's Cultural heritage City: contestation, conservation and planning. In: Rana PBS (ed) Heritagescapes and cultural landscapes. (Planet Earth & Cultural Understanding Series, Pub. 6). Shubhi Publs., New Delhi: pp 205–254

Singh RPB (2011b) Heritagescapes and cultural landscapes: an appraisal. In: Rana PBS (ed) Heritagescapes and cultural landscapes. (Planet Earth & Cultural Understanding Series, Pub. 6). Shubhi Publications, New Delhi: pp 7–56

Singh RPB (2013) Hindu tradition of pilgrimage: sacred space and system. Dev Publishers & Distributors, New Delhi

Singh RPB (2015) Banaras, the cultural capital of India: visioning cultural heritage and planning. SANDHI, A J Interfacing Sci-Herit Technol-Tradit India [IIT Kharagpur, India] 1(1):100–128

Singh, RPB (2017) Appraising the Indian Cultural Landscape: Envisioning Ecological Cosmology in the 21st Century. *North Eastern Geographer* (North East India Geographical Society, Gauhati University, Guwahati, India), vol. 39 (no. 1–2), 2016-17: pp 3–28

Singh RPB, Rana PS (2016) Indian sacred natural sites: ancient traditions of reverence and conservation explained from a Hindu perspective. In: Verschuuren B, Furuta N (eds) Asian sacred natural sites: ancient philosophy and practice in conservation and protected areas. Routledge, London and New York, pp 69–80

Singh RPB, Rana PS (2018) Kashi and cosmos, india: pilgrimage circuit of the panchakroshi yatra. In: Olsen D, Trono A (eds) Religious pilgrimage routes and trails: sustainable development and management. CABI Publishers, Wallingford, Oxon, pp 167–179

Singh RPB, Rana PS (2019) Visioning cultural heritage and planning: Banaras, the cultural capital of India. In: Bharne V, Sandmeier T (eds) Routledge companion of global heritage conservation. Routledge, London and New York, pp 152–175

Singh RPB, Rana PS, Kumar S (2020) Intangible dimensions of urban heritage: learning from holy cities of India. In: Silva KD (ed) The Routledge handbook on historic urban landscapes of the Asia-pacific. Routledge, London & New York: ca pp 241–258

Sinha A (compiled, ed.) (2014, December) Ghats of Varanasi on the Ganga in India: the cultural landscapes reclaimed. University of Illinois, Urbana IL, US

Chapter 14
Spatial Analysis and Mapping of Malaria Risk in Dehradun City India: A Geospatial Technology-Based Decision-Making Tool for Planning and Management

Ankita Sarkar, Vaibhav Kumar, Avtar Singh Jasrotia, Ajay Kumar Taloor, Rajesh Kumar, Rahul Sharma, Varun Khajuria, Girish Raina, Beena Kouser and Sagarika Roy

Abstract Land-use change emerged as one of the most rational component to the global environmental change, potentially has significant consequences on human health in relation to mosquito-borne blood diseases like malaria. Land-use change can influence mosquito habitat, and therefore the distribution and abundance of vectors and land use mediates human–mosquito interactions, including biting rate. Based

A. Sarkar
Department of Urban Planning, Indian Institute of Remote Sensing, Dehradun 248001, India
e-mail: ankitasrk26@gmail.com

V. Kumar
Centre for Urban Science and Engineering, Indian Institute of Technology, Bombay, India
e-mail: vaibhav.iirs@gmail.com

A. S. Jasrotia · A. K. Taloor (✉) · R. Sharma · V. Khajuria · G. Raina
Department of Remote Sensing and GIS, University of Jammu, Jammu 180006, India
e-mail: ajaytaloor@gmail.com

A. S. Jasrotia
e-mail: asjasrotia@yahoo.co.uk

R. Sharma
e-mail: rahul29453@gmail.com

V. Khajuria
e-mail: varunkhajuria8182@gmail.com

G. Raina
e-mail: girishrainaraina@gmail.com

R. Kumar
Department of Geology, Govt. Gandhi Memorial Science College, Jammu, India
e-mail: rajeshkumarjugis@gmail.com

B. Kouser
Department of Geology, University of Jammu, Jammu 180006, India
e-mail: beenajucryosphere@gmail.com

S. Sahdev et al. (eds.), *Geoecology of Landscape Dynamics*,
Advances in Geographical and Environmental Sciences,
https://doi.org/10.1007/978-981-15-2097-6_14

on a conceptual model linking the landscape, human, animal and mosquitoes, this study focuses on the impacts of changes in land use on malaria in Dehradun city of India. Health center wise data on malaria and land-use change data were prepared. Results of the different components of the study were integrated in the geographic information system (GIS) environment and linking land use to disease. The impacts of a number of possible scenarios for land-use changes in the region were delineated and also a risk map of the study area was prepared. Results indicated that land-use changes have a detectable impact on malaria. This impact varies according to the land use land cover (LULC) condition as well as the socio-economic condition but can be counteracted by the adoption of preventive measures.

Keywords Malaria risk · Land use land cover · GIS · Spatial analysis

14.1 Introduction

Malaria is one of the most widespread diseases affecting humans and it remains endemic in many parts of the world (Baeza et al. 2017). From the ancient history of man malaria has infected humans and may have been a human pathogen for the entire history of mankind. Tolle (2009) determines that malaria as a global concern and the World Health Organization (WHO) has estimated that about 40% of the world's population, mostly those living in the developing and underdeveloped countries are at high risk of malaria. Of these 2.5 billion people are at risk, more than 500 million become severely ill every year and more than 1 million die due to the effect of malaria. Malaria is a serious disease especially in Asia and Africa, where one in every five (20%) childhood death is due to the effects of this mosquito-borne infectious disease caused by parasitic single-celled microorganisms belonging to the Plasmodium group. On an average, an Indian child receives between 1.6 and 5.4 episodes of malaria fever each year and at every day five children dies from malaria. The vast majority of cases occur in children under the age of 5 years and pregnant women are also quite vulnerable (Kumar et al. 2007).

Since, the birth rule in India malaria has varied entomological, epidemiological, socio-economical and ecological determinants and thus remains a major public health problem. The interactive outcome of these disease determinants leads to various combinations of transmission risk factors at local and focal levels. The 1980s two million reported cases increased during the 90s both in terms of morbidity and mortality. In the last 5 years, about 40 epidemics including 1400 malaria deaths have been reported from nine states within the country (Srivastava et al. 2003). The urban centres in India are more prone to malaria as compared to rural India. Some of the reasons might be inadequate surveillance, poor reporting, and a time lag in reporting

S. Roy
Department of Civil Engineering, Indian Institute of Sciences, Bangalore 560013, India
e-mail: roy.sagarika@gmail.com

to decision makers and lack of georeferenced information to pin-point the trouble spots for a timely preventive action.

Geospatial technology is being widely used for identification, characterization, monitoring, surveillance, and managing of breeding habitats and mapping of malaria risk around the many parts of the world (Ahmad et al. 2018). Remote sensing imageries were used for identification and characterization of the habitats that produced potential Anopheles vector mosquitoes in the Republic of Korea (Sithiprasasna et al. 2005). Integrated use of remote sensing and GIS has been successfully demonstrated in many studies related to the mapping of malaria risk in different parts of Africa (Kleinschmidt et al. 2001). Malaria is a serious public health problem in Southeast Asian countries including India. After the implementation of Modified Plan of Operation (MPO) in 1976 in India, malaria incidence came down to 2–3 million from 7.5 million recorded cases. However, malaria continued to be an endemic in certain pockets particularly in the tribal regions. The emergence and wide-spread drug-resistance in malaria parasites in India initiated using cost-effective and reliable alternative method for the vector control (Dongus et al. 2007). Public funds for health services in India have been largely focused on medical services, and public health services have been neglected. This is reflected in the virtual absence of modern public health regulations, and of systematic planning and delivery of public health services. Various organizational issues also hinder the rational deployment of personnel and funds for disease control (Khaleghian and Gupta 2005).

Geographic information science (GI Science) and attendant technologies in remote sensing, geographic information systems (GIS), and spatial statistics can be used for the surveillance and control of disease vectors in general, and specifically mosquitoes in VBD (Pam et al. 2017). Geographic information systems (GIS), production of global environmental digital datasets, meteorological and remote sensing data collection and analysis at near real-time, development of methodologies, models and tools for data standardization, collection, spatialization, analysis and dissemination, networking and information sharing, development of integrated information management systems. Some experience has also been gained in the use of thematic digital datasets for spatial modeling of vector-borne diseases (Kazansky et al. 2016).

GIS-based malaria cases mapping can help health authorities to understand more about the spatial distribution of the disease in their area as well as its temporal occurrence. The provided information will provide a guideline for control programs and preparing health facilities based on the requirement of each area. GIS has been continuously used for the analysis of spatial health related data. It can be a useful tool for analyzing the spread of diseases in both developed and developing (Kazansky et al. 2016). This tool can also be useful for management strategy to allocate resources for preparing the needs for control of disease in high risk areas of disease. GIS also enable us to generate revised maps as soon as new data are available. In India, GIS-based malaria mapping as a decision support system to control malaria-a case study has also been made in the Koraput district in Orissa, where the problem is more severe due to minimal transport facilities and inadequate healthcare facilities. In rural India, due to lack of facilities people visit a hospital only if the illness is highly prolonged and if there is no response to the primary line of treatment (Daash et al. 2009).

During the rainy season, piedmont zone of the Indian Himalaya is highly prone to waterlogging. There are innumerable streams which flow into the river Ganga and Yamuna, the main river of Uttarakhand (Sidhu 2016; Jasrotia et al. 2018). The dense forest provides shelter and humidity for mosquitoes breeding which might cause an increase in vector density. As the population density is low and the breeding sites are very extensive, the ratio of the breeding sites to man is very high. Breeding sites are covered with dense aquatic vegetation which further makes it difficult to employ any control agent. Patches of swamps and seepages exist all along the streams and tributaries.

There are many factors which affect the spread of malaria in Dehradun city, India. Some of the major factors are anthropogenic, geo-climatic, socio-economic factors and way of living. It is essential to understand such factors that cause increase in the vector densities, and hence the transmission of diseases to prevent the emergence and resurgence of more diseases, as well as to serve as basis for effective control (Yadav et al. 2013). Dehradun has a good healthcare system which helps people from every socioeconomic class to get affordable treatment. Moreover, the cases are well documented and are updated in the form of spatial format by the responsible state agency.

14.2 Study Area

Dehradun city is situated in the newly formed state of Uttarakhand. It is one of the oldest cities in India. The city of Dehradun is situated in the south central part of Uttrakhand State located at an altitude of 640 m, lies at 30° 19′ N and 78° 20′ E covering a total area of 38.04 km^2 (Fig. 14.1). The site where the city is located slopes gently from north to south and southwest and is heavily dissected by a number of seasonal streams and nullahs. The drainage of the city is borne by the rivers Bindal and Rispana. Dehradun enjoys a salubrious climate due to its Dun location in the hilly part of the state. Temperature during the summer season ranges between 36 °C and 16.7 °C whereas, winter months are quite colder with the maximum and minimum temperatures ranges between 23.4 °C and 5.2 °C respectively (Matta 2014). During the monsoon season (June to September) Dehradun experiences most of the annual rainfall of about 2000 mm.

14.2.1 Soil Characteristics

The soil type which includes manly texture, organic matter content its infiltration capacity and permeability, greatly affects the soil loss and run-off. Fine texture soils are more susceptible to erosion than coarse texture soils, since rainwater enters in and passes through a dense clay much more slowly than through a porous sand or gravelly soil (Fischer et al. 2015). In the present study area, the alluvial soils at Vasad

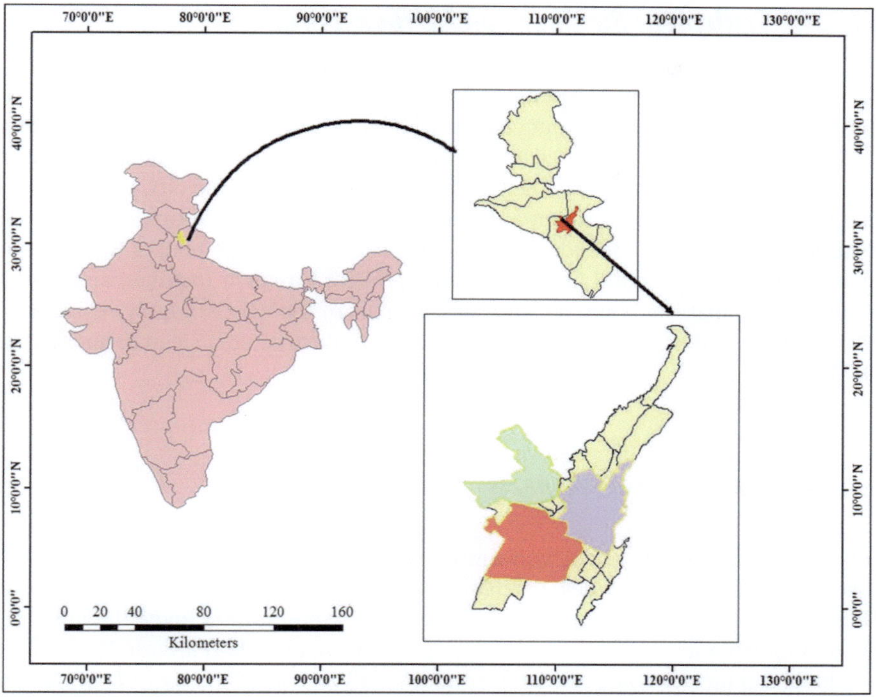

Fig. 14.1 Location map of the study area (*Source* Landsat-8, OLI)

and Dehradun have a very high rate of run-off; the black soils have an intermediate rate of run-off, but still the rate of run-off is high. The soil left in loose and pulverized condition is particularly liable to erosion through sheet-wash and gullying (Gupta and Panigrahy 2008).

14.2.2 Ground Cover and Land Use

When rain falls on a surface covered by a thick mantle of plants, its velocity and erosive power, reduces and most of the water either quickly percolates through the soil or moves over the surface with non-erosive velocity (Fischer et al. 2015). Areas not protected with a thick cover of plants are unable to absorb water effectively, because the dashing rains shatter the soil surface, the fine soil particles go into suspension and the thick mixture of water and soil quickly fills and closes the tiny interstices in the soil, reducing infiltration and consequently increasing run-off and soil loss.

14.2.3 Settlement Structure and Urban Form

The settlement structure of Dehradun depicts morphological expansion over a colonial structure. The eastern Rajpur canal was the most important feature in Dehradun during the British period that had served the needs of water for drinking and agricultural purposes (Parkash 2014). The central part consists of the old city, i.e., the colonial vestiges, and private residential areas. The prestigious educational and research institutions are situated outside the core city. The western side houses the Cantonment area, Oil and Natural Gas Corporation, Forest Research Institute, and Wadia Institute of Himalayan Geology. The eastern part of the city is largely residential. The southern part of the city is designated as an industrial area (Kumar and Thapa 2015).

14.2.4 Demographic Aspects

As a part of the past heritage, concentration of national and regional level institutions, and economic activities, availability of infrastructure, and the emergence of Dehradun as the state capital on November 2000, would further invite the influx of population from the rest of the valley as well as from outside (Parkash 2014). This would further increase the growth rate in addition to the natural increase of population within the city itself. Favourable climate, good regional linkages by rail and road, and feasibility of spatial expansion of Dehradun city would be therefore instrumental for further migration. However whether the city would support this population expansion on a sustainable basis also calls for a detailed insight into the socio-economic characteristics of population that in turn decides the quality of population as well as certain aspects of urban planning and urban environment.

14.2.5 Socio-Economic Function of Dehradun City

Cities come into existence due to the functions they perform as central places. A harmonious integration of functions and activities can lead to a healthy and orderly development of the city. The major town functions of Dehradun can be grouped under: Administrative: Dehradun is the capital of the newly formed state of Uttarakhand. Educational and Institutional: The city besides being the seat for prestigious educational institutions, and other technical institutes, are also famous for national level institutes as already stated. Commercial: Dehradun is the largest service centre within the hilly region of Uttaranchal. It meets the trade and commerce requirements of its region. With the expansion of national level institutes and offices, and the expansion of the cantonment area, the commercial activity had gained momentum. Industrial: Establishment of industries based mainly on limestone and forests have attracted ancillary industrial units and other industries. Development of industries is likely to play a vital role in building a sound economic base of the city. Tourism: Dehradun is endowed with immense potentialities for tourism industry besides being gateway to Mussoorie, the Queen of Hill stations. There are a number of tourist places and recreation spots within a short distance of the city that can be developed adequately. Defence: Dehradun is the headquarters of Indian Military Academy. A number of other defence establishments also are in Dehradun. The defence function has played a vital role in shaping the development and economy of the town (Suthar and Singh 2015).

14.3 Methodology

The Landsat 7 (ETM+Pan merged 15 m) data is used to prepare the spatial distribution map aided with ancillary data. The malaria data was collected from the Integrated Disease Surveillance Programme, National Centre for Diseases Control (IDSP 2013). Data consisted the entries from 15 health center of Dehradun city. The city witnessed highest cases of malaria (349) in 2015. Health center named Shri Mahant Indresh (SMI) reported highest number 61 Plasmodium falciparum (Pf) of malaria cases, while 23(pf) cases were reported in SPD Bhagat Singh colony. Lowest number of cases: 1(Pf), were reported by Synergy Institute of Medical Sciences. ArcGIS 10.0 software suit was used for mapping spatial distribution of malaria incidence at the micro level with the help of GPS survey. A 2 km buffer was created along each health location. Further, the wards falling in the buffer zone were divided into categories of high malaria incidence ward, moderate malaria incidence ward and low malaria incidence ward depending upon the number of malaria cases. Digitization of different data layers pertaining to the study area, rivers, water bodies, Income group population, location of health centres, Ward maps, etc. were carried out using Google earth explorer. Different attributes of the layers were added to the database for further analysis. The detailed methodology adopted in the present study is shown in the Fig. 14.2. The LULC for three area respectively high malaria cases, medium malaria cases and low malaria cases was prepared to find the spatial distribution of malaria in the Dehradun city. Supervised classification is done to prepare LULC map for different area in Dehradun municipal area. Various classes considered are built-up, vegetation, agriculture, and vacant land. The supervised classification approach was used to determine various the land cover types present in the scene. To do this, representative training sites of known land use land cover types are used to compile a numerical interpretation key that describes the spectral attributes for each type of interest the location and distribution of malaria vectors will be analysed through thematic presentation and mapping. To remove the bias and error field verifications for the training data has been done.

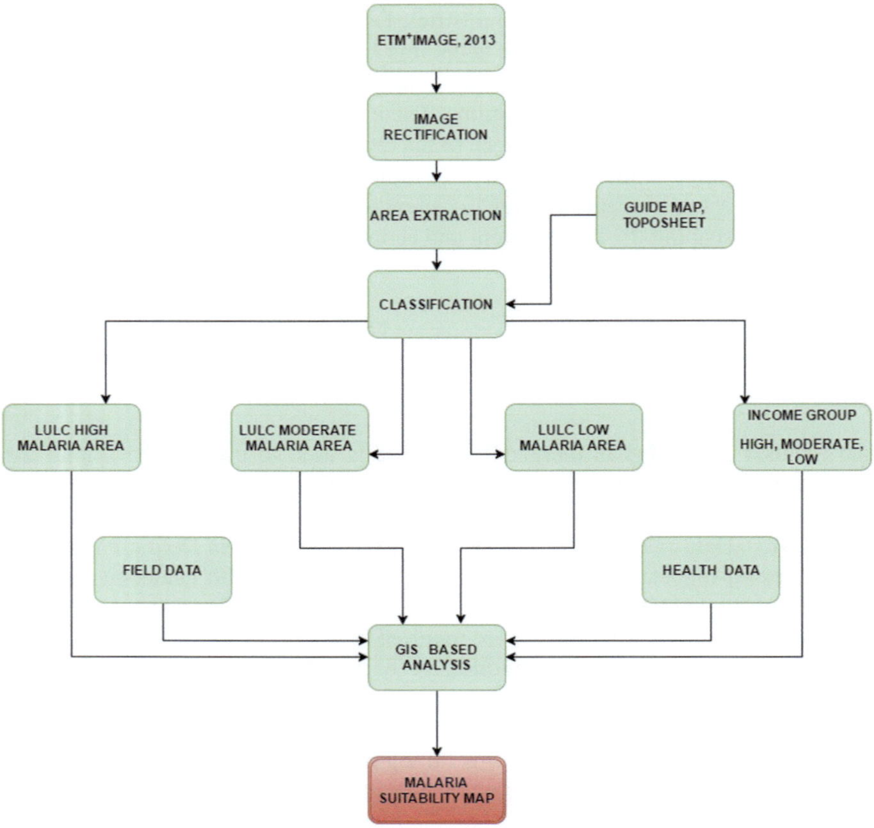

Fig. 14.2 Methodology adopted in the study

14.4 Results and Discussions

The section presents the results of the carried out analysis.

14.4.1 Temporal Distribution of Malaria Incidence

Temporal distribution of malaria incidence is considered, where 8 years (2006–13) data are taken from chief medical office, Dehradun. In 2013 total malaria examined case was 108521 (B/S exam) with total positive cases 346 persons (44Pf). During rainy season, the city is prone to water logging in the low lying area. There are innumerable streams which flow into the River Yamuna, the important river of Uttarakhand. The dense forest within the city provides shelter and humidity for mosquitoes

Table 14.1 Malaria cases in the Dehradun city (2006–2013) (*Source* National Centre for Diseases Control (IDSP 2013) Dehradun)

Year	2006	2007	2008	2009	2010	2011	2012	2013
Malaria cases	11	19	49	53	232	230	487	346

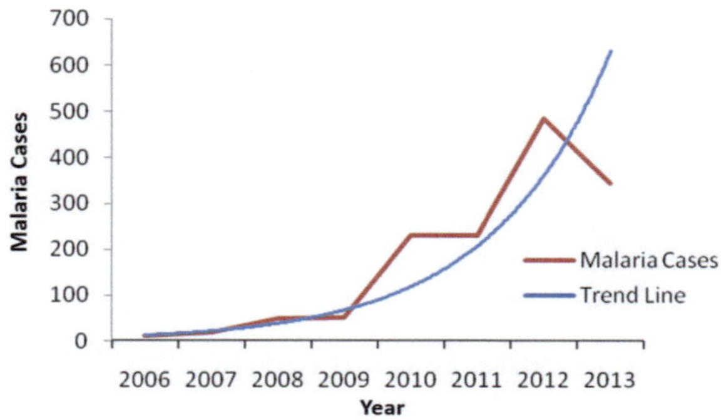

Fig. 14.3 Showing trend line of malaria growth from (2006–2013) (*Source* National Centre for Diseases Control (IDSP 2013) Dehradun)

breeding which may cause increase of vector density. The trend line shows gradually increasing malaria rate in 2009 but it increases exponentially after 2010 and get doubled in the 2014 (Table 14.1 and Fig. 14.3). This is a matter concern, thus, must be addressed by the decision makers.

14.4.2 Malaria and Rainfall Relationship

Rainfall data was collected from the year 2004 onwards from Indian meteorological department (IMD) website (www.imd.gov.in), which gives a picture about the rainfall pattern in the district. From the data yearly irregularities in the rainfall is evident. The average monthly variation of the rainfall shows that the rainfall is more prominent in the region from the month of June to September. Rainfall affects anopheles mosquitoes breeding site, a malaria vector, typically breeds in small water sources. In Dehradun city a relationship between rainfall and malaria incident in per year is shown in the graph below. The Correlation between rainfall and malaria is show a moderate positive relationship (r^2) = 0.42 (Fig. 14.4). Climatic factor like rainfall is not a single factor for malaria survival. Other factors like geographical and the socio-economic conditions also play an important role like LULC of a particular area. To understand this we analysed the LULC of the study areas with the malaria cases in the next section.

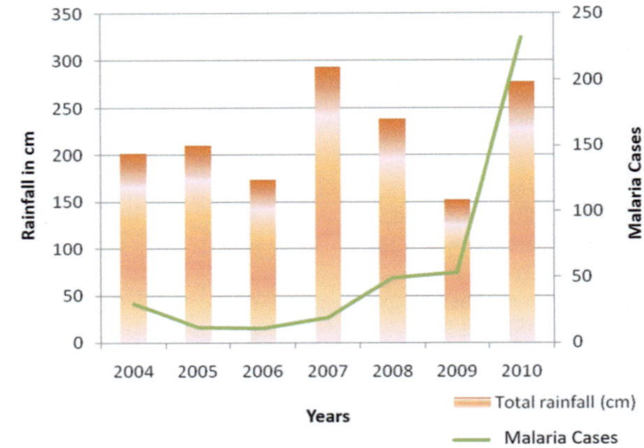

Fig. 14.4 Relationship between rainfall and malaria cases (*Source* National Centre for Diseases Control (IDSP 2013) Dehradun)

14.4.3 Land Use Land Covers Distribution

The spatial and temporal changes in the distribution of malaria and land use land cover changes (Table 14.2 and Fig. 14.5a–c) can influence malaria transmission intensity (Wagner 2017). The association between land cover type and occurrence of malaria is statistically significant (Himeidan et al. 2013). In high intensity area water is 2%, agriculture 12%, vegetation 18% and built-up 62% and river Bindal is flows almost vertically to the study area. In the moderate intensity area water level is similar (2%), but the vegetation area is 13% and a high built-up percentage (77%) is present which is higher than the high intensity area.

One of the important observations is high malaria incidence frequency prone area consist 11.74% of agricultural land of the total area which makes this area a high-risk malaria zone. Further analysis shows a positive correlation between high malaria cases and the forest cover. The correlation between each class is calculated and positive relationship between the land use/land cover and malaria has been found. Figure 14.6a shows the relationship graph of vegetation percentage with malaria cases

Table 14.2 LULC percentage of the study area (Hectare: ha) (*Source* Landsat-8, OLI)

Land use/Land cover classes	High malaria incidence area		Moderate malaria incidence area		Low malaria incidence area	
	(ha)	(%)	(ha)	(%)	(ha)	(%)
Water bodies	31.58	47.63	27.18	40.99	7.54	11.37
Agriculture land	166.86	11.74	0	0	0	0
Forest cover	254.99	47.89	140.10	26.31	137.25	25.78
Vacant land	81.52	35.30	83.91	36.34	65.45	28.35
Built-up	879.18	36.50	842.85	34.99	686.23	28.49

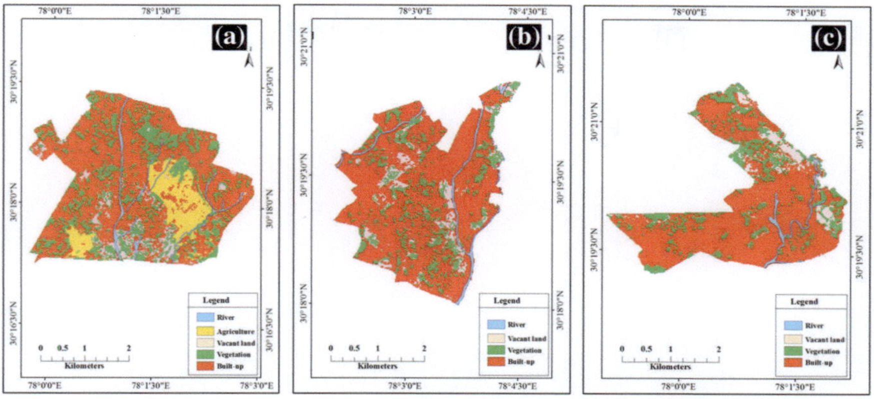

Fig. 14.5 a–c Land use land cover map of high malaria incidence area (**a**), medium malaria incidence area (**b**) and low malaria Incidence area (**c**) (*Source* Landsat-8, OLI)

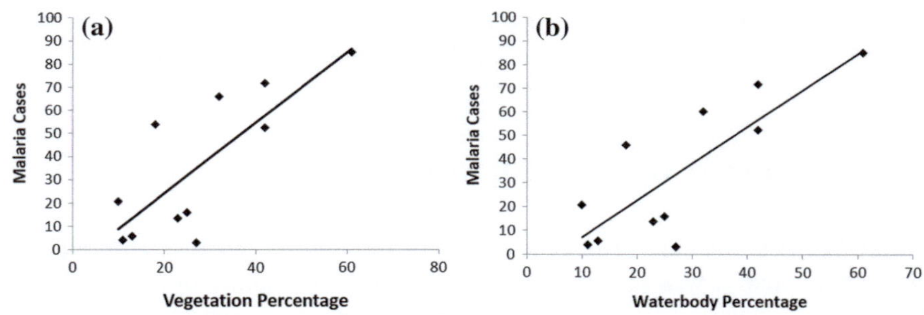

Fig. 14.6 a–b Relationship of malaria cases with vegetation area percentage (**a**), (**b**) Relationship of malaria cases with Waterbody area percentage (*Source* Landsat-8, OLI)

($r^2 = 0.6172$), and Fig. 14.6b shows the malaria case relationship with water body percentage ($r^2 = 0.6758$).

14.4.4 Socio-Economic Condition and Malaria

Socioeconomic, socio-cultural and behaviour patterns of the community plays an important role in disease transmission. To study the socio-economic conditions of the people the income group map for all the three areas have been prepared from Google earth and field verifications have been done to validate the datasets (Fig. 14.7a–c). We found that high income group (HIG) area is less vulnerable to malaria occurrences as compared to the low income group. Further, medium income group area is also found to be vulnerable to the malaria occurrence risks. The percentage of HIG people is

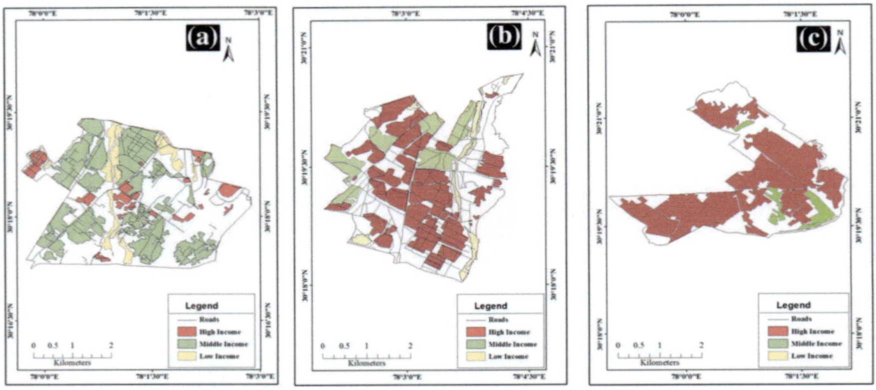

Fig. 14.7 **a–c** Income map of high malaria incidence area (**a**), Income map of medium malaria incidence area (**b**) and Income map of low malaria incidence area (**c**) (*Source* Field Survey)

low in high malaria incidence area but it is high in moderate and low area. The high malaria incidence area is predominant by the LIG group of people and vice versa.

14.5 GIS-Based Malaria Risk Analysis

Hazard is the probability of the occurrence of mosquitoes infective with malaria in a certain area. It was approached by assessing the suitability of environmental condition for malaria transmission based on environmental and physical factors. After preparing all the factor parameters compatible to risk analysis, estimating weights for risk parameters was defined. To illustrate, the malaria risk map health, hygienic condition, rainfall, land use land cover and income of the household were used to decipher the malaria risk map (Fig. 14.8).

14.6 Conclusion

GIS now days emerge as one of the most significant tool in the spatial analysis and demonstrates tremendous capabilities of the technology available to epidemiologists and researchers. Integration of GIS with remote sensing data helped us a lot in identification surveillance of breeding habitats and mapping of malaria risk areas in the study area. The global positioning system data in a GIS assisted format helped us in generating base map, mapping breeding habitats and analysis of areas of high disease prevalence. The GIS-based malaria incidence mapping assisted in risk mapping for analyzing past as well as present trends of malaria. Such kinds of

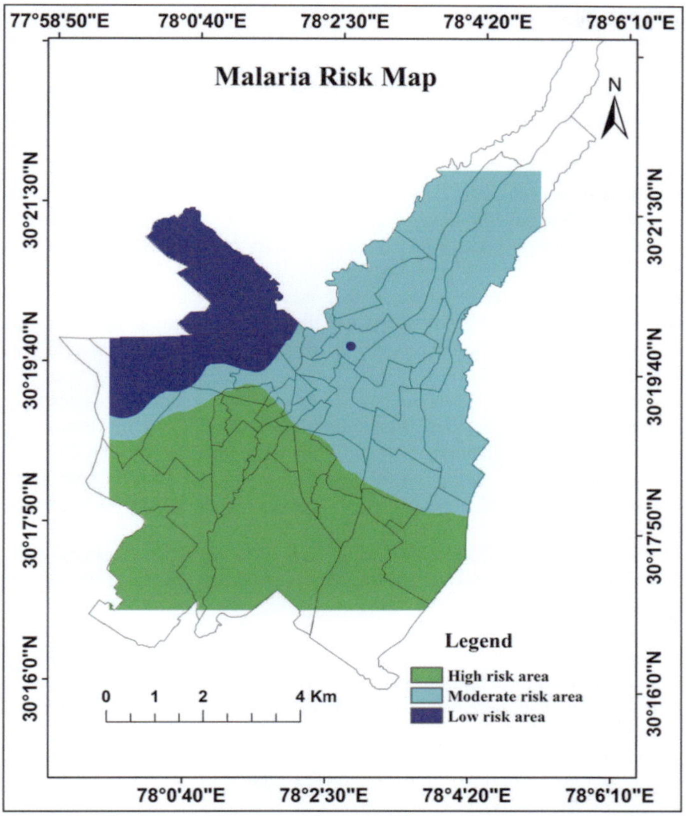

Fig. 14.8 Malaria risk map of the study area (*Source* National Centre for Diseases Control (IDSP 2013) Dehradun)

mappings have been extensively used in preparing maps of global malaria risk distribution in space and time. Spatial analyses using geo-processing tools have assisted in establishing relationship between malaria incidence and other potentially related factors. GIS mapping helped determine distribution of malaria vector mosquitoes along with monitoring and evaluation of malaria control activities in various countries. Three zones were identified for the malaria risks in the study area. It is found in the study area that less hygienic condition, poor health and economic standard are the major factor contributing the malaria in the study area. The areas which are delineated as high-risk area are low income area, poor hygienic condition are the major contributing factor to the high-risk malaria zone.

Acknowledgements The authors are thankful to the Health Department, State Government of Uttrakhand for providing us the necessary data. We are also thankful to the anonymous reviewers for helping us to update the manuscript.

References

Ahmad I, Arsin AA, Thaha RM (2018) A spatial analysis of malaria cases using geographic information system (GIS) application in gorontalo regency in 2016. In: Proceedings of the 2nd international conference on medical and health. ACM, pp 33–37

Baeza A, Santos-Vega M, Dobson AP, Pascual M (2017) The rise and fall of malaria under land-use change in frontier regions. Nat Ecol Evol 1(5):0108

Daash A, Srivastava A, Nagpal BN, Saxena R, Gupta SK (2009) Geographical information system (GIS) in decision support to control malaria-a case study of Koraput district in Orissa, India. J Vector Borne Dis 46(1):72

Dongus S, Nyika D, Kannady K, Mtasiwa D, Mshinda H, Fillinger U, Drescher AW, Tanner M, Castro MC, Killeen GF (2007) Participatory mapping of target areas to enable operational larval source management to suppress malaria vector mosquitoes in Dar es Salaam, Tanzania. Int J Health Geogr 6(1):37

Fischer C, Tischer J, Roscher C, Eisenhauer N, Ravenek J, Gleixner G, Attinger S, Jensen B, de Kroon H, Mommer L, Scheu S (2015) Plant species diversity affects infiltration capacity in an experimental grassland through changes in soil properties. Plant Soil 397(1–2):1–6

Gupta PK, Panigrahy S (2008) Predicting the spatio-temporal variation of run-off generation in India using remotely sensed input and Soil Conservation Service curve number model. Curr Sci 10:1580–1587

Himeidan YE, Temu EA, Rayah E, Amin E, Munga S, Kweka EJ (2013) Chemical cues for malaria vectors Oviposition site selection: challenges and opportunities. J Insects 1–9. Article ID 685182. http://dx.doi.org/10.1155/2013/685182

Integrated Disease Surveillance Programme (IDSP) National Centre For Diseases Control (2013). http://www.idsp.nic.in/index.php

Jasrotia AS, Taloor AK, Andotra U, Bhagat BD (2018) Geoinformatics based groundwater quality assessment for domestic and irrigation uses of the Western Doon valley, Uttarakhand, India. Groundw Sustain Dev 6:200–212

Kazansky Y, Wood D, Sutherlun J (2016) The current and potential role of satellite remote sensing in the campaign against malaria. Acta Astronaut 121:292–305

Khaleghian P, Gupta MD (2005) Public management and the essential public health functions. World Dev 33(7):1083–1099

Kleinschmidt I, Omumbo J, Briet O, Van De Giesen N, Sogoba N, Mensah NK, Windmeijer P, Moussa M, Teuscher T (2001) An empirical malaria distribution map for West Africa. Trop Med Int Health 6(10):779–786

Kumar A, Valecha N, Jain T, Dash AP (2007) Burden of malaria in India: retrospective and prospective view. Am J Trop Med Hyg 77(6):69–78

Kumar R, Thapa D (2015) Social media as a catalyst for civil society movements in India: a study in Dehradun city. New Media Soc 17(8):1299–1316

Matta G (2014) A study on physico-chemical Characteristics to assess the pollution status of river Ganga in Uttarakhand. J Chem Pharm Sci 7(3):210–217

Pam DD, Omalu IC, Akintola AA, Dan Azumi Y, Kalesanwo AO, Babagana M, Muhammad SA, Ocha IM, Adeniyi KA (2017) The Role of GIS and remote sensing in the control of Malaria. Online J Health Allied Scs 16(3):7

Parkash S (2014) Geo-environmental characteristics, natural resources and disasters in Uttarakhand State. Dolphin (PG) Institute of Biomedical and Natural Sciences, Dehradun (UK). ISBN: 978-81-907797-1-5.:15

Sidhu GS (2016) Soil conservation of Northwestern Himalayas (NWH): their constraints and potentials for sustainable hill agriculture. In: Conservation agriculture. Springer, Singapore, pp 315–338

Sithiprasasna R, Lee WJ, Ugsang DM, Linthicum KJ (2005) Identification and characterization of larval and adult anopheline mosquito habitats in the Republic of Korea: potential use of remotely sensed data to estimate mosquito distributions. Int J Health Geogr 4(1):17

Srivastava A, Nagpal BN, Saxena R, Eapen A, Ravindran KJ, Subbarao SK, Rajamanikam C, Palanisamy M, Kalra NL, Appavoo NC (2003) GIS based malaria information management system for urban malaria scheme in India. Computer Meth Prog Biomed 71(1):63–75

Suthar S, Singh P (2015) Household solid waste generation and composition in different family size and socio-economic groups: a case study. Sustainable Cities and Society 14:56–63

Tolle M (2009) Mosquito-borne diseases. Curr Probl Pediatr Adolesc Health Care 39(4):97–140

Wagner R (2017) The World Health Organisation: a time for reconstitution. In: Fighting the diseases of poverty. Routledge, pp 267–326

Yadav K, Dhiman S, Rabha B, Saikia PK, Veer V (2013) Socio-economic determinants for malaria transmission risk in an endemic primary health centre in Assam, India. Infect Dis Poverty 3(1):19. https://doi.org/10.1186/2049-9957-3-19

Chapter 15
Green Buildings: Building a Greener City, a Greener Future—An Indian Perspective

Kriti Kanaujia

Abstract Green building as a concept has become popular in recent years due to concerns of un-sustainability and climate change. It is premised on being energy and resource efficient from its inception to manifestation; in order to minimize environmental and energy consumption costs. Although, the concept may seem new, but the practice has been done since centuries in India in the form of white roofs of Jaisalmer to rain water harvesting methods of baolis (step-wells), hauz (water reservoir) and tals (lakes) in numerous old forts and palaces of India. Even in contemporary India, many sustainable techniques are being promoted by the Government of India in synchronization with State governments. Green buildings are the need of the hour because they will help reduce our ecological footprint by adapting to the existing climate and helping us to mitigate the effects of climate change to a considerable level. Green buildings are the building blocks of Smart Cities.

Keywords Green building · Old practices · New energy efficient techniques · Sustainability · Urban areas · Smart cities

15.1 A Need to Rethink Our Urbanization

It has been many centuries since Urbanization appeared and influenced the human lives. It has evolved through many stages and continues to evolve to keep in sync with the changing necessities of time and people. In the recent decades of urban history, a need for change in the approach to urban and its dynamics has been felt severely. It is in this light that a debate between development and regeneration emerged on the urban scene. While urban development acts as a mediator and facilitator for transformation, urban regeneration on the other hand, acts as a catalyst extending the process of transformation and as a means of sustaining urban development. An area may develop to be urban but it cannot continue to sustain itself beyond a point of time

K. Kanaujia (✉)
Department of Geography, Delhi School of Economics, University of Delhi, New Delhi 110007, India
e-mail: sak2lucy@gmail.com

S. Sahdev et al. (eds.), *Geoecology of Landscape Dynamics*,
Advances in Geographical and Environmental Sciences,
https://doi.org/10.1007/978-981-15-2097-6_15

without regeneration. This has been exemplified by western experiences especially that of London, post-reunified Berlin, Boston, Pittsburg, Scotland and many other European cities (Mitrojorgji 2003). Urban regeneration means differently to different people and contexts.

Broadly, urban regeneration is an attempt, a vision and a practice aimed towards achieving all encompassing, resilient as well as sustainable solutions towards myriad urban or urban related challenges while simultaneously providing improvements in physical, socio-economic and environmental conditions of a specific area that has been subjected to change(s). Regeneration encompasses these four aspects in entirety namely; community, infrastructure, employment, health, literacy and so on. Each aspect interacts with other aspects in an interrelated and complex manner along with their in/direct impact(s) on other aspects (Carrion and Hanley 2007). Therefore, it can be said that regeneration is not an individual oriented approach; rather is an all-encompassing perspective that is premised upon sustainable development. Urban regeneration is; therefore aimed at self-sustaining planned and regulated urban development (Kanaujia 2016).

Urban development of the world at present is extremely rapid and unsustainable with haphazard growth; especially in the developing world. Most of the West European countries and USA are highly urbanized due to industrialization and modernization. Urban regeneration has been accompanying urban development in the West; with major emphasis after the World War II period. It was only after many decades of urbanization experience as well as billions of dollars invested in urban development which led to the realization of the importance of urban image, quality of life and urban environment as necessary stimulants for economic development of cities (Lawless 2010). It was in this context that urban regeneration was adopted as an instrument of revival and rejuvenation of urban development in depopulating and degrading urban centres; with the aim of self sustaining development from that point forth. Urban Regeneration and not Redevelopment was adopted because the cities required more than spatial re-configuration and renovation (Brodies 2009). The cities were subjected to regeneration in order to revive economy, culture and a sense of community among its inhabitants. Unlike urban development, urban regeneration has the versatility to mould itself in order to meet the challenges of the time in a specific area aimed at specific section of society (Carrion and Hanley 2007). Urban regeneration is not a "one size fit all" approach, it needs to be modified and appropriated according to the needs of the area and its people in question (Brodies 2009). Urban regeneration as a concept works better and in synchronisation with the concept of Green Buildings since both are premised upon self-sustainability.

With recent drastic and changed weather phenomenon being witnessed; we need to relook at our current practices and norms and especially in urban areas because the world is rapidly progressing towards urbanization and urban population growth (refer to Fig. 15.1). Even though the urban population growth rate would register a declining trend in upcoming decades, it is important to note that the World urban population would continue to rise in absolute numbers implying that even a declining growth would imply huge population increments.

Fig. 15.1 Trend of Urban Population Growth (in percentage) from 2015 to 2030. *Data source* Urban Population Growth: 2015–2030 (World Health Organization 2014)

At present, 54% of the world's population resides in urban areas and it has been estimated that it will increase to 66% by the year 2050 (UN DESA 2014). It is in synchronisation with the estimates made in the year 2007 and 2014 (refer to Fig. 15.2) for the urban population growth by UN. There is an upward trend in urban population numbers across major regions of the world. Although the increment seems to be slowing, yet in absolute numbers the population added is large and will put immense burden on existing resources giving rise to mega cities (refer to Figs. 15.3 and 15.4) with multi-million urban populations. These cities would be characterised by high density, massive built-up area, unplanned growth and resource scarcity if actions and relevant measures are not taken up immediately. In this light, we need to question ourselves as to what kind of cities do we really want to live in or what kind of cities we ought to live in.

Many cities are already facing the problems of over-urbanization along with severe environmental impacts like water-logging during monsoons and Urban Heat Island effect. In this context, we really need to question the very notion of what we understood as Urban or as a City, should we really continue to conceive, plan and implement cities the way they are being planned and implemented or is it really the need of the

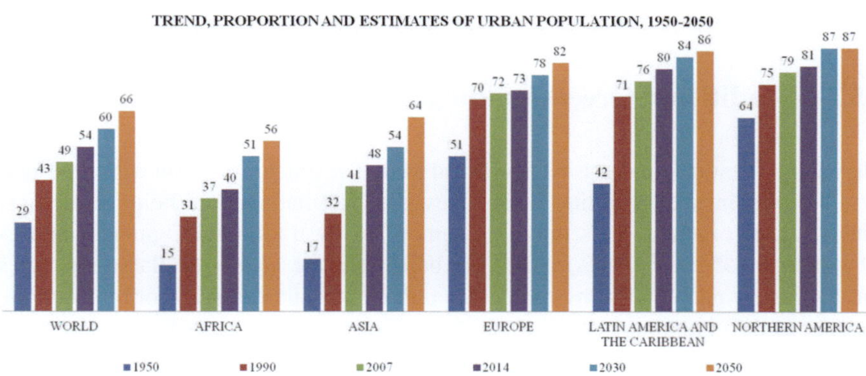

Fig. 15.2 Trend, Proportion and Estimate of Urban population growth (in percentage) from 1950 to 2050*. *Data source* World Urbanization Prospects: The 2005 Revision (UN DESA 2006), World Population Data Sheet (Haub 2007) and World Urbanization Prospects: The 2014 Revision (Highlights) (UN DESA 2014)
*[The data excludes Oceania's proportion of urban population (71, 71 and 74 for 1990, 2014 and 2050 respectively)]

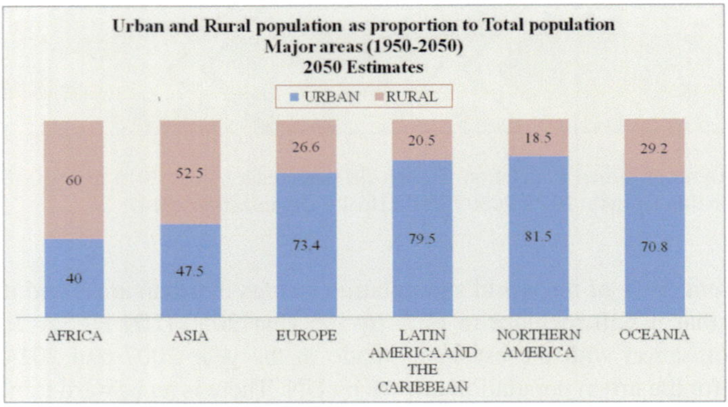

Fig. 15.3 Distribution of World Population according to Urban and Rural in Major regions. *Data source* World Urbanization Prospects: The 2014 Revision (Highlights) (UN DESA 2014)

hour to re-conceptualize and re-visualize the planning and implementation notions of and for a city.

It is in this context that, a need arises for sustainable urbanisms and urbanization which is centred upon living life in greener, smarter, sustainable, energy and resource-efficient ways. Green buildings or building in greener ways is a stepping stone towards the realization of Smart Cities' initiative and a relationship which needs to be pursued more rigorously. But the concept and usage of green buildings is a not a new phenomenon or idea, it has been a practice and an ideology that has been in existence since several decades and centuries premised on ingeniously living in-synchronization with the local geography.

15.2 Traditional Green Buildings: Living Smart in the Past

Human civilizations understood and learnt to live in synchronization with their natural surroundings by adapting themselves, their habitations and being ecological and resource efficient. This may be exemplified by traditional Ogimachi houses of Japan (wooden structures) which are built from the locally available resources and extremely adept at its environmental and seismic surroundings, or by the white facades of the numerous houses in Santorini, Greece or the locally suited stone tiled roofs of Apulia in Southern Italy. Perhaps the best examples of green buildings can be gathered in places like Ghana, Africa where the houses are designed to withstand extreme high day time temperatures and scarcity of building resources or the thatched or grass laden roofs of Portugal, Iceland and Norway respectively exemplifying being one with the nature, living in nature with nature. Perhaps it is important to point out one of the most famous examples of green buildings; Igloos. The Igloos are traditional houses in the northern frigid zone of our world, wherein the houses are

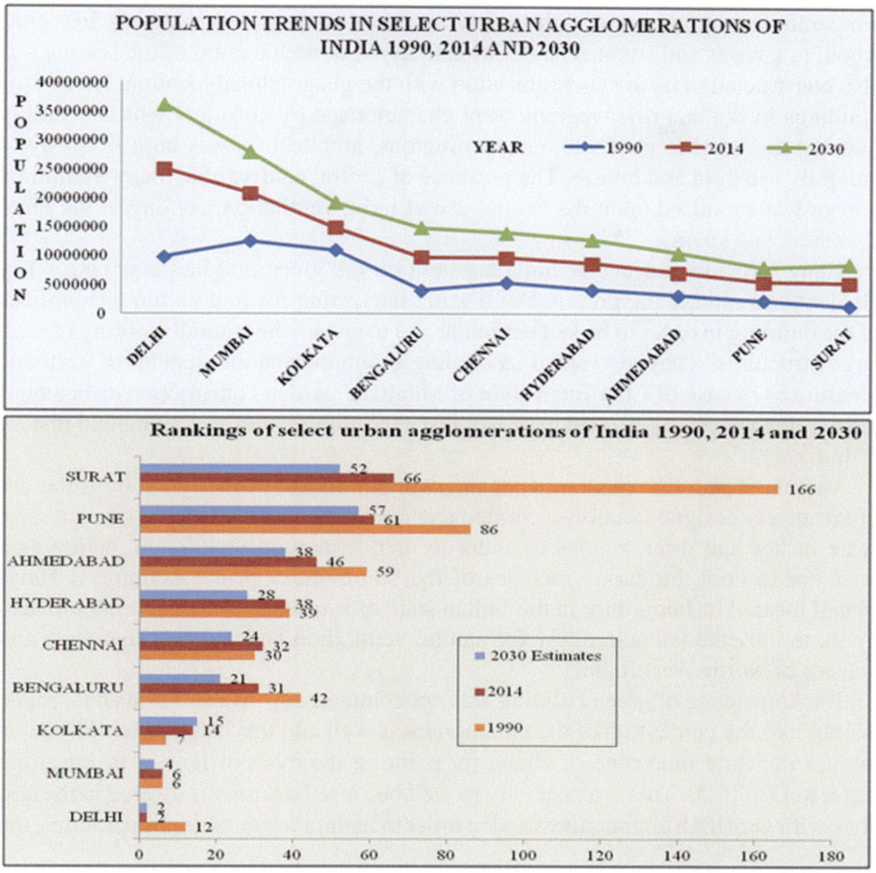

Fig. 15.4 Population Trends and Ranking of select Indian urban agglomerations with more than 5 million inhabitants as of 1 July 2014, for the years 1990, 2014 and 2030. *Data source* World Urbanization Prospects: The 2014 Revision (UN DESA 2014)

made by carving out bricks of ice. These houses are exceptionally environmentally sustainable since they cause no damage to their surroundings and are premised upon reuse–recycle.

Indians like their counterparts elsewhere; were aware of and makers of green buildings even in historical times. Traditional homes were built in accordance to the spatial and cosmological (the four cardinal directions, Vastu Shastra) location, local climatic regime, available building materials and the idea of utilising nature at its maximum.

Traditionally, Indian houses were made of baked roof tiles (red in colour) and clay for construction of walls which proved to be energy efficient as they were able to maintain cool ambience during summers and warmth during winters. Even till today, many of the traditional rural houses utilize this age old technique to maintain temperatures naturally and with locally available construction materials like clay, wood, jute ropes and so on. The traditional layout or architecture of the houses was also constructed to be in synchronisation with the geographical location. Hence, the buildings in hot and drier regions were characterised by corridors which naturally directed the wind to cool while in wet regions, architecture was built in a way to naturally use light and breeze. The presence of central courtyard in many traditional houses was premised upon the natural use of light, ventilation, cooling of the inner chambers and so on.

Many forts and palaces of India are built on the green building concept. A key element was always the presence of a water harvesting method within the confines of the building in order to be self sufficient and to ensure the smooth working of such large structures. They also acted as cooling fountains and an element of aesthetic pleasure as in case of Char Bagh style of Mughals' garden construction or Fatehpur Sikri or the various baolis and hauz that mark the cultural and environmental history of India.

Another aspect that predominates in green buildings of times gone by is the use of intricately designed Jaalis (decorated and carved windows of stone) on a massive scale in hot and drier regions of India to help in natural ventilation, acting as a mediator to cool the darker recesses of the buildings. A prime example is Hawa Mahal located in Jaipur city in the Indian state of Rajasthan which is predominated by these intricate jaalis to allow for natural ventilation and several other forts and palaces of North-West India.

The knowledge of green building was not confined only to the Royals but seeped deeply into the perception of the commoners as well and this can be exemplified by the use of white limestone or chuna for painting the roofs of houses in Jaisalmer (refer to Fig. 15.5). This is a necessity as well because Jaisalmer is located in the arid zone with very high temperatures and in order to maintain habitable temperatures, the

Fig. 15.5 Comparison of rooftops of Delhi and Jaisalmer respectively. *Data source* Author's illustration, compiled from Google Earth Imagery; Image ©2018 DigitalGlobe

Fig. 15.6 Baolis in Purana Qila and Humayun's Tomb Complex, respectively, New Delhi. *Data source* Author's illustration

roofs are white-washed to reflect and not absorb insolation. This practice is followed till today.

Delhi being a city located in the sub-tropics, experiences harsh summers with water scarcity. In order to sustain life in such a scenario, several rulers of Delhi and its various cities constructed and maintained; Smart and green structures to harvest water and provide a perennial source of water supply all year round for the needs of the Empire and its several citizens. These traditional water harvesters vary from tanks/kunds/hauz, baolis/step wells (refer to Fig. 15.6) or huge wells/kuan (like former Dhaula Kuan). These structures were self sustaining, and resource efficient in their complete building cycle. Few of the surviving and recorded traditional water harvesters of Delhi are listed in this section (refer to Figs. 15.7 and 15.8) wherein the status 'Exists' indicates that those structures could still be brought into water harvesting uses with few repairs.

On a national level in India, traditionally and even in contemporary times, different types of water harvesting and harnessing practices and systems are still used in accordance with the local environmental and anthropogenic conditions and cultures. The names vary according to the local dialects while the structures vary according to locally available materials and micro/regional geography.

Dams, tanks, wells, artificial lakes and ponds, reservoirs, systems and practices of irrigation, rain/water harvesting measures are few examples. Dams are stereotypically considered to be symbols of modern technology and progress but the idea and the practice is not so modern. In different parts of India, these were and till today being used to harness or conserve the precious resource of water. They are referred by many names across India such as Bhanadaras (check dams/diversion weirs) in Maharashtra,

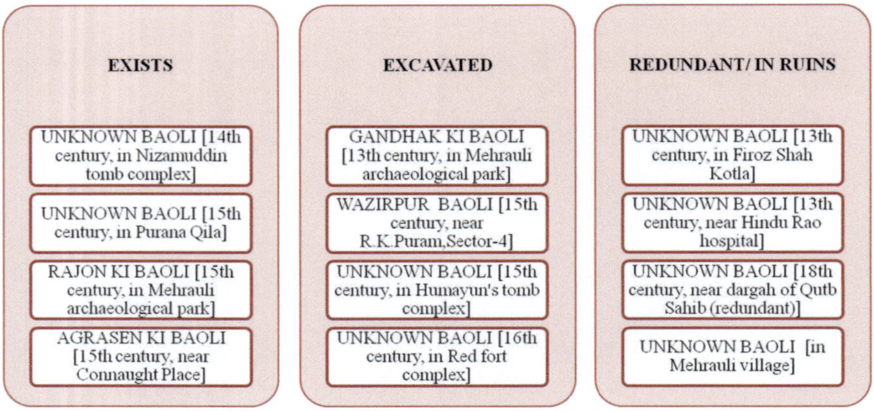

Fig. 15.7 Traditional green structures and smart water providers in Delhi: Step wells or Baolis. *Data source* Delhi: A thousand years of building (Peck 2009)

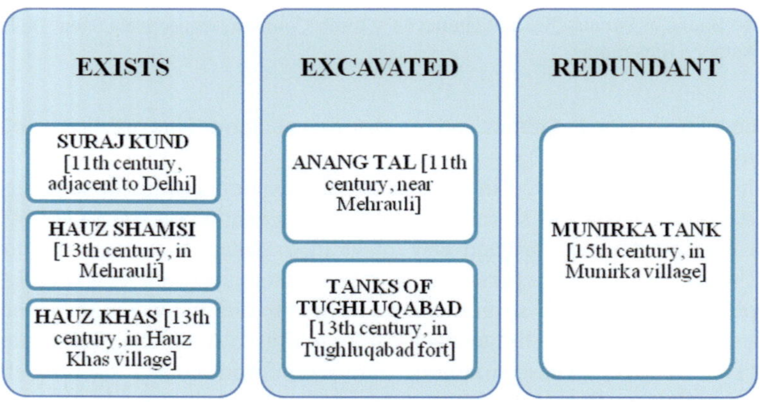

Fig. 15.8 Traditional green structures and smart water providers in Delhi: Tanks. *Data source* Delhi: A thousand years of building (Peck 2009)

Johad (small mud check dams) in Alwar district, Rajasthan, Naada/Bandha (stone check dam) in Mewar region of Thar Desert (Padigala 2017), Katas/Mundas/Bandhas (strong earthen embankment) in Odisha and Madhya Pradesh and Korambus/Chira (temporary dam) in Kasaragod and Thrissur district, Kerala.

Among other types of natural and artificial reservoirs are lakes which are referred as Sagar/Samand (bigger lakes), Talab/Bandhis (medium sized lakes/reservoirs), Talai (reservoir area of less than five bighas) and Pokhariyan (ponds) in Bundelkhand (natural) region and Udaipur (artificial), Rajasthan, Nadis (village ponds) in Jodhpur, Rajasthan and Dongs (ponds) in Assam. Tanks are far more prominent in South India due to topography of the region but are also found in north-western India as well.

They are named as Chandela (small sized tanks), Bundela (larger than Chandela tank), Tobas (natural ground depression) and Tankas (small home tank) in several areas, Rapat (percolation tank) in Rajasthan and Cheruvu (runoff storage reservoirs) in Chitoor and Cuddapah districts, Andhra Pradesh. In other parts of India, they are referred by several names such as Jhalaras in Rajasthan and Gujarat, Kohli in Bhandara, Maharashtra, Kere in Central Karnataka Plateau, Eri in Tamil Nadu and Ooranis in South Travancore, Tamil Nadu to name a few.

Wells are one of the most abundantly marked artificial reservoirs of water in Indian maps and especially in rural or village maps. Many a times, they have played significance in maintaining socio-cultural relations and power-politics in the village space. Their names and roles are signifier of their importance in community life. They may be referred as Baoris/Bers (community wells) in Rajasthan, Saza Kuva (open well with multiple owners) in Aravalli hills in Mewar, eastern Rajasthan, Virdas (shallow dug wells) in Great Rann of Kutch, Gujarat (Padigala 2017) and Vav/Vavdi/Baoli/Bavdi (stepwells) in Gujarat (refer to Fig. 15.9) and Rajasthan. It is also in the villages that agriculture and specifically traditional agricultural practices are utilised for crop production. Irrigation plays a major role in agricultural productivity especially in a sub-tropical monsoon dependent nation. Several types of irrigation practices are employed in India in different parts of the country; in-synchronization with the local geography and suitable conditions. These practices may be referred to as Bamboo drip irrigation (Bamboo pipes for irrigation) in Khasi and Jaintia hills of Meghalaya, Cheo-oziihi (channel irrigation) in Nagaland, Pat (irrigation system) in Jhabua district, Madhya Pradesh, Dungs/Jampois (small irrigation channels) in Jalpaiguri district, West Bengal and Apatani (wet rice cum fish farming system) in Arunachal Pradesh to name a few.

Fig. 15.9 Traditional Green building: Bavdi in Baroda, Gujarat. *Data source* Author's illustration

In several other parts of India, the abundance of water is utilised efficiently for times of scarcity. Several kinds of rain water and surface water harvesting systems are used to conserve, utilize and store abundant water availability rather than allowing it to be just run-off. Amongst the myriad rain water harvesting practices in India; Jack-wells in Great Nicobar Island, Andaman and Nicobar Islands, Khatri in Hamirpur, Kangra and Mandi districts, Himachal Pradesh, Kuis/Beris in western Rajasthan, Kunds/Kundis in Thar desert, Rajasthan and partially in Gujarat and Zabo/Ruza (impounding rainwater run-off system) in Nagaland are some significant few to mention. Ahar Pynes in south Bihar is a unique practice of harvesting flood waters while Kuhls in Jammu and Kul in Spiti valley, Himachal Pradesh are a practice revolving around the utilization of surface channels as well as Naula (surface water harvesting) in Uttaranchal. Khadin/Dhora is a surface run-off harvesting system found in Jaisalmer, western Rajasthan while Surangam/Thurangam/Thorapu/Mala is unique tunnel water harvesting system found in Kasaragod district of Kerala. Other names of water harvesting systems in different parts of India are Paar in western Rajasthan, Phad in Dhule and Nashik districts, Maharashtra and Zings in Ladakh to name a few.

It is in the light of this rich heritage of green buildings, that the relationship between green buildings and smart cities need to be rigorously pursued for attaining smarter cities with smarter solutions for maximum sustainable efficiency.

15.3 Green Buildings and Smart Cities: A Relationship to be Pursued

To understand Green buildings better, one can peruse this definition for a general understanding behind their conception. Green Building is:

Any built structure that is designed to judiciously and efficiently utilize its key components namely site, energy, water and materials (but not limited to) in an environment-friendly and sustainable manner along with the practical implementation of reuse and recycle ideology for improving quality of human life and environment. They are mediation between the best of both worlds that is; they are highly energy and water efficient as well as sustainable and economic in the longer duration.

Green Buildings are an integral part of a sustainable way of life especially necessitated in urban areas which are at present characterised by innumerable problems of over population, pollution, unplanned; haphazard, chaotic and congested urban spaces and energy scarcity with increasing economic and environmental costs over time. With so many problems and challenges, Green buildings provide an energy efficient, healthier and environmentally sustainable way of life. Green buildings are also paving way towards a greener, sustainable future by exhibiting resilience towards climate change, giving time for adjustments and for finding newer alternatives to adapt to climate change economically, socially, ecologically and even psychologically. Recently, in the Indian context, Green buildings and their key characteristics have found prominent place in the Smart City initiative.

To define a Smart City is difficult because it is not a "one size fit all approach", just like urban regeneration and green buildings. Hence in order to understand or explain the concept of smart city, there is no universally accepted definition for it; since it would imply different perceptions and understandings to different people, for instance in this case; it implies green buildings as an integral part of or a building block of a smart city.

The conceptualisation of a Smart City as stated in the Mission document released by the Ministry of Urban Development (MoUD), India in June, 2015, therefore defines it as a perceptual, spatio-temporal and resource-based approach. This implies that the concept itself would differ spatially in context of developmental levels, intention to change and reform and dependent upon the resource availability and aspirations of urban dwellers in a specific space and time. Even then, few parameters are required to act as guiding forces in this Mission which have been outlined as the list of essential urban infrastructures namely; the availability of sufficient as well as continuous: water and electricity supply, sanitation and solid waste management services, accessibility, affordable housing with special emphasis upon the urban poor, well networked information and communication technologies, effective governance along with citizen participation, safety and security of citizens, effective health and education system as well as environmental sustainability (MoUD 2015).

An urban area is typically premised upon inclusive and widespread development which can be broadly categorized into myriad physical, socio-economic and institutional infrastructure. It may be long or short term depending upon the needs and challenges of the situation, area and people. Smart cities are premised upon or imagined as the spaces providing a suitable quality of life to its inhabitants through comprehensive development and availability of essential infrastructure while simultaneously assuring hygienic, ambient and sustainable environment through Smart solutions' application. The comprehensive development through the addition of essential layers would progress towards incremental smartness. Towards the intention of smart city through green building, the strategies of retrofitting in and of existing buildings along with implementation of green building and energy efficiency codes would be implemented in Greenfield developments.

Green buildings are the need of the hour primarily because they are efficient in energy and water savings and provide a better alternative to harmful and wasteful utilization of resources especially in urban areas. Green buildings also fulfil the targets or goals set by the Urban Sustainable Development Goals and Smart Cities initiative. Even at the environmental level, Green buildings are a strong necessity given the present environmental scenario and the amount of non-renewable energy being consumed wastefully.

Green buildings like regeneration are not a "one size fit all" approach rather they are micro-based solutions that are better adept at tackling problems and resource scarcity at hand and appropriate to local conditions. It is high time that we relook at our present spate of urbanization and implement greener options to make our cities more sustainable and adept at tackling climatic changes along with bearing more resilience to urbanization related issues. Green buildings are an answer to many problems simultaneously and effectively and are also regenerative in the sense that they are based upon self-sustainability implying that much can be achieved in a positive sense if buildings are adopted, constructed or modified to be green. But all this can only be achieved or implemented if the people at their individual level become aware of the problems at present and the need for such measures to sustain a greener future and our future generations.

15.4 Green Buildings: Paving Way for a Better, Greener Future

In September, 2013, the campaign for Urban Sustainable Development Goal (USDG) was launched by Sustainable Development Solutions Network with the support of UN Habitat and many other organisations and agencies for Sustainable Cities and Regions (UNSDSN 2013). It is within the spirit and context of USDGs that the Green buildings are placed as an intention; to be pursued in solving and mitigating urban and urbanization challenges. Even in India, with Prime Minister Narendra Modi's vision of 100 Smart cities (India Today 2014), again due importance is premised upon the need for Green Buildings.

The concept of Green Buildings is somewhat rooted in the self-sustainability aspect of regeneration. To many, the concept of Green Building may appear to be new or modern or innovative but the fact is that the concept is as age old as human habitation. The practice to stay in synchronisation with our environment and seasons has been practiced since times gone by and to an extent are practised till today as part of rural traditions and techniques.

The concept of Green Buildings is premised upon Recycle and Reuse, Energy efficiency and Water management as key concepts from plan to implementation stage. Green buildings (refer to Fig. 15.10) are thought green, built green and use green to sustain and efficiently work. Even if they are not initially built green, simple measures and modifications can ensure greener use of the buildings. A green building must be in synchronisation with its environment and specifically with its local climatic regime because if this is not the case, then the added temperature controlling mechanisms will render the building non-green and unsustainable as it would consume energy to regulate an adequate temperature regime within the confines of the building. In such a scenario, the important role is played by the design of the building and the construction material.

At present in urban areas, the green building mechanisms employed range from using solar water heaters, solar panels for energy generation and solar cookers for cooking, rainwater harvesting systems for groundwater recharge, storage tanks, wastewater recycling system to minimize water loss and increasing efficient use, using motion sensors to switch off/on lights and air-conditioning, energy efficient appliances, CFLs or compact fluorescent light bulbs, using bio-degradable waste

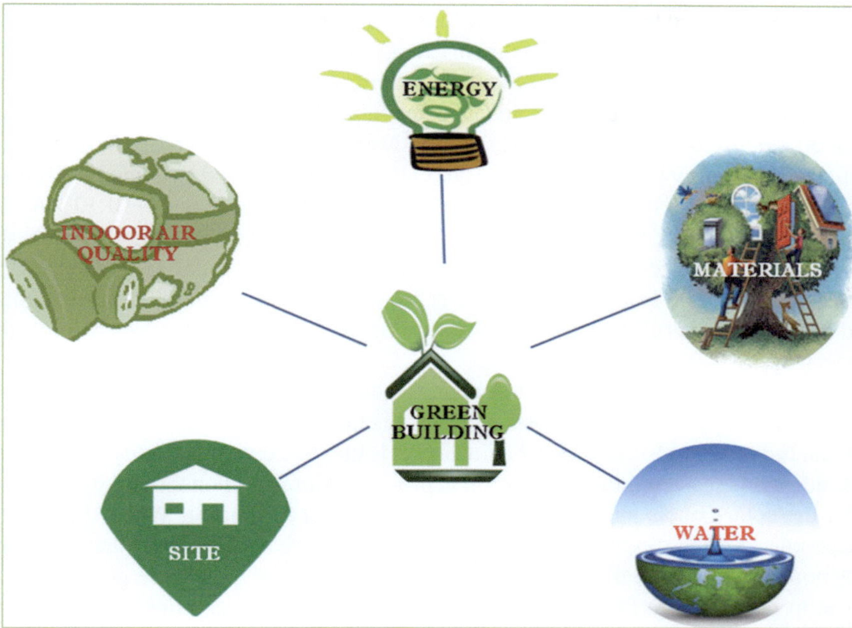

Fig. 15.10 Basis of Green building efficiency. *Data source* Based on LEED- India approach (2001)

from kitchen as compost manure for house garden and so on. Even the use of building material is also scrutinised to be in synchronisation with the local climate implying that glass is more appropriate for use in colder climates because it creates a greenhouse effect keeping the inside of the house/building warmer than the outside whereas wood, mud and lighter wall colours must be used in hot climates to keep the buildings/houses cooler without much dependence on artificial sources of ventilation and temperature control mechanisms. Green buildings are premised upon efficiency of water and energy saving and utilization which is extremely required in the present scenario.

15.5 Greening Smart Cities: Present Practices and Policies

Any green building in contemporary times resonates with Smart Technology. Smart technology helps in resource and energy conservation, efficiency and judicious utilization over time. To exemplify, Taipei 101, located in Taiwan can be observed as an energy efficient, smart-green building. Taipei 101 is the world's tallest green building standing at a height of 508 m above ground with smart technology (solutions provided by Siemens in 2008) such as motion sensors that detect and allow other technologies to function only in the presence of inhabitants, along with automatic ventilation control for air conditioning, improved insulation for heating and lighting controls utilizing natural light. It is (Leadership in Energy and Environmental Design) LEED-Platinum rated building. Further, Siemens has provided smart technology solutions to Crystal, London, UK in 2012; under its Sustainable Cities initiative. The Crystal is an all electric building that utilizes solar power and ground source heat pumps to fulfil its own energy requirements. It is an 'Outstanding' certified building by (Building Research Establishment Environmental Assessment Methodology) BREEAM (a Green building rating agency, similar to GRIHA, LEED and BEE). Apart from these two, a sustainable and green city is being constructed in Abu Dhabi, UAE referred to as Masdar—City of Tomorrow. The city is proposed to use Green building concepts and solar energy to fulfil its necessities to live in an environmentally sustainable and smart way.

In India, few preliminary steps have been taken towards green and smart buildings and one of the prominent examples at present is the Cochin International Airport in Kerala, India. Very recently, the Cochin airport achieved Power Neutrality, the first of its kind; at this scale in India. The airport at present is running upon the energy derived from the solar panels and works on principle of energy sharing. When they have excess energy, they supply to the grid, when they do not have adequate quantity, then they borrow from the grid, making them energy-neutral. They have emerged as a light house to guide others in adopting such cleaner, greener and smarter technological practices.

At present, India has a total of 752 certified green buildings, which are fully functional and operational (The Economic Times 2018) while their number is still increasing. To refer a building as a Green building in India, it is necessitated to apply for one of the following rating systems. The **GRIHA** or Green Rating for Integrated Habitat Assessment is an Indian rating system which has been jointly developed by TERI (The Energy and Resources Institute) and the Ministry of New and Renewable Energy, Government of India (www.glassisgreen.com, accessed 2018). It is an evaluation process comprising of three tiers. GRIHA rating system comprises of 34 criteria classified in four sections and includes such criteria as site selection and site planning, conservation and efficient utilization of resources, building operation and maintenance and innovation (www.grihaindia.org, accessed 2018). Prime examples of GRIHA buildings are Commonwealth Games Village, New Delhi, Fortis Hospital, New Delhi, CESE (Centre for Environmental Sciences and Engineering) Building, IIT Kanpur, Suzlon One Earth, Pune to name a few (www.greencleanguide.com, accessed 2018).

Another important rating system is the **IGBC** or Indian Green Building Council; which has licensed the LEED Green Building Standard from the USGBC or the U.S. Green Building Council. The Leadership in Energy and Environmental Design (LEED) is the rating system developed by U.S. Green Building Council (USGBC) for certifying Green Buildings. LEED is a framework for assessing building performance against set criteria and standard points of references. The benchmarks for the LEED Green Building Rating System were developed in year 2000 and are currently available for new and existing constructions (www.greencleanguide.com, accessed 2018).

LEED-INDIA approach for Green Buildings Confederation of Indian Industry (CII) formed the Indian Green Building Council (IGBC) in year 2001 (Madhumathi and Sundarraja 2014). IGBC facilitates Indian green structures to become green buildings. At present, nine Green Building rating systems are available under IGBC such as; LEED India 2011 for New Construction, LEED India 2011 for Core and Shell, LEED 2009 (V3) for Homes, LEED 2009 (V3) for Neighbourhood Development, LEED 2009 (V3) for Commercial Interiors to name a few (www.usgbc.org, accessed 2018).

Yet another green building rating system in India is the **BEE** or Bureau of Energy Efficiency. BEE rating is measured as a range of stars from one to five with more stars implying better energy efficiency. It also developed Energy Performance Index (EPI) specifically for rating air conditioned and non-air conditioned office buildings wherein one unit is measured as Kilo watt hours per square meter per year. The Reserve Bank of India's buildings in Delhi and Bhubaneshwar, the CII Sohrabji Godrej Green Business Centre are few examples that have received five star ratings from BEE (www.greencleanguide.com, accessed 2018).

At present, there is no nationwide initiative for building green buildings but the scenario is slowly changing in their favour. Around the year 2000, for instance in Haryana and Delhi, there was a rule for installing rain water harvesting systems on roofs of individual houses in order to get house completion certificates. At present, there are serious dialogues on benefits to be given to the owners of green buildings for reducing their carbon footprint. These green buildings may use solar water heaters and panels to fulfil their energy needs and use water harvesting and wastewater management systems to fulfil their water needs and so on. In several states across India, the need and importance of green buildings are being recognised and initiatives taken accordingly as in the case of Tamil Nadu; wherein the State government has targeted the rural poor to provide solar power run green houses while allocating Rs. 1,080 Crore for construction of 60,000 green houses (The Hindu 2011).

The Kerala State Housing Board initiated the work on sixty eco-friendly apartments (as limited scale commercial venture) located in Kochi and Thiruvananthapuram under the GRIHA system. This intention is further pursued under the Saphalyam Housing Scheme for the financially weak in the districts of Kottayam, Kollam, Idukki, Kozhikode, Malappuram, Palakkad and Thrissur. Several plans are also underway to implement GRIHA while constructing the new government residential flats as well as for rehabilitating slum dwellers in a green building way (The Hindu Business Line 2012). Furthermore, there are comprehensive plans by several states namely; Haryana, Rajasthan, Uttar Pradesh, West Bengal in north, Gujrat and Maharashtra is west, Odisha in east and Andhra Pradesh, Karnataka and Tamil Nadu in south along with the union territory of Delhi; to implement the Energy Conservation Building Code (ECBC) for all new commercial constructions (Institute of Building Efficiency 2012).

Recently in Green building context, the Delhi Development Authority or DDA has formally stated a need and initiation for talks of a policy concerning promotion of green buildings (The Economic Times 2014). Even at a personal level, people, communities and organizations are adopting for living and building green as in the case of Govardhan Eco Village in Thane, Maharashtra. This is a community which has built buildings with compressed stabilized Earth blocks, Rammed Earth Technique, Cob Houses (Adobe Bricks) with traditional thatched roofs. These buildings have received a five-star rating from GRIHA to be certified as Green buildings (www.teriin.org, accessed 2015). Another example is of Green One in Chittaranjan Park, New Delhi which became the first individual home to register for a green building rating from TERI in June 2011. This ultimately led to the initiation of the pilot project for Small Versatile Affordable Green Rating for Integrated Habitat Assessment or SVAGRIHA, TERI's adaptation of the GRIHA system for small homes. In January 2014, Green One became the first individual home in India to get a green rating, earning a five-star SVAGRIHA rating from TERI (www.teriin.org, accessed 2015). Along with this, the newly built TERI gram/village located on Gurgaon-Faridabad road is also rated as a green building.

In Delhi, institutions such as Centre for Science and Environment (Tughluqabad), Indian Spinal Injury Centre (Vasant Kunj) and Rashtrapati Bhawan (New Delhi), residential colonies namely Nizamuddin East and Defence Colonies (New Delhi) are prime examples of Rain water harvesting systems and thereby transformation (refer to Fig. 15.11) leading towards green buildings, green communities, green neighbourhoods (Singh 2015) and green city.

Further, the notion of green buildings is surpassing myriad activities, practices and processes; geared towards environmental and urban sustainability. Increasingly, industries are also being encouraged and mandated to go green, to exemplify; the recent instructions of the National Green Tribunal, 2015 (Times of India 2015) which state that all Textile Industries must fulfil Zero litre Discharge mandate wherein the concerned units are required to install Effluent Treatment Plants in conjunction with Reverse Osmosis plant to recycle and reuse waste/discharge water and minimise dependence upon fresh withdrawal of water for their processes. Essentially this system can be termed as individual wastewater treatment plants, a bold smart solution.

Green buildings are inclusive of myriad building and functional types and it is when all of them go green, become green and stay green, can we say that we can progress towards a future revolving around a smart city.

15.6 Building Greener Smart Cities

With the Smart Cities' initiative, the time has come to build and modify the buildings we live, work and access as Green for resource and environmental efficiency, embed smart solutions and technologies for efficient utilisations for an integrated system of smart-green buildings and neighbourhood. They also provide us with ways to mitigate climate change; for instance through the implementation of Green roofs (Castleton et al. 2010) which can have solar panels for power generation, rain water harvesting systems (Mentens et al. 2006), plant/grass cover for natural cooling or a roof top agricultural farm (Oberndorfer et al. 2007) and so on which may provide us

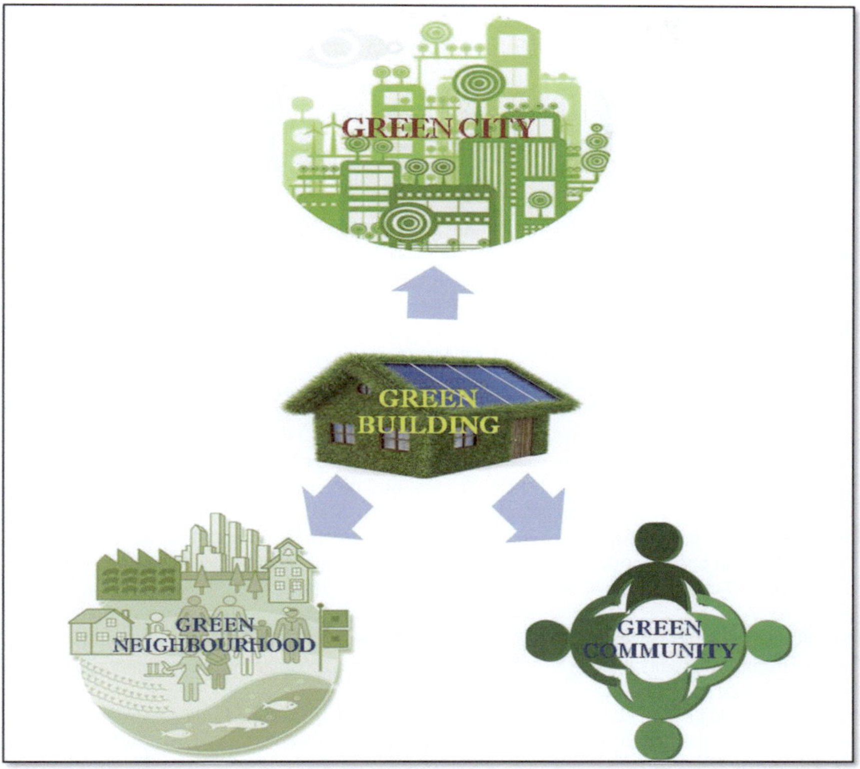

Fig. 15.11 The cyclic nature of building green: A step towards Smart City. *Data source* Author's illustration

with several opportunities and efficiencies. Some studies also indicate towards their positive role in tackling air pollution and climate change.

It is a fact and an initial desistance that building green buildings is slightly expensive than conventional buildings but this is true only in a very short run. In the longer run, green buildings have proven to be far more cost-effective than their conventional counter parts. Green buildings are a profitable venture in social, ecological and economic contexts and hence must be vigorously pursued. Green buildings have been built since time immemorial but in contemporary times riddled with several urban issues; they are needed with utmost urgency to sustain urban and urban ways of life but in an environment-friendly manner. Green buildings are a way to increase resilience towards climate change and to tap environmental degradation.

Green buildings have been made an integral part of Urban Sustainable Development Goals and Smart Cities' initiative because of their multiple uses and efficiencies in tackling so many challenges simultaneously along with providing an opportunity to live healthy, sustainably and pollution free lifestyle. Smart Cities are premised upon Green buildings without which the smartness can never be achieved completely.

References

Brodies LLP (2009) Urban regeneration–the great debate. Scotland. http://www.brodies.com. Accessed 28 Dec 2012

Carrion MF, Hanley LM (eds) (2007) Urban Regeneration and revitalization in the Americas: towards a stable state. Comparative Urban Project, Woodrow Wilson International Center for Scholars. https://www.wilsoncenter.org/sites/default/files/CUSP_Q.pdf. Accessed 20 Feb 2015

Castleton HF, Stovin V, Beck SBM et al (2010) Green roofs: building energy savings and the potential for retrofit. Elsevier Energy Build 42:1582–1591

Green Clean Guide (2012) Green buildings. http://greencleanguide.com/three-primary-rating-systems-for-green-buildings-in-india/. Accessed 22 Aug 2012

Green Rating for Integrated Habitat Assessment. http://www.grihaindia.org. Accessed 3 Jul 2018

Haub C (2007) World population data sheet. Population Reference Bureau, Washington, DC. http://www.prb.org/pdf07/07wpds_eng.pdf. Accessed 20 Feb 2015

India ranks third among top 10 countries for LEED green buildings. The Economic Times, 23 Jan 2018. https://economictimes.indiatimes.com/news/economy/indicators/india-ranks-third-among-top-10-countries-for-leed-green-buildings/articleshow/62619975.cms. Accessed 3 July 2018

India Today (2014) 5 key elements of PM Narendra Modi's 100 smart cities, New Delhi. http://indiatoday.intoday.in/story/pmnarendramodi100smartcitieskeyelements/1/382021.html. Accessed 15 Aug 2015

Institute of Building Efficiency (2012) Building efficiency: a key to fixing India's energy crisis. http://www.institutebe.com/energy-policy/Fixing-India%E2%80%99s-Energy-Crisis.aspx. Accessed 13 Apr 2016

Kanaujia K (2016) Urban regeneration: mediating geo-anthropogenic environment- a case study of urban villages of Delhi. In: VAV Raman (ed) Geoanthropogenic environment: an appraisal, A. K. Publications, New Delhi, pp 149–164

Lawless P (2010) Urban regeneration: is there a future? People, Place and Policy Online (4/1):24–28. http://extra.shu.ac.uk/ppp-online/issue_1_260410/documents/urban_regeneration_future.pdf. Accessed 22 Feb 2013

LEED in India. https://www.usgbc.org/help-topic/leed-in-india. Accessed 3 Jul 2018

Madhumathi A, Sundarraja MC (2014) Buildings in hot humid climatic regions using phase change materials as thermal mass in building envelope. Energy Environ 25(8):1405–1422

Mentens J, Raes D, Hermy M (2006) Green roofs as a tool for solving the rainwater runoff problem in the urbanized 21st century? Elsevier Landsc Urban Plan 77:217–226

Ministry of Urban Development (MoUD), Government of India (2015) Smart city: mission statement & guidelines. http://smartcities.gov.in/upload/uploadfiles/files/SmartCityGuidelines(1).pdf. Accessed 15 Aug 2015

Mitrojorgji L (2003) Urban regeneration in Berlin, Germany: new approaches at the neighbourhood level. Concentration paper, Department of Urban and Regional Planning, University of Illinois, Urbana-Champaign. https://www.ideals.illinois.edu/handle/2142/68736. Accessed 20 Feb 2015

Oberndorfer E, Lundholm J, Bass B et al (2007) Green roofs as urban ecosystems: ecological structures, functions and services. BioScience 57(10). http://www.bioscience.org. Accessed 15 Aug 2015

Padigala BS (2017) Traditional water management system for climate change adaptation in mountain ecosystems in reconsidering the impact of climate change on global water supply, use, and management. IGI Global

Peck L (2009) Delhi: a thousand years of building. INTACH Roli Guide, New Delhi
RB Singh's tweet (9.41 AM on 12 Sept 2015) stating a strong relationship between Green buildings and Green neighbourhoods
The Economic Times (2014) Delhi development authority to have incentive policy for green buildings soon. http://articles.economictimes.indiatimes.com/2014-10-30/news/55595601_1_green-buildings-delhi-development-authority-delhi-govt. Accessed 3 Mar 2015
The Energy and Resources Institute, TERI. http://www.teriin.org/. Accessed 15 Aug 2015
The Hindu (Madurai, India) (2011) Construction of solar-powered green houses for the poor to begin soon. http://www.thehindu.com/todays-paper/construction-of-solarpowered-green-houses-for-the-poor-to-begin-soon/article2625763.ece. Accessed 15 Mar 2015
The Hindu Business Line (2012) Kerala begins work on green buildings. http://www.thehindubusinessline.com/portfolio/kerala-begins-work-on-green-buildings/article3562966.ece. Accessed 13 Apr 2016
Times of India (2015) Green tribunal cracks whip on industries along Ganga. http://timesofindia.indiatimes.com/city/lucknow/Green-tribunal-cracks-whip-on-industries-along-Ganga/articleshow/46527347.cms. Accessed 15 Mar 2015
UN DESA. World urbanization prospects 2014. http://www.un.org/en/development/desa/news/population/world-urbanization-prospects-2014.html. Accessed 15 Aug 2015
United Nations (2006) World urbanization prospects: the 2005 revision. United Nations, New York.http://www.un.org/esa/population/publications/WUP2005/2005WUPHighlights_Final_Report.pdf. Accessed 20 Feb 2015
United Nations, Department of Economic and Social Affairs, Population Division (2014) World urbanization prospects: the 2014 revision, highlights (ST/ESA/SER.A/352). http://esa.un.org/unpd/wup/Publications/Files/WUP2014-Highlights.pdf. Accessed 20 Mar 2016
United Nations, Department of Economic and Social Affairs, Population Division (2015) World urbanization prospects: the 2014 revision (ST/ESA/SER.A/366). http://esa.un.org/unpd/wup/Publications/Files/WUP2014-Report.pdf. Accessed 20 Mar 2016
UNSDSN, Why the world needs an urban sustainable development goal (2013). http://unsdsn.org/resources/publications/why-the-world-needs-an-urban-sustainable-development-goal/. Accessed 15 Aug 2015
WHO, Urban population growth. http://www.who.int/gho/urban_health/situation_trends/urban_population_growth_text/en. Accessed 15 Aug 2015
WWW.GLASSISGREEN.COM. https://www.glassisgreen.com/knowledge-centre/glass-green-ratings/. Accessed 6 Aug 2018

Chapter 16
Conceptualising and Measuring Social Capital as a Sustainable Development Strategy in Mewat Region, Haryana

Naresh Kumar and B. S. Butola

Abstract This paper attempts to conceptualize and measure social capital as a sustainable development strategy in Mewat region, Haryana. Social capital resides in relations of the people that they can use for their benefits. Social capital here is reconceptualised as an instrumental use of social relations for the mutual benefit of the communities. Social capital is contextual and is best captured as a multidimensional variable. For lack of secondary data, this study relies on a primary field survey of ten sample Panchayat Villages in Mewat, Haryana. Sample Panchayats are identified through cluster random sampling and use measures of associatedness (economic, social and political), relations of production, institutions, social cohesion, and social pathologies as a set of variables. A mixed methods approach is used to analyze the variables. The empirical results show that socio-cultural activity has the highest level of associatedness followed by economic and political activities. The results show that self-help groups are an important source of social capital along with associatedness in socio-cultural activities, increase in mutually beneficial production relations, vibrant institutions and strong social cohesion. The paper concludes that social capital along with economic capital can be employed as a strategy of sustainable development in Mewat region, Haryana.

Keywords Social capital · Sustainable development · Meos · Other communities · Mewat region · Haryana

N. Kumar (✉) · B. S. Butola
School of Social Sciences, Centre for the Study of Regional Development, Jawaharlal Nehru University, New Delhi, India
e-mail: jnunaresh@gmail.com

B. S. Butola
e-mail: bsbutola@gmail.com

S. Sahdev et al. (eds.), *Geoecology of Landscape Dynamics*, Advances in Geographical and Environmental Sciences,
https://doi.org/10.1007/978-981-15-2097-6_16

16.1 Introduction

Fulfilment of basic human needs is the prime objective of development. The Neoliberal Model of Development has failed to meet these needs and even deprived them of the aspirations of having these needs in the future. It has led to the concentration of development in certain regions at the cost of many and in turn serious social and economic consequences for the rest of the world (Escobar 2011). It has led to stark regional and global inequalities all over the world. Consequent to this, doubts are being raised over the relevance of the current development model being pursued all over the world under the leadership of the Bretton Woods Institutions. In this pursuit, the United Nations Organisation (UNO) organized the Conference on Environment and Development (UNCED) in (1987) in Rio de Janeiro in Brazil to formulate 'a global agenda for change' to further a new development paradigm. The agenda proposed sustainable development as the only alternative to the current development paradigm. It defined Sustainable development as that "development that meets the needs of the present without compromising the ability of the future generations to meet their own needs" (World Commission on Environment and Development 1987).

There is significant evidence, which suggests that social capital adds a lot to the sustainable development process (Grootaert 1998). The idea of social capital has a dramatic rise in the social sciences in the last decade. The enthusiastic practitioners of the idea state that social capital has important implications for development policy especially community development (Defilippis 2001). The importance of the idea of social capital is due to its instrumental use, social action and civic value for sustainable development outcomes (Bourdieu 1986; Coleman 1988; Putnam et al. 1993). Social capital is being termed as the 'missing link' in development (Grootaert 1998). Village communities are socio-spatial units, which occupy social and cultural space and cannot be ignored altogether. They are the ideal units in the sense that they denote separate social and cultural boundaries and are easily recognizable from others. Several scholars (Portney and Berry 1997; Krishna and Uphoff 1999; Sampson and Graif 2009) have used the community as a scale of social capital.

Understanding new concepts and spatial information technologies is an integral component to the strategy of sustainable development. The conceptual framework, sampling, data collection, methodology, and the analytical tools are part of the decision support system to achieve sustainable development goals. The information on all of these aspects is crucial for this strategy to be successful. Following the statement of the problem, this paper has broadly two objectives: (1) To look at the problems in terms of conceptualization and measurement of social capital and (2) To look at the levels of social capital of Meos and other communities in Mewat, Haryana. This study relies on an extensive primary database to fulfill the objectives and follows a multidimensional methodological approach to study social capital as a sustainable development strategy.

16.2 Study Area

Mewat region of Haryana comprises of four tehsils of Mewat district and one tehsil of Palwal district (Fig. 16.1). The region is situated between 26° and the 30° north latitude and 76° and 78° East longitude (Ali et al. 1970). Mewat is surrounded by Gurgaon district on its north, Rewari from the west and Faridabad from the east. The demarcation of the Mewat is based upon the area for which the Mewat Development Board (MDB) works.

After the failure of the 'Gurgaon Experiment' (1928) and the Community Development Programme (CDP) in the 1950s, MDB was constituted in 1980 by the Haryana government 'with a commitment to deliver social and economic justice to the backward and under-privileged sections of society'. Its executive agency at the field level is Mewat Development Agency (MDA) headquartered at Nuh, the district capital of Mewat district. On its southern border, Mewat shares its boundary with the state of Rajasthan. It is largely a plane area although there are varied forms of landscapes.

Despite the relative prosperity of Haryana, Mewat has remained backward on all the parameters of development. This relative lack of development and the consequent repercussions are having adverse effects on the socio-cultural development of Mewat. Mewat is a unique ethno cultural tract providing glimpses of the mixing of both

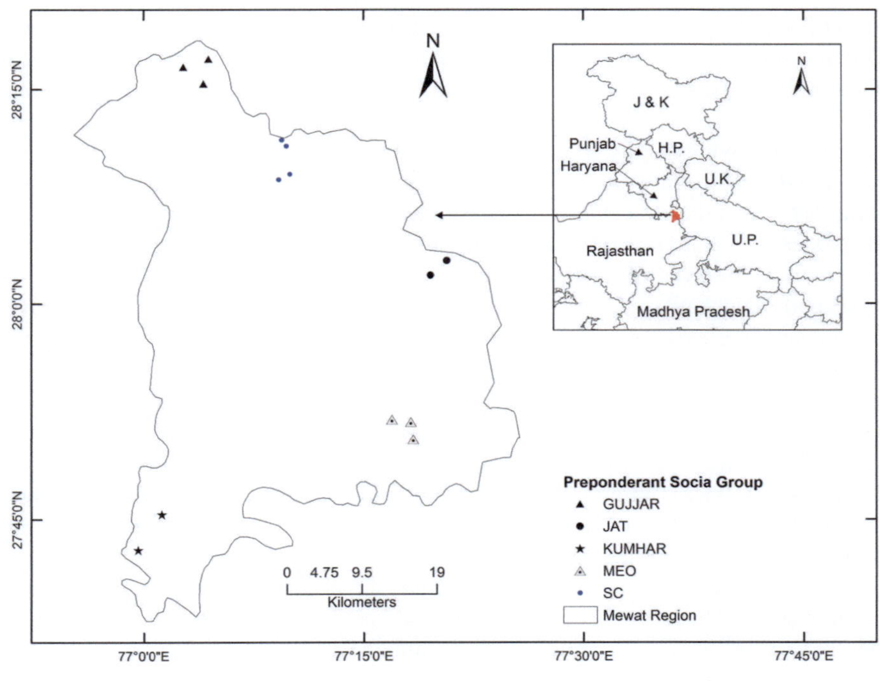

Fig. 16.1 Location map of Study area

Hindu-Muslim identities (Sikand 1995). The region is known for a strong sense of community among the Meos (Ali et al. 1970).

16.3 Methodology

The database is an important and integral part of any research. It is indeed difficult to locate the data sources of social capital, as there is no specific data set related to these aspects in India. Moreover, the secondary data sources are also scattered. To analyze the developmental problems of the Mewat region in terms of social capital at the community level, this research has relied on an extensive primary database consisting of community schedule, focus group discussion (FGD), key-informant interview and organization profile interview.

16.3.1 Survey Sampling

Ten Panchayat villages (fourteen revenue villages) were selected for the detailed survey based on cluster random sampling. Cluster sampling is used to select Meos and the other communities. This sampling is implemented in stages. It is a complex form of sampling in which two or more levels of units are embedded into the other. The first stage consists of constructing the clusters, which in turn was used to draw the sample. In the second stage, a sample of primary units is randomly selected from each cluster.

In the following stages, in each of those selected clusters, additional samples of units are selected, and so on. All ultimate units (individual unit, for instance) selected at the last step of this procedure are then surveyed. This technique, thus, is essentially the process of taking random sub-samples of preceding random samples. Following criteria were followed for the selection of the sample Panchayat villages:

(I) Only Panchayat villages were selected. In reality, some villages do not have Panchayat of their village. They are attached to the Panchayat of the other big village. The reason behind choosing the Panchayat village is that Gram Panchayat itself represents a form of social capital. It is a resource for the village community upon which they rely for solving their day-to-day problems and sustenance.

(II) Since one of the main objectives of the study looks at the social capital of the Meos and other communities, so a predominant village of either Meos or other communities (SC, Gujjar, Jat and Kumhar) was selected based on the inputs and advice of the local population.

(III) A Panchayat village and preponderant social group were preferred, which was located near the tahsil headquarters. It could not be possible in all cases because there was a problem in locating sample Kumhar Panchayat villages and even if a Gram Panchayat and sample social group preponderant village were located in the interior, it was selected.

16.3.2 Mixed Methods

Since most of the quantitative data is categorical in nature, therefore, cross tabulation was done to analyze the various aspects of social capital(s). Mixed methods approach was used to analyze both data sets. The qualitative data is used here to supplement the quantitative data. The analysis of qualitative data poses certain problems of coding and quantification. To solve these problems, numerical weights have been attached to make the data scale-free as per the advice and discretion of the local population.

The weights have been attached keeping in mind the importance attached to that associational activity by the respondents in the field area (Table 16.1). There are three methods of giving weightage. These include discretionary measures, equal weightage and statistical method (Mahmood 1977). In a complex and heterogeneous society, it

Table 16.1 Weightage to economic, socio-cultural and political activities

	Weights
(i) Economic activities	
(a) Agriculture	5
(b) Government job	4
(c) Self-employed	3
(d) Private contract jobs	2
(e) Casual labour	1
(ii) Socio-cultural activities	
(a) Id-ul-fitr	13
(b) Bakrid	12
(c) Holi	11
(d) Teej	10
(e) Rakshabandan	9
(f) Bhandara	8
(g) Havan	7
(h) Diwali	6
(i) Dussehra	5
(j) Gugga Pir	4
(k) Dargah Shariff Ajmer	3
(l) Durga mata	2
(m) Jai Ambe Puja	1
(iii) Political activities	
(a) Panchayat	3
(b) State assembly	2
(c) Lok Sabha	1

Source Community Field Survey Mewat 2012

Table 16.2 Weights to organizations in the sample communities

Organizations	Weights
(i) Informal organizations	
(a) Non-government organizations	5
(b) Self-help groups	4
(c) Tablighi Jamaat	3
(ii) Formal organizations	
(a) Gram Sabha	2
(b) Gram Panchayat	1
(iii) Which persons or organization help or support these community-based organizations	
(a) Non-government organizations	5
(b) Gram Panchayat	4
(c) Religious organization	3
(d) Zamindar	2
(e) Any other (specify)	1
(iv) Which place the members of the community regularly use for meetings?	
(a) Panchayat Bhawan	4
(b) Dharamshala	3
(c) Village school	2
(d) Personal homes	1

Source Community Field Survey Mewat 2012

is not possible to give equal weightage to all since it would distort the whole picture. The whole purpose of the exercise would be defeated.

There is always some sort of subjective reasoning involved in the weightage of the variables. Keeping in mind the limitations of the quantitative methods, they are not able to measure, which is most likely to be measured. However, it has been taken care of by the logical reasoning as implied by the respondents in the field area. The views and responses of people of the field have lessened the subjectivity inherent in the weights. The weights being attached are provided above.

Associatedness includes economic activities, socio-cultural activities, and political activities. Economic activities include agriculture, government job, self-employment, private contract jobs, and casual labour. Socio-cultural activities include festivals and ceremonies observed by the village communities. Political activities include participation in Gram Panchayat, state assembly, and Lok Sabha elections.

Two types of organizations are found, formal and informal. There are also community organizations in the village. The members of the community regularly meet to resolve the issues facing the community. Here also, the weightage is given as per the information given by the respondents in the study area (Table 16.2).

16.4 Conceptualising Social Capital

Social capital 'while not all things to all people, is many things to many people' (Narayan and Pritchett 1999). It has important implications for development policy and has a dramatic rise in the social sciences in the last decade of the twentieth century (Woolcock 1998). This rise has been accompanied by enthusiasm in the first instance and skepticism in the second. The enthusiasts of the idea state that social capital has important implications for development policy especially community development while the critics assert that until now there is disagreement about its definition and methodology. The literature on social capital is divided into many lines and presents a confused scenario. There are contrasting views to conceptualize and measure social capital in the available literature. Most scholars agree with Putnam on the need for a 'lean and mean' definition. Following Bourdieu (1986), this paper tries to reconceptualise social capital as instrumental use of social relations for mutual benefit and cooperation.

There are two dimensions of social capital, which includes both cognitive and structural dimensions. Both the dimensions need to be looked at understanding this idea. There is a need to develop a general theory of social capital, which can devise some basic principles on which the empirical investigations can be carried out. However, the current literature is not helpful in this direction. The literature is divided on how to operationalise the concept in an empirical setting. Practitioners of the idea assert that to operationalise social capital; there is a need to follow a multidimensional, coherent and contextual methodology (Krishna and Uphoff 1999; Bullen and Onyx 2005; Krishna 2002, 2007).

The earliest attempt at conceptualization and measurement was by the World Bank under the Social Capital Initiative project (SCI 1998) before Bourdieu (1986), Coleman (1988, 1994)) and Putnam et al. (1993), Putnam (2000) made initial efforts in this direction. Stone (2001) argues that there is a need for a measurement framework that takes into account the theoretical understanding of the concept and recognizing that it is a multidimensional construct. Paxton (1999) states that to measure social capital, single measures are not enough and variables and outcomes need to be separated for better measurement. Krishna and Uphoff (1999) look at the empirical correlates of social capital in Rajasthan in India. Brehm and Rahn (1997) look at the individual level evidence for the causes and consequences of social capital. Bullen and Onyx (1998, 2005) provide a measure of community social capital in New South Wales, Australia.

Further, there is a macro versus micro-debate in the literature on social capital. There are some studies, which support the macro social capital (Knack and Keefer 1997). These studies look at the international perspective and work within the broader political economy framework of development. Bjornskov (2006) looks at the determinants of generalized trust in a cross-country investigation. Svendsen and Bjornskov (2007) attempt to construct measures of social capital of twenty-five countries in eastern and western Europe. Most of the studies under this framework rely on cross-national data. All of these studies are undertaken in the West since they have a robust institutional network of data collection and dissemination. Apart from the macro-perspective, there is also an inclination towards meso studies like community development. Brisson and Usher (2012) attempt to conceptualize and measure social capital in low-income neighbourhoods. While advocating community studies, Saegert et al. (2001) write

> Making use of social capital as an analytical construct requires a shift from the individual to the community as the unit of analysis for strategies to combat poverty. Social capital is a collective asset, a feature of communities, rather than the property of an individual. As such, individuals both contribute to it and use it, but they cannot own it.

However, the importance of social capital lies in the fact that it looks at non-material resources and how these resources can be accessed at the meso and micro levels. Therefore, it is important to look at community levels to understand its effects on the macro level phenomena.

16.5 Results and Discussion

16.5.1 Associatedness

The history of human beings is the history of embedded association (Bourdieu 1986). Because of living in close proximity to each other, the village communities have joined each other in times of need and distress (Kropotkin 1998). By associating with each other, they try to add economic value to the social relations. The addition of value to social relations results in the accumulation of (social)capital. Table 16.3 looks at the associatedness of male for Meos and other communities in Mewat. The activities include economic, socio-cultural and political. The table records weighted scores for each social group in sample Panchayats.

Figure 16.2 looks at the level of associatedness of Meos and other communities in economic activities. Scheduled castes report the highest level of associatedness in total economic activities followed by Gujjars, Jats, and Kumhars. Meos reports the lowest level of economic activities. Agricultural has the highest level of associatedness among all the social communities followed by government job, self-employment, and casual labour. Private contract jobs have the least level of associatedness.

In socio-cultural activities, Gujjars have the highest level of associatedness followed by Jats and SCs (Fig. 16.3). Meos here also have the lowest level of associatedness except for the Kumhars, which reports the extremely low level of associatedness and hence may be considered as an outlier.

Holi festival among the Hindus has the highest level of associatedness followed by Teej, Rakshabandan and Diwali. What is interesting to note that the role of local deities is slowing declining in the lives of the Hindus day by day as is evident from this data? The Gugga Pir festival and Havan ceremonies report the lowest level of associatedness.

Table 16.3 Level of associatedness in Meos and other communities (Weighted Scores)

Economic	Gujjar	Jat	Meo	SC	Kumhar	Total
Agriculture	10	20	20	20	0	70
Casual labour	2	2	2	0	4	10
Private contract jobs	2	0	0	0	0	2
Self-employed	0	3	0	0	12	15
Government job	12	0	0	18	8	38
Total	26	25	22	38	24	135
Diwali	24	14	0	24	0	62
Holi	48	37	0	48	0	133
Dussehra	20	20	0	20	0	60
Teej	32	32	0	42	0	106
Gugga Pir	4	0	0	0	0	4
Rakshabandan	36	36	0	0	0	72
Durga mata	8	0	0	0	8	8
Jai Ambe Puja	4	0	0	0	4	8
Dargah Shariff Ajmer	0	0	12	0	0	12
Id-ul fitr	0	0	50	0	0	50
Bakrid	0	0	46	0	0	46
Havan	0	7	0	0	0	7
Bhandara	9	32	0	0	0	41
Total	185	178	108	134	12	617
Panchayat election	12	12	12	12	12	60
Assembly election	8	8	8	8	8	40
Lok Sabha election	4	4	3	3	4	18
Total	24	24	23	23	24	118
Sum total	235	227	153	195	60	870

Source Community Field Survey Mewat 2012

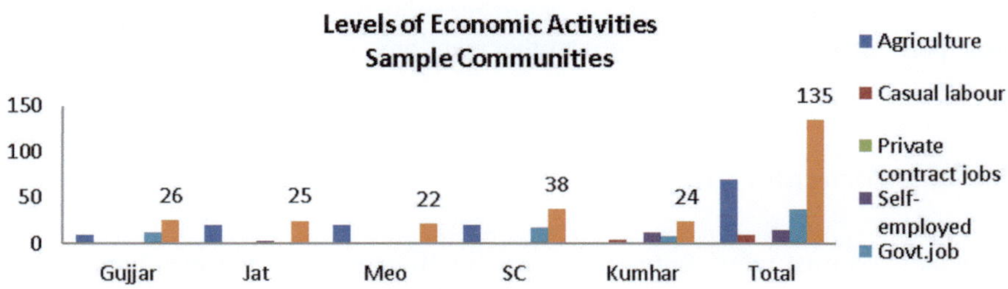

Fig. 16.2 Levels of Economic Activities *Source* Community Field Survey Mewat 2012

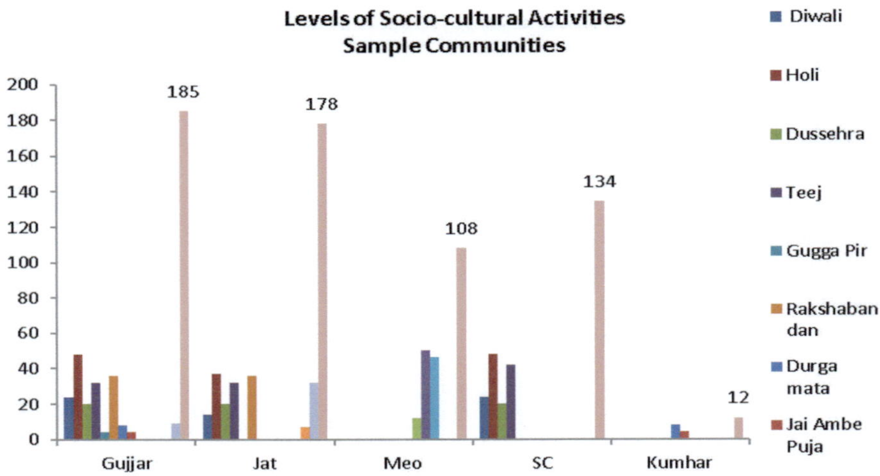

Fig. 16.3 Levels of Socio-cultural Activities *Source* Community Field Survey Mewat 2012

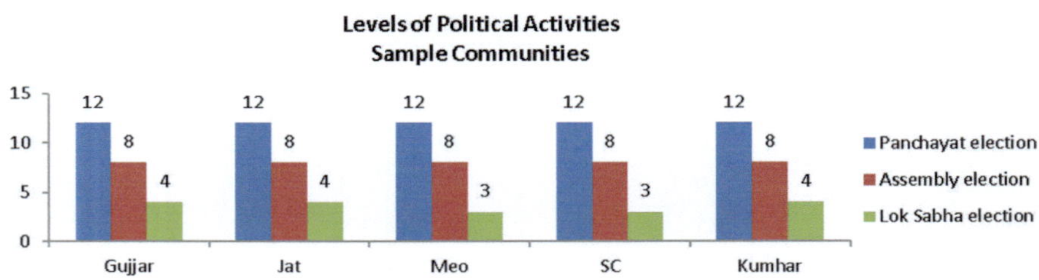

Fig. 16.4 Levels of Political Activities *Source* Community Field Survey Mewat 2012

Political activities show the highest level of similarity among the sample communities (Fig. 16.4). Gujjars, Jats, and Kumhars show the highest level of associatedness followed by Meos and SCs. Panchayat election shows the highest associatedness followed by Assembly and Lok Sabha election meaning thereby that local elections are more participatory and directly affect other elections.

16.5.2 Particularized Trust

Particularized trust refers to that trust which is found among the members known to each other. The members are known to each other, and this similarity leads to what is known as reputation among the members of the community (Uslaner 2008). Opposite to the particularized trust is the generalized trust. Generalized trust is trust towards strangers whom you do not even know (Bjornskov 2006). It is based on mutual reciprocity and behaviour by the individual members.

Table 16.4 Particularized trust in Meos and other communities

	Gujjar	Jat	Meo	SC	Kumhar	Total
Do people in this community trust one another in matters of lending and borrowing?						
Yes	2	2	2	2	2	10
No	0	0	0	0	0	0
In the last five years, what is the level of trust among the community members?						
Improved	2	2	2	2	2	10
Worsened	0	0	0	0	0	0
Remained the same	0	0	0	0	0	0
Compared with other community groups, how much do people in this community trust each other in matters of lending and borrowing?						
More trust than in other communities	2	2	2	2	2	10
Same as in the other communities	0	0	0	0	0	0
Less trust than in other communities	0	0	0	0	0	0
Do you agree or disagree with the following statement: "People here look out mainly for the welfare of their families, and they are not much concerned with community welfare"						
Strongly agree	1	1	1	0	0	3
Agree	0	0	0	0	0	0
Strongly disagree	0	0	0	0	0	0
Disagree	1	1	1	2	2	7

Source Community Field Survey Mewat 2012

There are obligations upon the members in the particularized trust, and they fear social boycott in case trust conditions are violated and infringed upon. If a stranger breaks trust or cheats another stranger, then it is highly unlikely that he would be subjected to the same treatment as the member who is known among the members of the community.

Table 16.4 shows the particularized trust among Meos and other communities in Mewat. The first variable looks at the trust regarding lending and borrowing in the community members. The entire sample communities report that they trust each other regarding lending and borrowing.

The second variable looks at the level of trust among the community members in the last five years. The entire communities report that the level of trust among the community members has improved. The third variable looks at how much people in this community trust each other in terms of lending and borrowing as compared with the other communities. The entire communities report that the people in their community trust each other more than other communities in terms of lending and borrowing.

The fourth variable puts forward a statement, 'people here look out mainly for the welfare of their families, and they are not much concerned with community welfare'. The Gujjar, Jat and Meo communities strongly agree that people here look mainly for the welfare of their families and they are not much concerned with community welfare. The same communities also disagree that people here look out mainly for the welfare of their families and they are not much concerned with community welfare. The entire SC and Kumhar communities disagree that people here look out mainly for the welfare of their families and they are not much concerned with community welfare.

16.5.3 Relations of Production

The need to trade goods, commodities, and services arise from the basic fact that human needs are numerous and to satisfy these, it is impossible to meet these or satisfy from local produce. Therefore, there is always a need to bring the goods, commodities, and services from other places. Due to this, there is a constant rise in the exchange of goods, commodities, and services from other places. This process of exchange has travelled a long distance over human civilization. Barter or exchange through money and virtual trade are some of the important features of the same. It is understood that these progressions in the form of trade have transcended the limitations imposed by various geographical and sociological obstacles and added new dimensions into trade including convenience and comfort.

However, when one looks at these conveniences and comforts, it is an undeniable truth that they have been at the cost of social capital meaning thereby that though barter and direct exchanges are inconvenient and cumbersome processes, yet they are rich in social capital. As opposed to this, the virtual trading is most convenient and very poor in social capital.

Table 16.5 shows the relations of production of Meos and other communities in Mewat. The first variable looks at where the inhabitants of Meos and other communities sell their livestock products and other produce. All the Gujjar, Jat, Meo and SC communities report that they generally sell their livestock products and produce at the local *mandi.* The members of the Kumhar community report local mandi as well as nearby city market for their products and produce. Only the Kumhars of Patkhori Panchayats sell their products in the village.

The second variable looks at the visit of agricultural extension workers to advise the farmers in their areas. The members of the Gujjar, SC and Kumhar communities report that the agricultural extension workers sometimes visit and sometimes, do not visit the area and provide advice to the farmers. The members of the Meo community report that agricultural extension workers do not visit the area and provide advice to the farmers.

The third variable looks at the facility of the agricultural cooperative for credit and loan services by Meos and other communities. The entire Gujjar, Jat, Meo, and SC communities report that they avail of the facility of the agricultural cooperative for credit and loan services. The members of the Kumhar community do and do not avail of the facility of agricultural cooperatives for credit and loan services. The fourth variable looks at those institutions, which provide credit to the Meos and

Table 16.5 Relations of production in Meos and other communities

	Gujjar	Jat	Meo	SC	Kumhar	Total
Where do the inhabitants of this community generally sell their livestock products and produce?						
Local *mandi*	2	2	2	2	1	9
District *mandi*	0	0	0	0	0	0
Cooperative society	0	0	0	0	0	0
Any other (specify)	0	0	0	0	1	1
Do the agricultural extension workers visit the area and provide advice to the farmers?						
Yes	1	2	0	1	1	5
No	1	0	2	1	1	5
Does this community avail of the facility of the agricultural cooperative for credit and loan services?						
Yes	2	2	2	2	1	9
No	0	0	0	0	1	1
What are the three main institutions that provide credit to farmers in this community?						
Moneylenders	2	0	2	0	1	5
Agricultural cooperatives	0	1	0	1	1	3
Development banks	0	0	0	0	0	0
Commercial banks	0	1	0	1	0	2
In the last three years, the yield per acre has						
Increased	1	1	2	1	2	7
Decreased	1	0	0	0	0	1
Remained the same	0	1	0	1	0	2
In the last three years, the sale of agricultural/livestock products in this community have						
Increased	1	1	2	1	2	7
Decreased	1	0	0	0	0	1
Remained the same	0	1	0	1	0	2

Source Community Field Survey Mewat 2012

other communities. The entire Gujjar and Meo communities report that moneylenders provide credit to the farmers in this community while the Kumhar community avails of the credit facility from both moneylenders and agricultural cooperatives. The members of the Jat community report agricultural cooperatives and the commercial banks provide credit to the farmers in their community.

The fifth variable looks at yield per acre in the sample Panchayats. The members of the Gujjar community report that yield per acre has increased as well as decreased over the last three years. The members of the Jat and SC communities report that the yield per acre has increased over the last three years but lately have remained the same. The members of the Meo and Kumhar communities report that yield per acre have increased over the last three years.

The sixth variable looks at the role of agricultural/livestock products in the community. The members of the Gujjar community report that the sale of agricultural/livestock products has increased and decreased in the last three years. The members of the Jat and SC communities report that the sale of agricultural and livestock products have increased and remained the same in the last three years. The members of the Meo and Kumhar communities report that the sale of agricultural and livestock products have increased in the last three years.

16.5.4 Types and Role of Institutions

Institutions are arrangements by which people come together for fulfilling certain needs. Institutions are instrumental in the sense that they provide service to the people. Institutions are essential for the proper functioning of society. They also provide the necessary social and cultural sanction to social groups and communities.

Table 16.6 looks at the institutions of Meos and other communities in Mewat. The entire sample communities attach highest weightage to the Gram Sabha followed by statutory Gram Panchayat, self-help groups, and NGOs. The members of the Meo communities attach highest weightage to the Tablighi Jamaat, a religious Islamic

Table 16.6 Types and Role of Institutions in Meos and Other Communities

Organizations	Gujjar	Jat	Meo	SC	Kumhar	Total
Gram Sabha	2(4)	2(4)	2(4)	2(4)	2(4)	10(20)
Statutory Gram Panchayat	2(2)	2(2)	2(2)	2(2)	2(2)	10(10)
Self-help group (SHG)	1(5)	2(10)	0	2(10)	2(10)	7(35)
Non-government organization (NGO)	0	0	0	0	1(4)	1(4)
Tablighi Jamaat	0	0	2(6)	0	0	2(6)
Which persons or organization help or support these community-based organizations?						
Gram Panchayat	2(8)	2(8)	2(8)	2(8)	2(8)	10(40)
Zamindar	0	0	0	0	0	0
Non-government organization (NGO)	0	1(5)	0	0	0	1(5)
Religious organization	0	0	2(6)	0	0	2(6)
Any other (specify)	0	0	0	0	0	0
Which place the members of the community regularly use for meetings?						
Village school	0	0	0	1(2)	0	1(2)
Dharamshala	1(3)	1(3)	0	0	2(6)	4(12)
Panchayat Bhawan	0	0	0	1(4)	0	1(4)
Personal homes	1(1)	1(1)	2(2)	0	0	4(4)

Note Figures outside the parenthesis indicate number of institutions and inside the parenthesis total score of the institutions
Source Community Field Survey Mewat 2012

organization, for the upliftment of their community. The second variable looks at which persons or organizations support the community-based organizations (CBOs). The members of the entire sample communities report that they help and support the Gram Panchayat. The members of the Meo community report that along with Gram Panchayat; they also support the Tablighi Jamaat.

The third variable looks at the meeting place of the members of the community. The members of the Gujjar and Jat communities report that they regularly use *dharmshala* and personal homes for meeting members of the community. The members of the Meo community report that they use personal homes for meeting members of the community. The members of the SC community report that they use village school as well as Panchayat Bhawan for meeting members of the community. The members of the Kumhar community report that they use *dharmshala* for meeting members of the community.

16.5.5 Community Cohesion

Community cohesion is belongingness among the members of the society for achieving shared values and goals. It looks beyond diversity and strives to bridge social inequalities in the society. It advocates positive relations among the community members. A sense of belonging has to be there in the community so that all sample communities feel safe and secure. Table 16.7 looks at the community cohesion of Meos and other communities in Mewat.

The first variable looks at gendered roles of the community members in solving the issues facing the community. The entire sample communities report that male members of the community participate most in solving the issues facing the community. Women have no say in solving the issues facing the community. Even males do not work in association with females in solving the disputes. Patriarchy is so much rampant that women are not allowed to come out of the four walls of the house in these matters. The second variable looks at which age groups participate most in solving the issues facing the community. Old persons participate most in solving the issues facing the community.

The third variable looks at the occupational status of those, which participate most in solving the issues facing the community. The village zamindars participate most in solving the issues facing the community. The fourth variable looks at whether the community has organized itself to address a problem or not in terms of initiatives. The members of the Gujjar, Jat, Meo and Kumhar communities report that they have organized themselves to address a problem (drinking water, electricity) and the same also report that they have not organized themselves. The members of the SC community report that they have organized themselves to address a problem (electricity) in the village.

The fifth variable looks at whether there are specific affirmative action programmes for the welfare of the communities. The members of the Meo and the SC community report that there are specific assistance programmes for the welfare

Table 16.7 Community cohesion in Meos and Other Communities

Organizations	Gujjar	Jat	Meo	SC	Kumhar	Total
Which members of the community participate most in solving the issues facing the community?						
Gender						
Male	2	2	2	2	2	10
Female	0	0	0	0	0	0
Both male and female	0	0	0	0	0	0
Age						
Youth and adolescents	0	0	0	0	0	0
Adults	0	0	0	0	0	0
Older persons	2	2	2	2	2	10
None participates	0	0	0	0	0	0
Occupation status						
Village zamindars	2	2	2	2	2	2
Businesspersons	0	0	0	0	0	0
Govt. jobholders	0	0	0	0	0	0
Any other	0	0	0	0	0	0
In the last five years, has the community organized to address a need or problem?						
Initiatives one						
Yes	1	1	1	2	1	6
No	1	1	1	0	1	4
Initiative two						
Yes	0	0	0	0	0	0
No	0	0	0	0	0	0
Are there any specific assistance programmes for this community?						
Yes	0	0	2	2	0	4
No	0	0	0	0	0	0

Source Community Field Survey Mewat 2012

of their community which include Multi-sectoral Development Programe (MsDP) and lending schemes of National Minority and Development Finance Corporation (NMDFC).

16.5.6 Social Pathologies

The importance of social capital lies in its consequences. It can be used for both the positive and the negative outcomes in the society. Most of the literature on social capital has focused on the positive aspects of social capital. The 'other side' of aspects or the dark side of the social capital has been ignored in most of the literature (Portes and Landolt 2000). While focussing on the positive aspects, it is also very important to highlight the negative aspects, keeping in mind that the same social capital can be used for the negative outcomes.

Table 16.8 looks at social pathologies of Meos and other communities in Mewat. One Panchayat of Meo and Kumhar communities report that there are burglaries/vehicle theft in the community and they have been affected most by these in the last one year. The entire members of the SC community also report burglaries/vehicle theft in the village. As one SC member puts it: ‘Recently one of our neighbours attended a marriage. After the bestowing ceremony, he found that his

Table 16.8 Social Pathologies in Sample Communities

	Gujjar	Jat	Meo	SC	Kumhar	Total
Do any of the following problems exist in this community? If yes who is the most affected or at risk group during the last one year?						
Burglaries/vehicle theft						
Yes	0	0	1	2	1	4
No	2	2	1	0	1	6
Robberies						
Yes	2	0	0	1	0	3
No	0	2	2	1	2	7
Assaults						
Yes	1	0	2	0	0	3
No	1	2	0	2	2	7
Gangs						
Yes	0	0	2	0	0	2
No	2	2	0	2	2	7
Vandalism						
Yes	0	0	0	0	0	0
No	2	2	2	2	2	10
Violent disputes						
Yes	0	0	0	0	1	1
No	2	2	2	2	1	9
Alcohol abuse						
Yes	1	1	1	0	0	3
No	1	1	1	2	2	7
Drug abuse						
Yes	2	0	0	1	0	3
No	0	2	2	1	2	7

Source Community Field Survey Mewat 2012

motorcycle was gone. The thieves just broke the lock'. This proves the hypothesis that a given form of social capital at the individual level that is valuable in facilitating certain actions may be useless or even harmful to others at the community level. The entire members of the Gujjar community Panchayats and one SC Panchayat report robberies (Jewellery). The entire members of the Meo Panchayat and one Gujjar Panchayat report assaults. The entire members of the Meo community report small gangs. The entire sample communities report that there is no vandalism in the sample Panchayats. The members of the Kumhar community report violent dispute resulting in a murder in the Pathkhori Panchayat. The members of the Gujjar, Jat and Meo communities report rampant alcohol abuse. The entire members of the Gujjar community and the lone SC Panchayat report drug abuse.

16.6 Conclusion and Policy Implications

It can be concluded from the above discussion that a consensus is fast emerging towards a 'lean and mean' definition of social capital, which points to an instrumental use of resources for mutual benefit and cooperation. It also points to a deep theoretical understanding of the concept in historical setting particularly of Pierre Bourdieu. The first step in measuring social capital is to recognize that it cannot be measured by just one variable and is multidimensional. The second step relates to contextual underpinning, which brings out its relevance and the utility for development policy outcomes. The selection of variables and indicators is an important exercise in understanding and analyzing social capital and its implications for local development. Variables have to be contextualized and made relevant to the situation under consideration. Along with the quantitative variables, qualitative information is also processed for a coherent understanding of social capital. In the backdrop of all these considerations, this paper proceeds to analyze the levels of social capital of Meos and other communities in Mewat, Haryana.

The empirical results show that in terms of embedded association, socio-cultural activity has the highest level of associatedness followed by economic and political activities. It shows higher informal socio-cultural interaction among the communities. The entire sample communities report an increase in yield per acre except for the Gujjar community as the surface soil is badly affected by termites. While the Meos report an increase in the sale of agricultural and livestock products, the others (Jats and SCs) report no change in sale in the last three years. The self-help groups are the most prominent informal organizations of the women and contributing to the welfare of the households and communities. The formal organizations include Gram Sabha and statutory Gram Panchayats. The Gram Panchayats support the CBOs followed by religious and non-government organizations. Members of the community regularly use dharmshala as a meeting place.

The women Sarpanch do not play an active role in resolving the issues facing the community as in most of the cases the Sarpanch-pati continues to run the Gram Panchayat. The members of the SC community have organized community members to address problems like drinking water and electricity as compared to Gujjar, Jat and Kumhar Panchayats. In this regard, they have organized *bandhs* and submitted a memorandum to the public representatives. The results also show that self-help groups are an important source of social capital along with associatedness in socio-cultural activities, increase in mutually beneficial production relations, vibrant (informal) institutions and strong social cohesion. The paper concludes that social capital may prove important in improving the decision support system for community development outcomes as it allows the instrumental use of social relations for mutual benefit and this in association with economic capital in the form of income support can be employed as a strategy of sustainable development in Mewat region, Haryana.

Acknowledgements The author certifies that no funding was received for this paper. At the time of writing this research, he had no affiliation with any organization with any financial interest or non-financial interest in the subject matter or materials discussed in this manuscript. All procedures performed were in accordance with the 1964 Helsinki declaration and its later amendments and ethical approval was obtained from all sources. Informed consent was obtained from all individual participants included in this article. The author further declares that there is no conflict of interest involved in the preparation of this article.

References

Ali AH et al (1970) Meos of Mewat: old neighbours of New Delhi. Oxford University Press, New Delhi

Bjornskov C (2006) Determinants of generalized trust: a cross-country comparison. Public Choice 130(1):1–21

Bourdieu P (1986) The forms of capital. In: Richardson JG (ed) Handbook of theory and research for the sociology of education. Greenwood, New York, pp 46–58

Brehm J, Rahn W (1997) Individual level evidence for the causes and consequences of social capital. Am J Polit Sci 41(3):999–1023

Brisson D, Usher CL (2012) Conceptualizing and measuring bonding social capital in low-income neighbourhoods. J Soc Sci Res 34(1):1–11

Bullen P, Onyx J (1998, 2005) Measuring social capital in five communities in NSW: a practitioner's guide. Management Alternatives, NSW

Coleman SJ (1988) Social capital in the creation of human capital. Am J Sociol 94(1):S95–S120

Coleman SJ (1994) Foundations of social theory. Harvard University Press, Massachusetts, pp 300–324

Defilippis J (2001) The myth of social capital in community development. Housing Policy Debate 12(4):781–806

Escobar A (2011) Encountering development: the making and unmaking of the third world. Princeton University Press, New Jersey, pp 21–54

Grootaert C (1998) Social capital: the missing link. Social Capital initiative, Working paper No. 3. World Bank, Washington D.C

Knack S, Keefer P (1997) Does social capital have an economic payoff? A cross-country investigation. Q J Econ 112(4):1251–1288

Krishna A (2002) Active social capital: tracing the roots of development and democracy. Columbia University Press, New York, pp 55–84

Krishna A (2007) How does social capital grow: a seven year study of villages in India. J Polit 69(4):941–956

Krishna A, Uphoff N (1999) Mapping and measuring social capital: a conceptual and empirical study of collective action for conserving and developing watersheds in Rajasthan, India, Social Capital Initiative, Working paper No. 13. World Bank, Washington D.C
Kropotkin P (1998) Mutual aid: a factor of evolution. New York University Press, New York, pp 76–77
Mahmood A (1977) Statistical methods in geographical studies. Rajesh Publications, New Delhi, p 91
Narayan D, Pritchett L (1999) Cents and sociability: household income and social capital in rural Tanzania. Econ Dev Cult Chang 47(4):871–897
Paxton P (1999) Is social capital declining in the United States? A multiple indicator assessment. Am J Sociol 105(1):88–127
Portes A, Landolt P (2000) Social capital: promise and pitfalls of its role in development. J Lat Am Stud 32:529–547
Portney KE, Berry MJ (1997) Mobilizing minority communities: social capital and participation in urban neighbourhoods. Am Behav Sci 40(5):632–644
Putnam DR (2000) Bowling Alone: the collapse and revival of the American community. Simon & Schuster, New York
Putnam DR, Leonardi R, Nanetti YR (1993) Making democracy work: civic traditions in modern Italy. Princeton University Press, New Jersey
Saegert S, Thompson JP, Warren MR (2001) Social capital and poor communities. The Russel Sage Foundation, New York
Sampson JR, Graif C (2009) Neighbourhood social capital as differential organization: resident and leadership dimensions. Am Behav Sci 52(11):1579–1605
Sikand Y (1995) Meonis of Mewat. Econ Polit Wkly 30(10):490–492
Stone W (2001) Measuring social capital: towards a theoretically informed measurement framework for researching social capital in family and community life, Working paper No. 24. Australian Institute of Family Studies, Melbourne
Svendsen GT, Bjornskov C (2007) How to construct a robust measure of social capital: two contributions. J Comp Policy Anal Res Pract 9(3):275–292
Uslaner EM (2008) Trust and consequences. In: Gupta KR, Svendsen GLH, Maiti P (eds) Social capital, vol 2, Atlantic Publishers, New Delhi, p 58–87
Woolcock M (1998) Social capital and economic development: towards a theoretical synthesis and policy framework. Theory Soc 27(2):151–208
World Commission on Environment and Development (1987) Our common future. United Nations, Oxford University Press, New York

Chapter 17
Monitoring and Modelling of Urban Sprawl Using Geospatial Techniques—A Case Study of Shimla City, India

Pawan Kumar Thakur, Manish Kumar and Vaibhav E. Gosavi

Abstract Urban Sprawl, a burning global phenomenon, refers to that extent of city or urbanisation, which may driven by industrialisation, population growth, settlement on the periphery of the city and large-scale migration. India, with population of over one billion, which is one-sixth of the world's total population, phenomenon of Urban Sprawl is affecting its natural resources like never before. This results in multiple problems such as unmanageable transportation, unemployment which results in poverty, illegal housing colonies, slums, etc. This study illustrates the use of Geospatial and Statistical Techniques to highlight the extent of Urban Sprawl in Shimla Municipal Corporation, Himachal Pradesh, India; at a detailed level. In the present work, four temporal satellite images of Landsat Thematic Mapper have been used over a period of nearly two decades (i.e. 1991–2011). Landsat imageries of two time periods (Landsat Thematic Mapper (TM) of 1991 and 2011; and ETM+ (Enhanced Thematic Mapper) 2001), were used and quantified for studying the decadal change in land use/land cover. Supervised classification methods have been employed using Support Vector Machine in ENVI 5.0 and ERDAS 2014. The study area is categorized into five different classes, viz. water bodies, forest area, built-up area, agriculture and open space. The results indicate that built-up area is the major land use in study area. During the period 1991–2011, the area under built-up land has increased by 529.65 ha (26.48)% due to construction of new buildings on forest land, agricultural land and open spaces. As a result, the area under vegetation (forest), open space and water bodies decreased by −361.71 ha (−18 09%), −178.02 ha (−8.90%), 0.36 ha (0.02%) respectively. Urban built-up density was calculated in percentage (%) for 25 wards

P. K. Thakur (✉) · V. E. Gosavi
G. B. Pant, National Institute of Himalayan Environmental & Sustainable Development, Himachal Regional Centre, Mohal-Kullu 175 126, Himachal Pradesh, India
e-mail: thakurpawankumar0431@gmail.com

V. E. Gosavi
e-mail: vaibhavgosavi8@gmail.com

M. Kumar
Department of Geography, Kalindi College, University of Delhi, East Patel Nager, New Delhi, India
e-mail: manish.ks1@gmail.com

S. Sahdev et al. (eds.), *Geoecology of Landscape Dynamics*,
Advances in Geographical and Environmental Sciences,
https://doi.org/10.1007/978-981-15-2097-6_17

of the city for 1991, 2001 and 2011. Depending on the density levels, it categorized as low, medium and high density. The process of urban expansion in the Shimla City during 1991, 2001 and 2011 are further studied by examining a distance decay concept from a major road. The study also highlights the nature, rate and location of change; and the importance of digital change detection techniques in land use planning for sustainable growth of the study area. The Shannon's Entropy Index and Landscape Metrics have been computed in order to quantify the urban growth using built-up area as a spatial unit. Further, Stepwise Regression techniques were used to explore the relationship between the urban growth and its contributory factors. Further, Stepwise Regression was used to study the impact of Independent Variables (population density, female literacy rate, road density, total workers, sex ratio, α-population, number of HH, β-population) and Dependent Variables (percentage of built-Up area) factors on urban growth. The result shows that the growing population triggers the increase in built-up area in study area. This study also demonstrates the potentials of geospatial data in mapping, measuring and modelling the urban sprawl, which could be helpful to decision makers to form/make particular decision support system for sustainable growth of hilly urban areas or in making Smart City planning.

Keywords Urban sprawl · SVM · Digital change detection · Built-up density · Shannon's entropy · Stepwise regression · Shimla city

17.1 Introduction

The world urban centers are growing very faster in terms of the extent geographical area and density of population. Moreover, the urban areas change a rapid rate in developing countries and are experiencing stress. According to the United Nations (2014) report, the trend of global urbanization shows that 66% of the world's population is projected to be in urban area by 2050, with 90% of this expansion being anticipated in developing countries. In 2050, most of the urban population living of the world will be combination in Asia (52%) and Africa (21%). In India, the percentage of urban population to the total population has been steadily increasing since Independence. India is the 2nd largest urban organization in the world with more than 30% of urban population. The half of India's population will become urban has been estimated by 2025. The urban population of Indian cities is highly concentrated in large cities which are growing rapidly compared to small urban centres. The most Indian cities population is highly concentrated in large which are growing rapidly compared to small CBD (Central Business District) centers.

Urbanization drives the change in land use/land cover pattern. Actual information on the extent of urban growth is of great interest for the various purposes such as developmental activities like urban planning and management, land and water resource management, service and marketing analysis, etc. Urban local bodies are required to dedicate more time, effort and attention to able to use of land resources in order to accommodating the growing population or other urban land uses (Jat et al.

2008). The traditional surveying and mapping techniques are time-consuming and also expensive for the estimation of urban sprawl. Therefore, more research interest is being focused on mapping and monitoring of urban growth using Remote Sensing and GIS techniques (Epstein et al. 2002). Several researches on urban sprawl have been attempted by many scholars (Batty et al. 1999; Torrens and Alberti 2000; Barnes et al. 2001; Hurd et al. 2001; Jantz and Scott 2005; Yang and Liu 2005).

Urban sprawl usually takes place on the border area of the city or along the highways. Some studies of urban sprawl (The Regionalist 1997; Sierra Club 1998) was carried out in developed countries (Batty et al. 1999; Torrens and Alberti 2000; Barnes et al. 2001, Hurd et al. 2001; and Epstein et al. 2002) as well as in developing countries such as China (Yeh and Li, 2001; Cheng and Masser 2003) and India (Jothimani 1997; Lata et al. 2001; Sudhira et al. 2003). To quantify the urban sprawl, built-up is considered as a defining parameter (Torrens and Alberti 2000; Barnes et al. 2001; and Epstein et al. 2002), which used to be delineated using standard toposheets. Remote Sensing and GIS systems played important role in quantifying, monitoring, and subsequently predicting urban sprawl phenomenon. On spatial scale, GIS assist in determining the landscape properties in terms of structure, function and change (ICIMOD 1999; Civco et al. 2002). Modelling of the spatial dynamics mainly cover the LULC change studies (Lo and Yang 2002). In predicting the situation of Ipswich watershed, USA, in terms of its land use change, Pontius et al. (2000) carried out future land use changes prediction based on validated model for 1971, 1985 and 1991. Further, the Cellular Automata (CA) technique was extensively used in the urban growth models (Clarke et al. 1996) and in simulating urban sprawl (Torrens and O' Sullivan 2001; Waddell 2002).

The Geo-spatial techniques, such as Remote Sensing and GIS plays vital role in quantifying and modelling urban landscape which is otherwise not possible or time consuming by mapping through traditional methods. Remote Sensing have proven its usefulness in mapping urban areas and as a data sources (spatially and temporarily) for modelling of urban sprawl as well as in studying LULC (Batty and Howes 2001; Clarke et al. 2002; Donnay et al. 2001; Herold et al. 2001; Jensen and Cowen 1999). Therefore, more often, it is used for the analysis of urban sprawl (Sudhira et al. 2004; Yang and Liu 2005; Haack and Rafter 2006). The integration of Remote Sensing with GIS and database management systems helps in quantifying, monitoring and modelling the urban sprawl phenomenon. This has reflected in extensive research efforts made in last three decades for urban change detection studies using remotely sensed images (Gomarasca et al. 1993; Green et al. 1994; Yeh and Li 2001; Yang and Lo 2003; Haack and Rafter 2006).

Sudhira et al. (2004) used geospatial techniques in modeling simulation of future urban sprawl. Kumar et al. (2007) carried out similar study in Indore city of India using three temporal satellite Remote Sensing data (1990–2000). In Ajmer city (India), Jat et al. (2008) analyzed the urban sprawl using eight multi temporal satellite imageries of duration 1977–2005. They found that the built-up development in Ajmer city is more than three times the population growth. The spatial extent and patterns of systematic/random changes in urban development in Haridwar city (India) was studied by Jha et al. (2008) for developing future plans. In applying Entropy approach, Punia and Singh (2012) conducted study on Jaipur city (India) to quantify urban sprawl. This study reveals that the rate of built-up growth in Jaipur city have outstripped the rate of population growth. Based on study using satellite data from 1990 to 2010, Rawat and Kumar (2015) found that there is sharp increase in built up area in Almora town of Uttarakhand state in India, which is attributed to construction of new buildings on agricultural and vegetation lands.

Apart from the research work of land use/land cover change detection, there are studies in which uses statistical techniques with geospatial technology to quantify, estimate, map & model urban sprawl (Jat et al. 2008). Shannon's Entropy is based on information theory. One of such statistical techniques is Shannon's Entropy which used as a mathematical estimation of Urban Growth that occurs in a disorganized way. Shannon's Entropy (Hn) is used to measure the degree of spatial concentration or dispersion of geophysical variables (Xi) among 'n' spatial units/zones. In quantifying urban forms, Shannon's Entropy has been used by Joshi et al. (2006).

Changes in urban structures and pattern can be obtained using landscape metrics, which gives information about the configuration of the land cover classes and the landscape, are algorithms used to describe and quantify the spatial characteristics of patches, class areas and the entire landscape (Cabral et al. 2005; Herold et al. 2002). GIS helps in calculating the urban landscape metrics, like patchiness and density in order to characterize landscape properties in terms of spatial distribution and change (Trani and Giles 1999; Yeh and Li 2001; Civco et al. 2002, Sudhira et al. 2004). However, the analysis of the sprawl pattern and the configuration of landscape metrics correlate with each other, and cannot be seen apart. With the identification of the pattern a first impression of the pattern is obtained, the landscape metrics provide statistical insight of the configuration of the urban sprawl.

In present study, an attempt has been made to investigate the usefulness of the Remote Sensing and GIS techniques along with Shannon's Entropy (Hn) and landscape metrics for urban sprawl dynamics and monitoring of spatial and temporal variability of Shimla city, Himachal Pradesh, India from 1991 to 2011. Variation of urban sprawl (spatially and temporally) is considered for identifying the relationship between urban sprawl and some its contributory factors such as (independent variables) population density, female literacy rate, road density, total workers, sex ratio, α-population, number of household's (HH), β-population, using Stepwise Regression Analysis. In order to quantify and measure the urban growth, Shannon's Entropy, Built-up Density, Landscape Metrics and Spatial Dynamic of urban sprawl were calculated. The landscape metrics were used to enhance understanding of the urban forms. Computation of these parameters helps in understanding the process of urbanization at a landscape level. An attempt has also been made in the study for mapping the status of land use/land cover of Shimla Municipal Corporation, Himachal Pradesh (India) for detecting the land use rate and the change that have taken place during 1991–2011 using Remote Sensing and Geographic Information System (GIS) technique.

17.2 Study Area

Shimla is a capital of Himachal Pradesh. The population of Shimla city is 1,69,578 (Census of India 2011). Shimla city, located between 31°3′ and 31°8′ N latitudes and between 77°7′ and 77°13′ E longitudes (Fig. 17.1). It has an average altitude of 2195 m above mean sea level and extends along a ridge with seven spurs. The city is spread over an area of 30 km^2. Population of Shimla city has increased from 1,42,555 (in 2001)–1,69,578 (in 2011), accounting for 18.96% decadal growth. Population growth in percentage (%) has increased from 39.51% (in 2001) and 18.95% (in 2011).The climate in Shimla is mostly cool during winter months and moderately warm during summer months (Cwb), under Koppen climate classification, with three distinct seasons—summer, rains and mild winter. Shimla has an average temperature of 19 °C–28 °C during March to June (in summer) and average temperatures in winter season (November–February) ranges from 03 °C–09 °C. As a result of north-west monsoon, the city receives an annual rainfall of 1577 mm between July–October. The warmest month of the year is June.

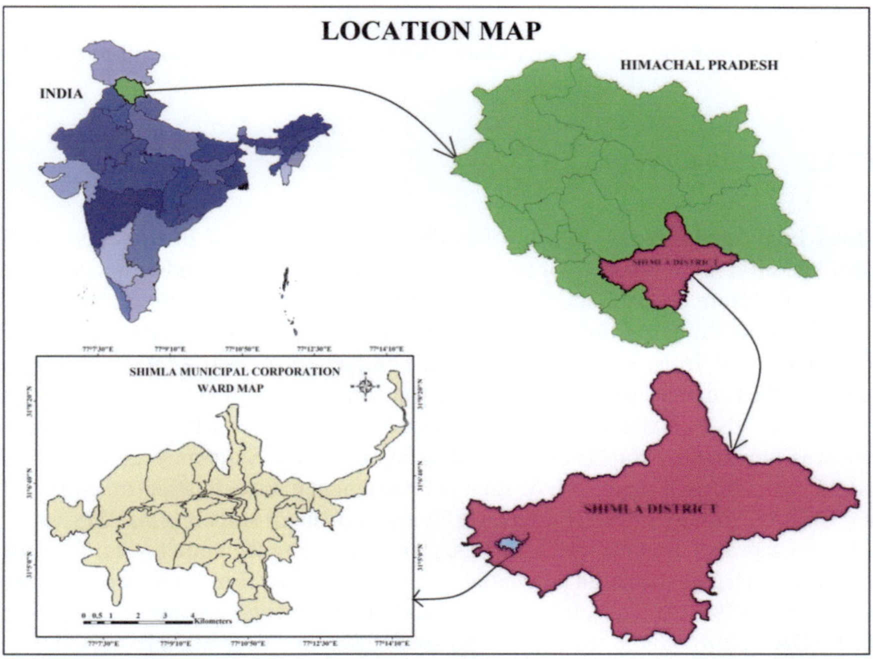

Fig. 17.1 Location Map of the study area

Table 17.1 Data used for study

Data	Date of acquisition	Spatial resolution & scale (m)	Sources
A. Satellite images			
Landsat TM	16–11–1991	30 m	Earth Explorer
Landsat ETM+	18–10–2001	28.5 m	Earth Explorer
Landsat TM	22–10–2011	30 m	Earth Explorer
B. Additional data			
Toposheet (53E/4)	–	1:50,000	Survey of India
SMC Map (Ward)	2011	–	SMC
Demographic data	1991, 2001 and 2011	–	Census of India
Road layers	2011	–	Google Earth

17.3 Database and Methodology

17.3.1 Data Used

Primary and secondary data sources have been used for present study (Table 17.1).

17.3.2 Methodology

Cloud Free Landsat satellite data of multi-temporal image of three years: 1991, 2001 and 2011 were selected for analysis. The freely available Landsat images were downloaded from Global Land Cover Facility (GLCF) and Earth Explorer website of United States Geological Survey (USGS). All the data are preprocessed and projected to the Universal Transverse Mercator (UTM) projection system. Another reason for selecting these images was their availability at similar resolution.

Using bands 2, 3 and 4 of the preprocessed images, map of land use/cover pattern was prepared by Supervised Classification using the Support Vector Machine (SVM) classification algorithm of Envi (Environmental Visualization Imagine) 5.1 and ERDAS 2014 (Earth Resource Data Analysis System) software. ArcGIS 10.2.2 software was also used for spatial analysis and generating thematic layers. To study the built-up change detection especially in the land use/cover over the period of two decade (from 1991 to 2011), unsupervised and supervised classification was used. In the ERDAS 2014, indices such as Normalized Difference Vegetation Index (NDVI), Normalized Difference Water Index (NDWI) and Normalized Difference Built up Index (NDBI) were also applied to classify the Landsat images at a spatial resolution of 30 m of 1991 and 2011, ETM+ at a spatial resolution 28.5 m of 2001. The Error Matrix and Kappa coefficient method were used to assess the mapping accuracy. Five Land use/cover types have been identified and used in this study, namely (i) Water Body(ii) Forest Area (iii) Built up Area (iv) Agriculture Area (v) Open Space

Table 17.2 Description of land use/land cover classes

Land use/land cover categories	Description
Urban or built-up	Residential, commercial, services and industrial, roads and playground, others means of transportation and other urban feature and Recreational areas also included
Forest area (Vegetation)	Vegetation areas, grass land and green parks and Forest, scattered trees
Agriculture Area	Production of food and fiber, This includes cropped areas, fallow lands and plantation (Apples and pears orchards and nurseries)
Water bodies	River, lakes, tank, ponds, reservoirs etc
Open space/land	Exposed soil, landfill sites, polder area of active excavation, open space in the built-up land etc

(Table 17.2). A change matrix (Weng 2001) was build using ERDAS 2014 and ArcGIS 10.2.2 software. The quantitative areal data of land use/land cover changes and its gains and losses in each categories from 1991 to 2011, were compiled.

In determining qualitative as well as quantitative aspects of the changes from 1991 to 2011, classified image pairs of respective time frame were compared using cross-tabulation. In order to produce the change information on pixel basis and interpreting the changes more efficiently, a pixel-based comparison was performed. A change matrix (Weng 2001) and Land Encroachment map was produced using ERDAS Imagine 2014. The matrix showed the respective gains and losses in land use/cover category during 1991–2011. Field data were collected for two reasons: first, to acquire GPS data for ground verification of doubtful areas; and secondly, to calculate the mapping accuracy of the classified images (Table 17.3). The ground control points were used to correct the misclassified areas using recode option in ERDAS Imagine 2014 software. To assess the classification accuracy of all the produced outputs a reference dataset of randomly selected 200 pixels were used. Land use/land cover for these pixels was determined with the help of urban settlement map, and data collected from other secondary maps. In order to avoid the errors in the reference dataset for sensitive classes such as vegetation, which is temporal in nature; the original satellite data was used for better assessment in its accuracy. Some of the ground truth data that have been used in present study essentially includes urban settlement map and

Table 17.3 Description of land use/cover mapping accuracy

Land use/land cover	Overall accuracy (%)	Kappa coefficient
1991	92.00	0.714
2001	97.00	0.926
2011	96.00	0.920

Sources: Landsat TM and ETM+ satellite data for 1991, 2001 and 2011

geographical locations of some of the features, location-wise type of vegetation, buildings, play grounds, permanent water bodies, and road network.

Urban sprawl over a period of two decade (1991–2011) is obtained from the classified images. To understand and model the urban sprawl pattern, different Landscape Metrics (Shannon's Entropy, Stepwise Regression, Map density and Buffering along major roads) were calculated using the demographical and built-up area statistics. Shannon's Entropy (H_n) is mostly used for ascertaining the urban growth phenomena. This method is basically used from information theory which acts as a measure of uncertainty of conveyed information over a noisy channel (Jat et al. 2007; Bailey 2009). The greater value of Shannon entropy imparts higher uncertainty in the information conveyed (Bailey 2009). High Entropy leads to high occurrence of phenomena that contributes to disorder. This is probably same for urban sprawl. Thus, Shannon entropy has been computed to quantify the decay of sprawl. The Shannon entropy H_n is given by (Yeh and Li 2001).

$$H_n = \sum_{i}^{n} P_i \log_e \frac{1}{P_i} \tag{17.1}$$

where, P_i is the proportion of occurrence of a phenomenon ith spatial zone out of n zones (wards). This is given by

$$P_i = \frac{x_i}{\sum_i^n x_i} \tag{17.2}$$

where, x_i is the area of built up at the ith zone.

The value of entropy varies between 0 to $\log_e n$. The value closer to 0 indicates concentrated or compact distribution of the built up while value closer to $\log_e n$ revels dispersed distributed pattern. However, high value of entropy indicates the occurrence of sprawl.

Apart from the whole study area, entropy model has been also applied to different administrative wards in the study area to detect the form and type of urban growth phenomena (Punia and Singh 2011). For ward wise calculation of Shannon entropy, each ward has considered as an individual spatial unit, and is given by

$$H_i^n = \sum_{i}^{n} P_i \log_e \frac{1}{P_i} \tag{17.3}$$

where, P_i represents proportion of land use/land cover class in ith ward, n denotes total number of land use/land cover classes. The range of Entropy values is minimum of 0 to $\log_e n$ representing upper limit of entropy which is equal to log 25 = 1.3413.

In developing the relationship of percentage built-up (dependent variable) with causal factors (independent variables) of sprawl, Stepwise Regression analysis was carried out. In this study, independent variables are Population Density (PD), α-Population Density (α-PD), β-Population Density (β-PD), Total workers (TW), Road density (RD), Female literacy rate (FLR), Sex ratio, (SR) and No of Household's (HH). In order to identify the probable relationship of percentage of built-up (PB) (dependent variable) and individual causative factors, different distributions (Linear, Polynomial, Logarithmic, and Exponential) have been calculated and documented.

In present study, eight spatial metrics indices, i.e., Class Area (CA), Number of Patches (NP), Largest Patch Index (LPI), Edge Density (ED), Euclidian Mean Nearest Neighbour Distance (ENN_MN), Area Weighted Mean Patch Fractal Dimension (AWMPFD) and Contagion (Table 17.10), were used for quantify the spatial characteristics of urban landscape. The changes in urban landscape were measured and analysed using the FRAGSTATS software. The process of urban expansion in the study area during the period of two decade (1991–2011) was examined by evaluating a distance decay concept from major roads. In an obvious phenomenon like density of land development declines rapidly as the distance from roads increases, the National Highway No.—NH 22, National Highway No.—NH 88, summer hill roads, Bharari road, Mall road and Ridge road, etc., have been considered for present urban sprawl analysis, where four buffer zone were created along the roads with 0–50 m each.

17.4 Result and Discussion

17.4.1 Image Analysis

The spatial analysis was performed beyond the Shimla Municipal Council's administrative boundary. The classified images for land use/land cover in 1991, 2001 and 2011 are shown in Figs. 17.2, 17.3 and 17.4 respectively. The extent of land use/land cover during 1991, 2001 and 2011 comprised built-up area with 361.05 ha, 646.92 ha, 890.7 ha respectively. Similarly, the extent of non-built-up area comprising open land, forest area and water bodies were 2329.47 ha, 2106.72 ha and 1968.12 ha during 1991, 2001 and 2011 respectively. The LULC change analysis was carried out based on the differences in land use/land cover on spatial scale and temporal variation. During 1991–2011, it was observed that the extent of built-up area has increased by 26.48%. Here, the extent of increase in built-up area only describes the magnitude of change but without the pattern of this transition. The cross-tabulation for the studied classified images was performed in order to analyze the probable LULC change from various non-built-up classes to Built-up class. It was found that during 1991–2011, the land use change from open land into Forest area was significant. The open land-use class was found to be a major land use that contributed to the increase of built-up area. Similarly, during 1991–2011, the Forest Area was the major land use/land cover that was lost (by almost 361.71 ha) due to changes into built-up areas.

The overall accuracy and Kappa coefficient for all the classified images are presented in Table 17.3. For better classification results, random sets of 200 samples were generated and classification results were compared with the true information classes in the reference image. The high level of accuracy was obtained in the present

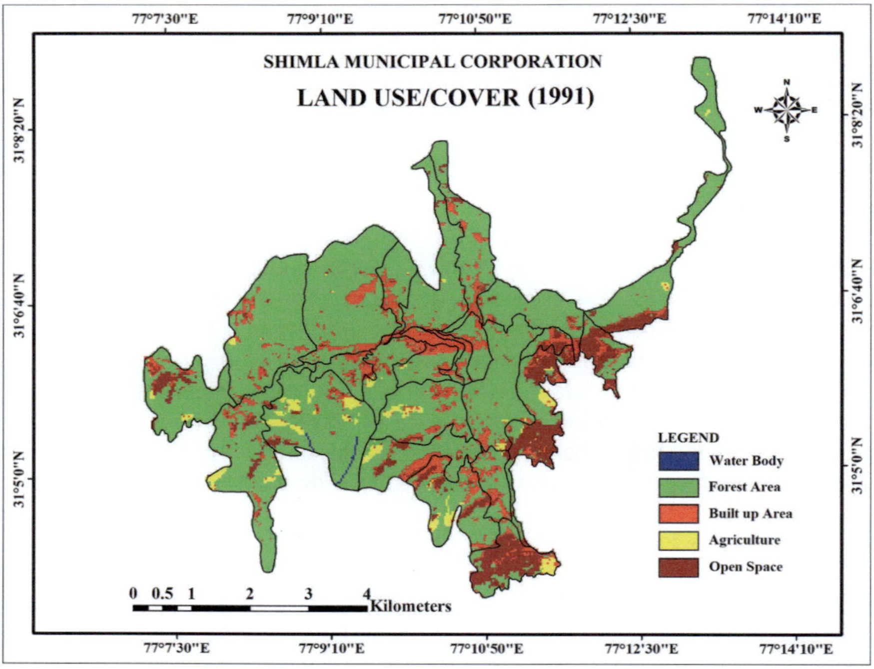

Fig. 17.2 Land use/cover classification of Shimla City (*Source* Landsat TM 1991)

study could be the result of the coarse classification, since only five classes are used. The classified image of original bands using Support Vector Machine (SVM) of Landsat TM (1991 and 2011) and Landsat-7 ETM+ (2001). From the image it is seen that most of the forest area is classified as vegetation due to similar spectral reflectance.

Table 17.4 and Fig. 17.5 reveal that both positive and negative changes occurred in the land use/cover pattern of Shimla city. According to the land use/cover maps produced, the total built-up area for 1991 was 361.05 ha (11.78%). This has increased to 646.92 ha (21.16%) by 2001, and finally reached 890.7 ha (29.06%) in 2011. These figures represent about 26.48% growth in built-up area over a period of time from (1991 to 2011). With respect to forest cover, it is clear that in 1991 it was 2325.6 ha and decreased to 1963.89 ha by 2011. Likewise, open space was about 296.19 ha in 1991 and decreased by 118.17 ha in 2011. The total decrease in forest area and open space during the study period was 18.09% and 8.90% respectively. The decrease in both the land use/land cover is mainly due to over exploitation of land for built-up purpose.

During this period, forest area decreased from 2325.6 ha in 1991–1963.89 ha in 2011 which accounts for −18.09% change (−361.71 ha) of total land cover area. The open space decreased from 296.19 ha in 1991 to 118.17 ha in 2011 which accounts for −8.90% change (−178.02 ha) of the total area. Forest area is decreasing due to

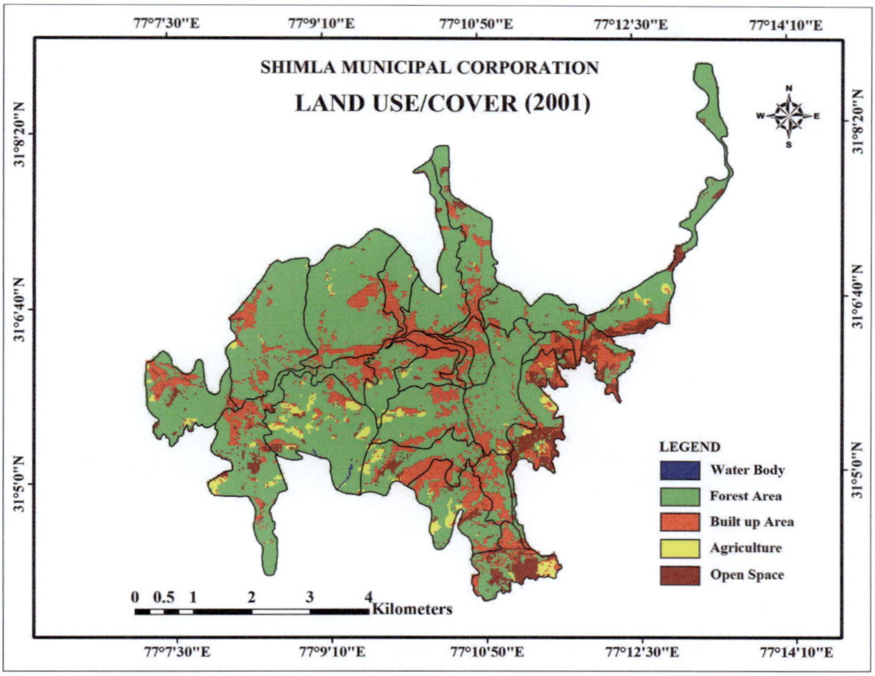

Fig. 17.3 Land use/cover classification of Shimla City (*Source* Landsat ETM+ 2001)

increasing population of the city which required more land for new settlers. Forest land which included both land of different vegetation and also included the grass park and green grass land. The built-up area has increased from 361.05 ha in 1991–890.7 ha in 2011 which accounts for +26.48% (529.65 ha) of the total area (Table 17.5). Built-up area which included both playground and stadium. This dramatic increase in built-up area is due to migrated population problem, rapid growth of local population, tourism development, continuous establishment of national/multinational companies, and development of major and minor's roads, etc. During the study period, it is observed that the fringe area, forest area and open space has change to built-up area in 1991–2011, decrease the forest area and open space is due to increase in the built-up area.

In order to understand the land conversion into different land categories in two decade (i.e., 1991–2011) a change detection matrix (Table 17.6) was prepared that reveals about 362.52 ha area of forest cover has been converted into built-up area, 48.06 ha into agriculture area, 30.69 ha into open space and 3.78 ha into water body. Similarly, about 55.71 ha area of open space has been converted into forest area, 148.77 ha area converted into built-up area, and 6.21 ha area converted into agriculture area. Likewise, about 24.48 ha area under agriculture area has been converted into forest area, 18.36 ha area converted into built-up area, 1.98 ha area converted into

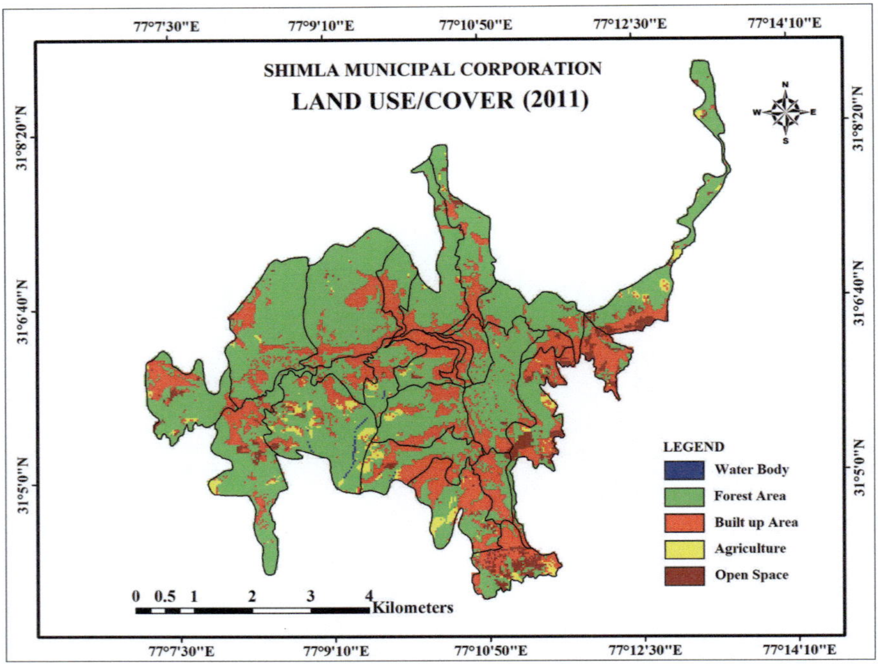

Fig. 17.4 Land use/cover classification of Shimla City (*Source* Landsat TM 2011)

Table 17.4 Status of land use/land cover in all categories in Ha

Land use classes/years	1991(Ha)	2001(Ha)	2011(Ha)
Water bodies	3.87	1.98	4.23
Forest area	2325.6	2106.72	1963.89
Built-up area	361.08	646.92	890.73
Agriculture area	77.85	96.66	87.57
Open space	296.19	212.22	118.17

Sources Landsat TM and ETM+ satellite data for 1991, 2001 and 2011

open space. And, about 3.15 ha area open water body has been converted into forest area and 0.27 ha area has been converted into agriculture area.

17.4.2 Population Growth and Built-up Area

The urban impervious area such as built-up land has increased from 361.05 ha in year 1991–890.7 ha in year 2011. Results in Table 17.7 and Fig. 17.6 reveals that

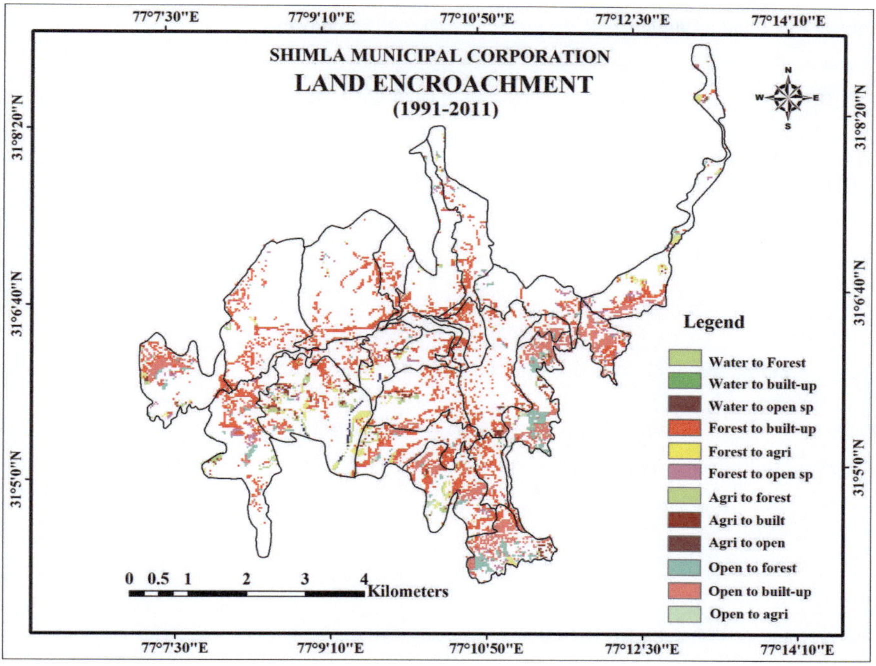

Fig. 17.5 Land Encroachment of Shimla City (*Sources* Landsat TM and ETM+ satellite data of 1991–2011)

Table 17.5 Area and per cent changes in land use/cover categories (1991–2011)

Land use/cover categories	Area (1991)	Area (2011)	Change (1991–2011)	Area (1991)	Area (2011)	Change (1991–2011)
	Hectares	Percent	Hectares	Percent	Hectares	Percent
Water body	3.87	0.13	4.23	0.14	0.36	0.02
Forest	2325.6	75.89	1963.89	64.08	−361.71	−18.09
Built up	361.05	11.78	890.7	29.06	529.65	26.48
Agriculture	77.85	2.54	87.57	2.86	9.72	0.49
Open space	296.19	9.67	118.17	3.86	−178.02	−8.90

Source Landsat TM and ETM+ satellite data for 1991, 2001 and 2011

the rate of land development in Shimla has exceeded the rate of population growth, from the year 1991–2011. A compound annual growth rate for built-up area for two decade (1991–2011) and average annual exponential growth rate for population of Shimla city has been calculated with considering the logic of population grows exponentially while area does not. In the description of population growth and built-up area from, 1991 and 2001, the highest annual growth rate was observed in both built-up area and population. However, the built-up area growth is much higher

Table 17.6 Land use/land cover change matrix (Ha) of Shimla city

Land use/land cover categories		Year 1991					2011 Total
		Water body	Forest	Built Up	Agriculture	Open Space	
Year 2011	Water body	0.45	3.78	0	0	0	4.23
	Forest	3.15	1880.55	0	24.48	55.71	1963.89
	Built up	0	362.52	361.05	18.36	148.77	890.7
	Agriculture	0.27	48.06	0	33.03	6.21	87.57
	Open space	0	30.69	0	1.98	85.5	118.17
1991 Total		3.87	2325.6	361.08	77.85	296.19	3064.56

Source Landsat TM and ETM+ satellite data for 1991, 2001 and 2011

Table 17.7 Urban growth statistics for Shimla city 1991–2011

Years	Built-up area (Ha)	Percentage increase in built-up area (%)	Projected Population	Percentage growth in population (%)
1991	361.05	–	102,186	–
2001	646.92	79.18	142,555	39.51
2011	890.73	37.69	169578	18.95

Source Landsat TM and ETM+ satellite and census data for 1991, 2001 and 2011

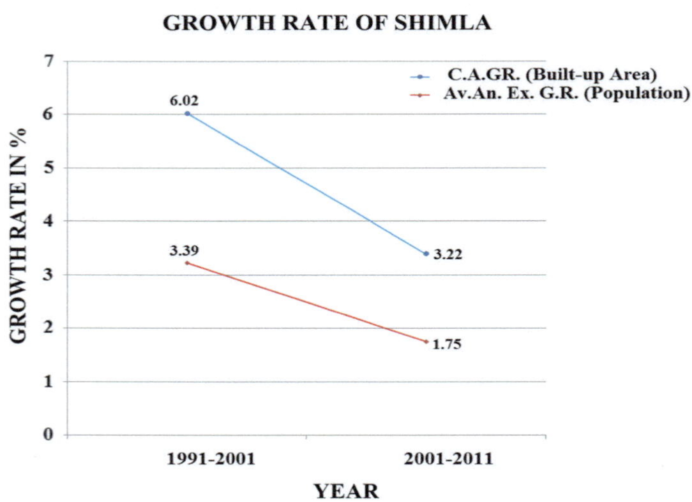

Fig. 17.6 Growth of population and built-up area of Shimla city in last 20 years, 1991–2011 (*Sources* Landsat TM and ETM+ satellite data and Census of India, 1991, 2001 and 2011)

(6.02%) than the growth rate of population (3.39%). This denotes the higher per capita land consumption and the city has been expanded physically with high rate of expansion. In the next period of 2001–2011, it is also observed that built-up area is growing much higher (3.22%) than the growth rate of population (1.75%). It is noteworthy here that with this high base, growth rate of the city is very significant that indicates spatial and temporal growth of the city.

Shimla city is growing fast in the pattern of construction of new building and others factors. Since the inception of Shimla Municipal Corporation (SMC), small pockets of land in parts of the villages were added increasing the area within the jurisdiction of SMC. The border area growth has resulted into the increased residential areas and other facilities such as transportation nodes. The population of Shimla city has increased manifold in the last 20 years and the rate of increase has been very speedy after 1991. The continually increasing population pressure has led to the growth of the adjoining areas and the city has extended outwards filling in spaces between it and the suburbs. The urban expansion has taken place in all directions but more extensively in the eastern, southern and south-western directions. Spatial distribution of ward-wise urban sprawl in last 20 years from 1991 to 2011 is shown in Fig. 17.7. It was found that urban sprawl is faster in outer area of (ward number 20–23) and inner area of (ward number 9–13) along the major roads of the city as compared to central

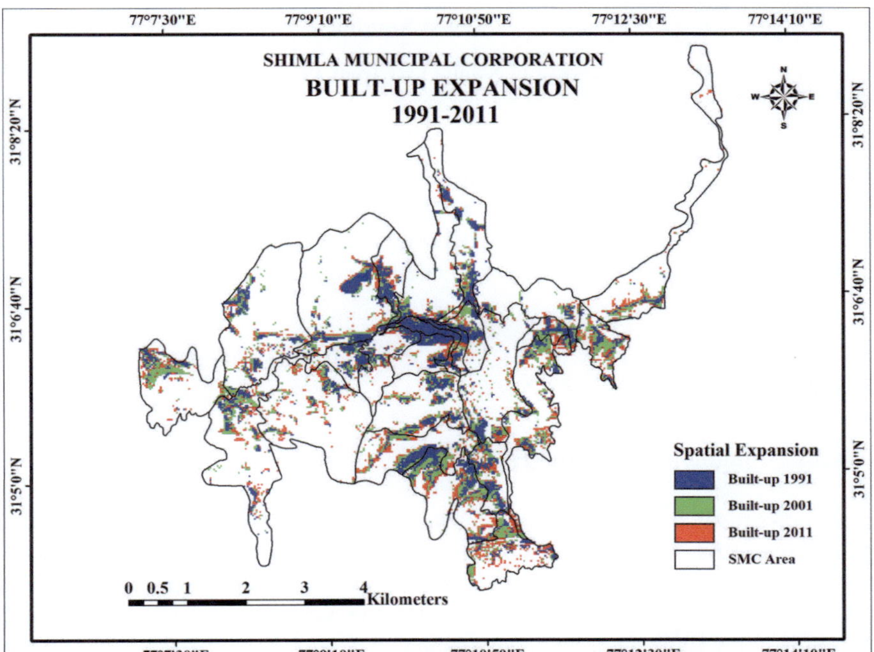

Fig. 17.7 Urban growth of Shimla city from 1991, 2001 and 2011 (*Sources* Landsat TM and ETM+ satellite data for 1991, 2001 and 2011)

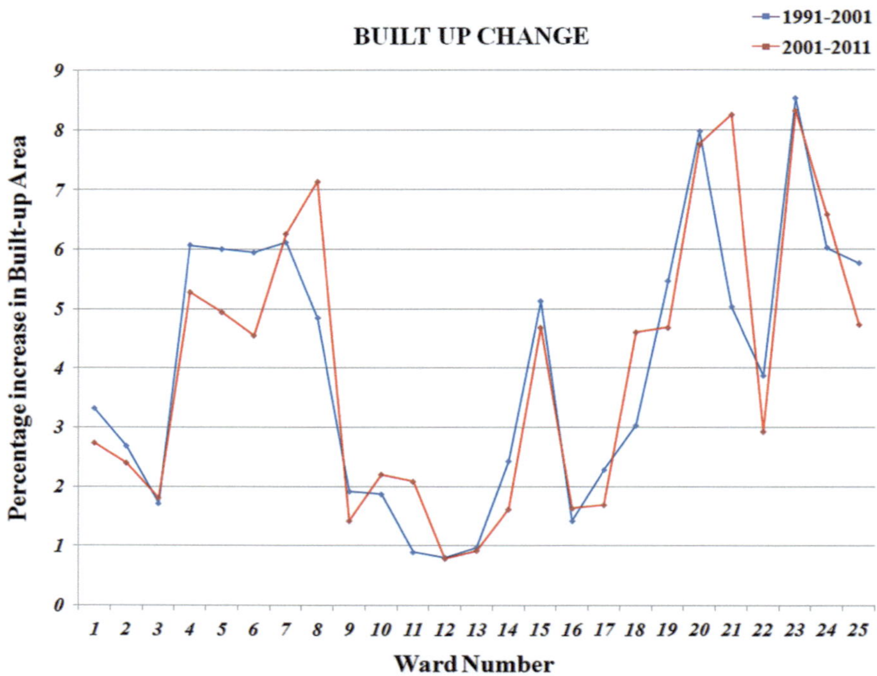

Fig. 17.8 Ward-wise urban growth of Shimla City from 1991 to 2011 (*Sources* Landsat TM and ETM+ satellite data for 1991, 2001 and 2011)

portion, which is also validated by landscape metrics. This is in accordance to the hypothesis which stated that increase in economic conditions among the population and development relates to urban sprawl (Census of India 2011). Highest built up growth was observed in ward number 23 and lowest in ward number 12 (Fig. 17.8).

17.4.3 Metrics of Urban Sprawl

17.4.3.1 Shannon's Entropy (H_n)

In the present study, Shannon's entropy was used in order to measure the extent of spatial concentration or dispersion (homogeneity) of a geophysical variable, impervious (built-up area) among 'n' spatial units/zones (wards).The entropy indicates that there was substantial internal variation in the patterns of urban sprawl among the wards of the study area. Shimla municipal area grew in all the wards has almost reached the threshold value of growth (log (n) = log, (25) = 1.34). The value of entropy varies between 0–log n. Entropy closer to zero correspond to compact distribution or homogeneity of urban growth; whereas, entropy closer to log n signify the

Table 17.8 Values of Shannon's entropy: Shimla city (1991–2011)

Years	Values of Shannon's entropy
1991	1.30
2001	1.32
2011	1.34

Source Landsat TM and ETM+ satellite data for 1991, 2001 and 2011

dispersed distribution of urban sprawl (specially built-up area). Entropy values have been calculated across all wards, and were summed-up to represent the entropy for the whole urban area. The change in entropy helps identifying whether land development is towards the more dispersal or in compact pattern. For ward wise calculation each ward has considered as an individual spatial unit.

$$H_n i\text{th} \Sigma p_i \log(1/p_i) \tag{17.4}$$

P_i represents the proportion of land use/cover class in ith ward.

n denotes total number of land use/cover classes.

Log n represents the upper limit of entropy, i.e., equal to log 1.3413

Relatively lower value of Shannon's entropy (1.3092) in the year 1991 indicates the compact and homogeneous distribution of the built-up area. Entropy value has increased from 1.3092 in year 1991 to 1.3276 in year 2001. Further, value of Shannon's entropy has increased from 1.3276 in year 2001 to 1.3413 in year 2011. This increase in value of entropy indicates increase in dispersion of built-up area, which indicates urban sprawl. The entropy values obtained during 3 years (1991, 2001 and 2011) are closer to the upper limit of log n, i.e., 1.3413, showing the degree of dispersion of built-up area in the region. Higher value of overall entropy for the whole urban area represents higher dispersion of built-up area, which is an indication of urban sprawl. Lower entropy values of wards (given in table), during 2011 shows an aggregated or compact growth in the whole region. However, the wards under Shimla city, grew phenomenally with dispersed growth during the study period (1991, 2001 and 2011) and reached higher value of entropy (Table 17.8) during 2011.

17.4.3.2 Urban Map Density

In present study Urban map density parameter was used to determine the compactness/dispersion of spatial occurrence of urbanization. Distribution of built-up areas, as an indicator of urban sprawl, has been studied using density metrics. Map density values were estimated by determining the number of built-up area pixels out of the total number of pixels in a 3 × 3 kernel. In this analysis, again size of kernel has been used as per the maximum number of land use/cover categories available in a specific classified image. There are five land use/cover categories available in most

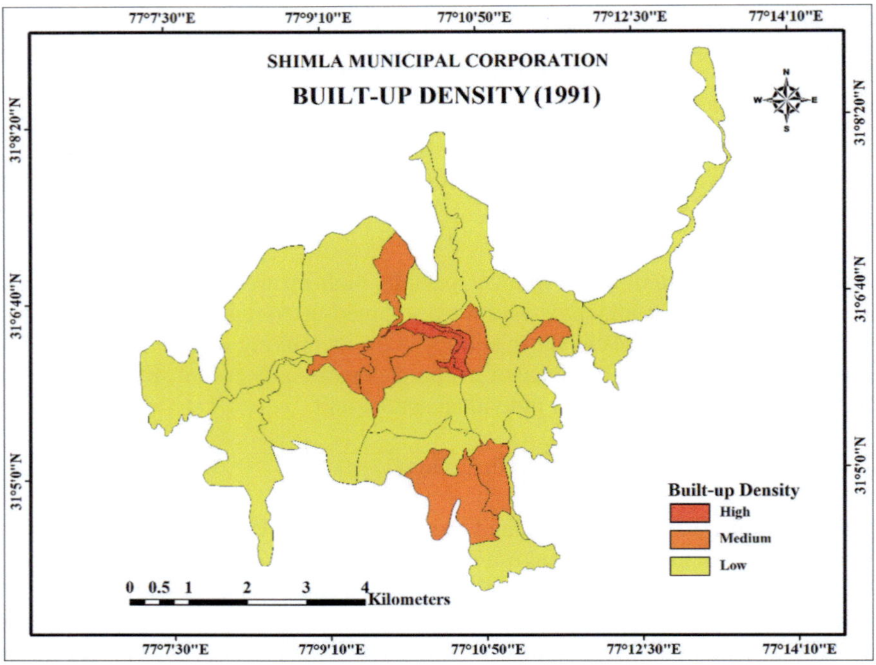

Fig. 17.9 Density of Urban Area of Shimla for 1991 (*Source* Landsat TM 1991)

of the classified images, which is less than the number of pixels in a 3 × 3 size kernel. Therefore, kernel of 3 × 3 size has been use in the present analysis. This when applied to a classified satellite image of 1991, 2001, and 2011, converts land cover classes to density classes. Depending on the map density levels, it was categorized as Low, Medium and High density (Figs. 17.9, 17.10 and 17.11). Further, the relative share of each category was computed (area in percentage). This helps in identifying the different urban growth centres which then subsequently assist in correlating the results with Shannon's Entropy. The calculation of built-up density gave the distribution of the High, Medium and Low-density built-up clusters as categorized in the study area. High density of built-up refers to compact nature of the built-up theme, while medium density denotes relatively lesser compact built-up; whereas, low density referred to sparsely spread built-up areas. This revealed that the study area comprised of a smaller area (of more compact or highly dense in nature); and more dispersed or least dense built-up area. Results showed that the high and medium map density was found all along the highways and in the city centres. High map density was found within and closer to the cities; whereas, medium density was found mostly around the city periphery and along the highways. Further, the distribution of low density was observed in the study area due to the higher dispersion of the built-up in the study area. This further confirms the results of Shannon's Entropy, which denotes a high dispersion of the built-up feature in the study area.

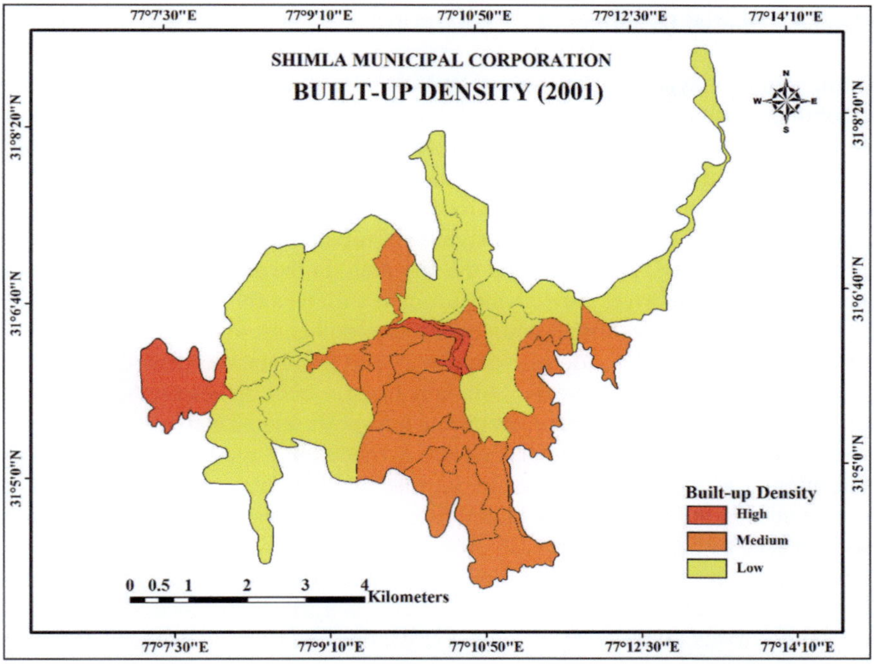

Fig. 17.10 Density of Urban Area of Shimla for 2001 (*Source* Landsat ETM+ 2001)

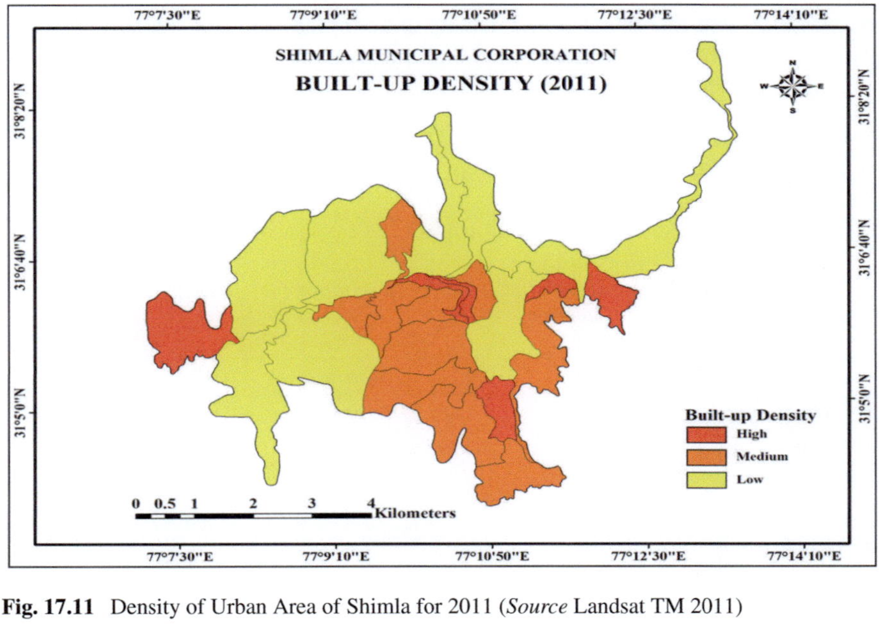

Fig. 17.11 Density of Urban Area of Shimla for 2011 (*Source* Landsat TM 2011)

17.4.3.3 Spatial Dynamics of Urban Sprawl

The modeling of dynamic urban sprawl and its future prediction is a greater challenge rather than its quantification; which gives ways to policy planning for city development. Although, different sprawl types were identified here, still there is inadequacy in developing mathematical relationships to define them. This necessitates the characterization of urban sprawl. This exercise helps in city planning and development of water resources projects and/or designing of urban drainage infrastructure along with other facility within and around the city area. In the present study, population and related densities were used as independent variables for simulation modelling of the urban sprawl. The Shimla Municipal Corporation is known to have undergone extremely fast area expansion in recent years due to an unprecedented population growth over a period of 20 years. Shimla is heavily built up across the hill slopes and characterized by mixed commercial and transports related activities. The public, semi-public, residential and other land use activities in the city have been mostly concentrated in the South. On comparing the satellite data sets of 1991–2011, it was found that the built up area in and around the Shimla Municipal Corporation has increased by 529.65 hectares (26.48%) over the period of 20 years (1991–2011). Urban expansion processes in the Shimla Municipal Corporation during the period of 20 years (1991–2011) were further evaluated by analyzing a distance decay concept from major roads (Table 17.9 and Fig. 17.12). For this purpose National Highway No.—NH 22, National Highway No.—NH 88, Summer hill road, Bharari road, Mall road, Ridge road, etc., have been taken for analysis. Urban expansion and four buffer zones were created along these roads with 0–50 m each. The result of the present analysis indicated that the area and density under urban land was decreasing while going the away from major roads.

Table 17.9 Urban Expansion along a major road of Shimla city in 1991–2011

Proximity to road (mts)	1991		2011		Change During 1991–2011	
	Urban area (km^2)	Urban density (urban area km^2)	Urban area (km^2)	Urban density (urban area km^2)	Urban area (km^2)	Rate of expansion of urban area (km^2/Year)
0–50	1.25	0.21	2.74	0.46	1.49	0.07
50–100	0.79	0.18	1.80	0.41	1.01	0.05
100–150	0.44	0.14	1.09	0.34	0.65	0.03
150–200	0.30	0.12	0.70	0.28	0.40	0.02

Sources Landsat TM and ETM+ satellite data (1991, 2011) and Google Earth Engine for roads (2011)

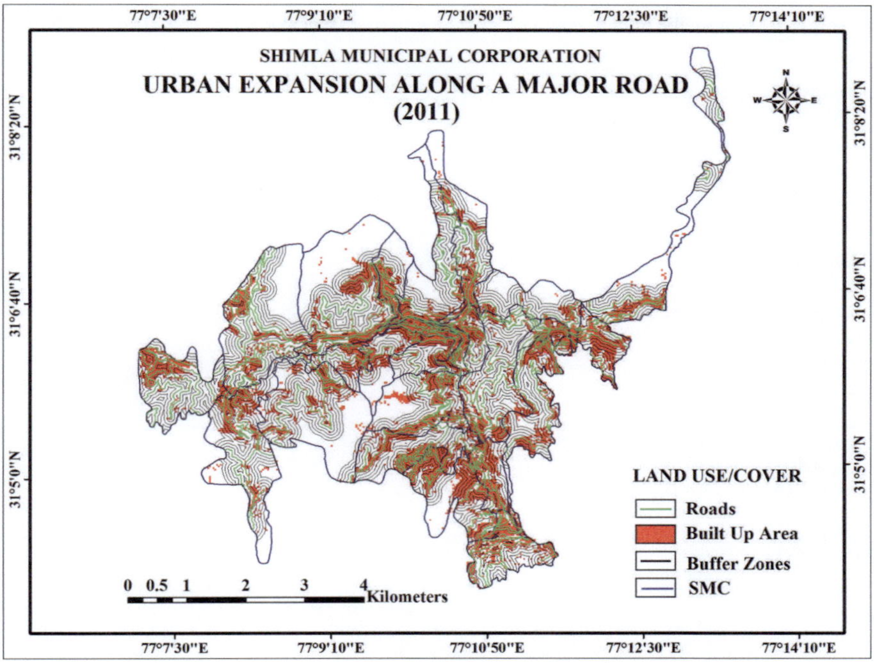

Fig. 17.12 Buffering along Major Roads of Shimla City (*Source* Landsat TM and Google Earth Engine of 2011)

17.4.4 Analysis of Landscape Metrics

Landscape metrics were used to quantify and measure the spatial patterns and structures of urban sprawl. Eight landscapes indices were used to quantify urban structure and form (Table 17.10). Class Area (CA) measures the total built-up growth of an area. It basically measures the amount of built-up pixels per time step. It is also possible to measure the percentage of the total area covered by the built-up class, by dividing the class area by the total area (Alexakis et al. 2011; Araya 2009; Herold et al. 2002; Xi et al. 2009). Spatial metrics and their variation were calculated for the built-up area (urban impervious area), which increased by 79.16% between 1991 and 2001. Similarly, Built-up growth was also increased by 37.68% from 2001 to 2011; however, the rate of growth was slower compared to previous decade. The Number of Patches (NP) is a simple quantification of the amount of individual built-up patches. It does not contain any information about the area, or the disparity of patches, or the density. However, it does provide information on the amount of new created patches in time (Alexakis et al. 2011; Araya 2009; Herold et al. 2002; Xi et al. 2009; Yu and Ng 2007). The NP in a landscape analysis indicates the aggregation or disaggregation in the landscape, while Large Patch Index (LPI) measures the proportion of total landscape area comprised by the largest urban patch. The NP (*i.e.*, built-up

Table 17.10 Landscape indices and their percentages of changes of Shimla city from 1991–2011

Metrics	1991	2001	2011	Δ%1991–2001	Δ%2001–2011	Δ%1991–2011
Class Area (CA)	361.080	646.920	890.730	79.162	37.687	146.684
Number of patches (NP)	696.000	932.000	784.000	33.908	−15.879	12.643
Largest patch index (LPI)	38.2263	38.227	38.226	0.002	−0.002	0
Edge density (ED)	34.1856	42.449	42.153	24.174	−0.698	23.307
Euclidian mean nearest Neighbour distance (EMN_MN)	117.791	104.045	108.588	−11.669	4.366	−7.813
Area weighted mean patch fractal dimension (FRAC_AM)	1.135	1.138	1.141	0.246	0.2811	0.528
Contagion	69.863	68.044	68.155	−2.604	0.1638	−2.445

blocks) increased (33.90%) between 1991 and 2001, which depicts the dispersed pattern of urban growth. The development of a number of isolated, fragmented, or discontinuous built-up areas occurred in the first period while in the second (i.e., 2001–2011) it was decreased by −15.87%. This suggests that small patches were transformed into larger patches and urbanization is made in agglomerated form.

Largest Patch Index is the ratio of the area of the largest patch in the landscape and the total landscape area. When the largest patch in the landscape is increasingly small, LPI approaches 0%. It reaches 100% when the complete landscape consists of a single patch. LPI increased by 0.002% between 1991 and 2001, thus indicating significant growth within the historical city core. Edge Density (ED) does in contrast to number of patches and patch density take into account the shapes of patches, the complexity of the shape of a patch. When new built-up areas emerge the edge density of the urban land use class should increase, a decrease is possible when areas agglomerate (Araya 2009; Alexakis et al. 2011; Eiden et al. 2012; Herold et al. 2002). ED increased by 24.174% and −0.698% between 1991–2001 and 2001–2011 period respectively, thus indicating an increase in the total length of the edge of the urban patches, as due to disintegration of land use pattern. The increasing trend in first two periods demonstrated the strong urban sprawl phenomena.

A decreasing trend in Euclidian Mean Nearest Neighbour Distance (ENN_MN) was −11.669% seen from 1991, which shows a reduction in the distance between the built-up patches, thus suggesting coalescence. The Area Weighted Mean Patch Fractal Dimension (AWMPFD) measures the shape of urban patches. Whereas, the edge density takes the complexity of the shapes of the patches into account. Relatively higher values were estimated during first phase of urban development (i.e., 2001–2011). Contagion index measures the extent to which a landscape is aggregated or clumped (dispersed or interspersed). The value of the contagion index ranges between 0 and 100. A value of 0 is obtained when patch types are well interspersed, scattered and not aggregated, referring to many small and dispersed patches. A value of 100 is obtained when maximum aggregation of patch types is witnessed, i.e., only a few large contiguous patch types do exist (Weijers 2012). Relatively high Contagion index values during first decadal period were obtained which depicts the landscape with few large, and contiguous patches (contagion is high because many cells are adjacent to each other in large and contiguous patches (many internal cells)) while lower values was obtained during the second decadal period. During this period the development was taken place in scattered and not aggregated form, referring to many small and dispersed patches. But during second period again relatively high contagion index values was seen because of conversion of small patches into large and contiguous patches.

17.4.5 Modelling of Urban Sprawl

Urban sprawl dynamics was modelled using the causal factors (Independent variables) such as Population Density (PD), α-Population Density (α-PD), β-Population Density (β-PD), Total workers (TW), Road density (RD), Female literacy rate (FLR), Sex ratio, (SR) and No of Household's (HH). The percentage built-up is the proportion of the built-up area to the total area of the ward. The α-population density denotes the proportion of the population in each ward to the built-up area of the respective ward. The β-population density (often referred as population density) is the proportion of population in each ward to the total area the respective ward. Since the present study mainly concentrated on built-up area, the percentage built-up, α and β population densities were computed and analyzed ward-wise. Ward-wise population data includes the information of some of the key factors of urban sprawl like total workers in secondary and tertiary sectors, and female literacy rate. The physical growth of city is mostly driven by its secondary and tertiary sectors. Hence, ward wise total worker in secondary and tertiary sectors were calculated. When the population of any region grows and become urbanized. Urban growth follows the major transportation nodes. So as to calculate the road density, ward wise road network has been digitized in GIS platform. With the help of these underlying factors, modelling of urban sprawl was carried out. To explore the possible relationship of percentage built-up (dependent variable) with causal factors of sprawl, regression analysis was carried out. Various regression analyses (Linear, Polynomial, Exponential and Logarithmic) were undertaken to determine the nature of significance of the causal factors (independent variables) on the sprawl, and quantified in terms of percentage built-up. The Regression analyses disclose the individual contribution of causal factors on the nature of urban sprawl. A variety of relationships and corresponding statistical parameters have been presented in Appendices 1–5. Some of the significant relationships are,

Population

$$y = -(0.098)\,x^2 + (1.436)\,x + 0.100 \tag{17.5}$$

y = percentage built-up, x = Population

Road Density

$$y = -(0.207)\,x + 4.826 \tag{17.6}$$

y = percentage built-up, x = RD

α-population density

$$y = (0.123)\,x^2 - (1.686)\,x + 8.408 \tag{17.7}$$

y = percentage built-up, x = α − opulation density

β – population density

$$y = (0.029)\,x^2 - (0.551)\,x + 5.473 \tag{17.8}$$

y = percentage built-up, x = β − P

Female Literacy

$$y = -(0.125)\,x^2 + (1.742)\,x - (0.601) \tag{17.9}$$

y = percentage built-up, x = Female Literacy

Total Workers

$$y = -(0.037)\,x^2 + (0.819)\,x + 1.407 \tag{17.10}$$

y = percentage built-up, x = Total Workers

Sex Ratio

$$y = -(0.086)x^2 + (1.546)x - 0.775 \tag{17.11}$$

y = percentage built-up, x = Sex Ratio

No of household's (HH)

$$y = -(0.072)\,x^2 + (1.127)\,x + 0.909 \tag{17.12}$$

y = percentage built-up, x = No of household's (*HH*)

Linear regression model evaluated here has been found to be the lower correlation coefficient Eq. (17.6) ($R^2 = 0.101$) to show the relationship between % built-up and road density as compared to linear, polynomial, logarithmic, exponential, distributions. Relationships between % built-up and β-population density have been found to be polynomial. Polynomial regression results show higher correlation coefficient ($R^2 = 0.178$). Relationships between percentage built-up and α-population density have been found to be linear with higher correlation coefficient ($R^2 = 0.349$). Polynomial regression model has higher correlation coefficient ($R^2 = 0.325$) to explain the relationship between percentage built-up and female literacy rate. Relationships between percentage built-up area and total works have been found to be polynomial regression with higher correlation coefficient ($R^2 = 0.191$). Relationships between percentage built-up area and population have been found to be linear with highest correlation coefficient ($r^2 = 0.059$). The linear and polynomial regression analyses reveal that the road density and population density have considerate influence on percentage built-up. To estimate the collective effects of causal factors, stepwise multivariate regression analysis was also carried out (Table 17.11). In the stepwise multivariate regression, it is assumed that the relationships between variables are linear. Considering all the causal factors in the stepwise regression, Appendix 5 indicates the highest correlation coefficient ($R^2 = 0.928$) which collectively explains the 92.08% variation in the urban growth. The α-population density of 90.04% represents the distinctive characteristic of Shimla city development.

Table 17.11 Model summary of stepwise regression

Model	R	R square	Increase in R square	Adjusted R square	Increase in adjusted R square	Std. error of the estimate
1	0.540	0.291	–	0.260	–	1.497
2	0.953	0.908	0.617	0.900	0.640	0.551
3	0.957	0.915	0.007	0.903	0.003	0.542
4	0.960	0.921	0.006	0.905	0.002	0.535
5	0.963	0.928	0.007	0.908	0.003	0.527
6	0.963	0.928	0	0.904	−0.004	0.540
7	0.963	0.928	0	0.899	−0.005	0.554
8	0.964	0.928	0	0.893	0.006	0.571

a. Predictors: (Constant), α-PD
b. Predictors: (Constant), α-PD, P
c. Predictors: (Constant), α-PD, P, No HH
d. Predictors: (Constant), α-PD, P, No HH, SR
e. Predictors: (Constant), α-PD, P, No HH, SR, FLR
f. Predictors: (Constant), α-PD, P, No HH, SR, FLR, β-PD
g. Predictors: (Constant), α-PD, P, No HH, SR, FLR, β-PD, RD
h. Predictors: (Constant), α-PD, P, No HH, SR, FLR, β-PD, RD, TW

17.5 Summery and Conclusion

Shimla city had experienced a tremendous rate of growth over the period of two decade. Satellite Remote Sensing is crucial in dealing the dynamic event, like urban sprawl phenomena. Without past remote sensing data, one may not be able to monitor past urban growth events and compare with the recent trend and predict the future growth pattern effectively over a time period at less time, in low cost and with better accuracy which is otherwise not possible to attempt through conventional mapping techniques. The application of Remote Sensing, GIS, spatial metrics and urban growth models provides novel approach for the study of spatio-temporal dynamics of urban sprawl. Considering the availability of past data sources from Landsat archive at medium resolution, demonstrated the capabilities for improving the knowledge, understanding and modelling of urban sprawl dynamics of Shimla city. This paper has presented a detailed analysis of 20 years of spatial urban growth pattern of Shimla city. The result shows that built-up area has increased from 361.05 ha in year 1991 to 890.7 ha in year 2011. This continuous increase in built-up area in Shimla city has outstripped the rate of population growth. The Shannon's entropy index and landscape metrics have quantified the urban sprawl. The entropy values obtained during the study period are closer to the upper limit of log n, i.e. 1.3413, showing the degree of dispersion of built-up area in the region, which is an indication of urban sprawl. All the parameter of spatial metrics has shown their inclination towards overall urban growth pattern in the study area. Multivariate regression analysis was performed in order to establish a relationship between urban sprawl and some of its contributory factors. Analysis of the contributory factors of urban growth collectively explains the 92.08% variation in the urban growth. Regression model developed here reveals that the most reasonable variable for the variation in built up area is α population density. These measures of spatial sprawl may even help in refining our understanding of this phenomenon. In addition, future development of Shimla city would also influenced by other contributory factors such as tourism initiatives, physical constrains, socio-economic change, population migration, future government investments corridors, smart city plan, development of small and medium towns around vicinity, scope of industrialization, and distances from major locations. These factors can be considered for urban growth modeling in future research work.

Acknowledgements The authors are thankful to the Vice Chancellor, Kumaun University, Nainital, Uttarakhand for providing facilities in SSJ Campus of University at Almora, where the present work has been done for Master Dissertation. Authors are also like to acknowledge the Director, G. B. Pant National Institute of Himalayan Environment and Sustainable Development, Kosi-Katarmal, Almora, Uttrakhand for providing facilities in Himachal Regional Centre of the Institute where we could modify and finalize the present work.

Appendix 1: Coefficients of Causal Factors and Percentage Built-Up by Linear Regression Analyses

Dependent variable (y)	Independent variable (x)	Equation ($y = ax + b$)	R	R^2	Adjusted R^2	Std. error of the estimate
Percentage Built-up	Population	$y = 0.460x + 2.161$	0.46	0.21	0.17	1.58
Percentage Built-up	α-population density	$y = -0.530x + 6.118$	0.54	0.29	0.26	1.50
Percentage Built-up	β-population density	$y = -0.228x + 4.910$	0.40	0.16	0.12	1.63
Percentage Built-up	Road Density	$y = -0.207x + 4.826$	0.32	0.10	0.06	1.69
Percentage Built-up	Total Workers	$y = 0.452x + 2.193$	0.43	0.19	0.15	1.60
Percentage Built-up	No. of Household's	$y = 0.398x + 2.408$	0.45	0.20	0.17	1.59
Percentage Built-up	Sex Ratio	$y = 0.834x + 0.662$	0.23	0.08	0.04	1.71
Percentage Built-up	Female Literacy	$y = 0.834x + 0.662$	0.49	0.24	0.21	1.55

Sources Landsat TM and ETM+ satellite data and Census of India (1991, 2001 and 2011)

Appendix 2: Coefficients of Causal Factors and Percentage Built-Up by Polynomial (Order = 2) Regression Analyses

Dependent variable (y)	Independent variable (x)	Equation ($y = ax^2 + bx + c$)	R	R^2	Adjusted R^2	Std. error of the estimate
Percentage Built-up	Population	$y = -0.098x^2 + 1.436x + 0.100$	0.51	0.26	0.19	1.57
Percentage Built-up	α-population density	$y = 0.123x^2 - 1.686x + 8.408$	0.59	0.35	0.29	1.45
Percentage Built-up	β-population density	$y = 0.029x^2 - 0.551x + 5.473$	0.42	0.18	0.10	1.65
Percentage Built-up	Road density	$y = -0.018x^2 + 0.016x + 4.351$	0.33	0.11	0.02	1.72
Percentage Built-up	Total workers	$y = -0.037x^2 + 0.819x + 1.407$	0.44	0.19	0.12	1.64
Percentage Built-up	No. of household's	$y = -0.072x^2 + 1.127x + 0.909$	0.49	0.24	0.17	1.59
Percentage Built-up	Sex ratio	$y = -0.086x^2 + 1.546x - 0.775$	0.28	0.08	−0.01	1.75

(continued)

(continued)

Dependent variable (y)	Independent variable (x)	Equation ($y = ax^2 + bx + c$)	R	R^2	Adjusted R^2	Std. error of the estimate
Percentage Built-up	Female literacy	$y = -0.125x^2 + 1.742x - 0.601$	0.57	0.31	−0.26	1.49

Sources Landsat TM and ETM+ satellite data and Census of India (1991, 2001 and 2011)

Appendix 3: Coefficients of Causal Factors and Percentage Built-Up by Logarithmic Regression Analyses

Dependent variable (y)	Independent variable (x)	Equation ($\log y = \log(a) + b \log x$)	R	R^2	Adjusted R^2	Std. error of the estimate
Percentage Built-up	Population	$y = 2.105 \ln(x) + 1.250$	0.48	0.23	0.20	1.56
Percentage Built-up	α-population density	$y = -2.265 \ln(x) + 6.934$	0.57	0.32	0.29	1.47
Percentage Built-up	β-population density	$y = -0.935 \ln(x) + 5.032$	0.42	0.18	0.14	1.61
Percentage Built-up	Road density	$y = -0.805 \ln(x) + 4.984$	0.26	0.07	0.03	1.72
Percentage Built-up	Total workers	$y = 1.915 \ln(x) + 1.483$	0.45	0.17	0.14	1.62
Percentage Built-up	No. of household's	$y = 1.777 \ln(x) + 1.723$	0.47	0.22	0.19	1.57
Percentage Built-up	Sex ratio	$y = 3.322 \ln(x) - 0.571$	0.28	0.08	0.09	1.71
Percentage Built-up	Female literacy	$y = 2.268 \ln(x) + 1.043$	0.53	0.28	0.25	1.51

Sources Landsat TM and ETM+ satellite data and Census of India (1991, 2001 and 2011)

Appendix 4: Coefficients of Causal Factors and Percentage Built-Up by Exponential Regression Analyses

Dependent variable (y)	Independent variable (x)	Equation ($y = be^{ax}$)	R	R^2	Adjusted R^2	Std. error of the estimate
Percentage Built-up	Population	$y = 2.276e^{0.118x}$	0.45	0.20	0.17	0.41

(continued)

(continued)

Dependent variable (y)	Independent variable (x)	Equation ($y = be^{ax}$)	R	R^2	Adjusted R^2	Std. error of the estimate
Percentage Built-up	α-population density	$y = 6.455e^{-0.143x}$	0.56	0.32	0.29	0.38
Percentage Built-up	β-population density	$y = 4.722e^{-0.065x}$	0.44	0.19	0.16	0.41
Percentage Built-up	Road density	$y = 4.644e^{-0.061x}$	0.36	0.13	0.09	0.43
Percentage Built-up	Total workers	$y = 2.328e^{0.112x}$	0.41	0.17	0.13	0.42
Percentage Built-up	No. of Household's	$y = 2.421e^{0.102x}$	0.45	0.20	0.17	0.43
Percentage Built-up	Sex ratio	$y = 1.465e^{0.228x}$	0.29	0.09	0.05	0.44
Percentage Built-up	Female literacy	$y = 2.209e^{0.125x}$	0.45	0.24	0.21	0.40

Sources Landsat TM and ETM+ satellite data and Census of India (1991, 2001 and 2011)

Appendix 5: Stepwise Regression Equations

S.N.	Stepwise Regression Equations	R^2
1	pc built-up = 6.118 − 0.530 α-PD	0.29
2	pc built-up = 4.131 − 0.921 α-PD + 0.888 P	0.91
3	pc built-up = 3.835 − 0.929 α-PD + 1.553 P − 0.583 HH	0.92
4	pc built-up = 2.790 − 0.907 α-PD + 1.563 P − 0.599 HH + 0.245 SR	0.92
5	pc built-up = 0.819 − 0.905 α-PD + 2.320 P − 0.375 HH + 0.766 SR − 1.010 FLR	0.93
6	pc built-up = 1.010 − 0.892 α-PD + 2.231 P − 0.360 HH + 0.717 SR-0.937 FLR − 0.012 β-PD	0.93
7	pc built-up = 0.783 − 0.869 α-PD + 2.256 P − 0.342 HH + 0.740 SR-0.968 FLR-0.040 β-PD + 0.028 RD	0.93
8	pc built-up = 0.867 − 0.874 α-PD + 2.395 P − 0.352 HH + 0.724 SR-1.024 FLR-0.041 β-PD + 0.030 RD-0.075 TW	0.93

Sources Landsat TM and ETM + satellite data and Census of India (1991, 2001 and 2011)

References

Alexakis D, Hadjimitsis D, Agapiou A, Themistokleous K, Retalis A (2011) Monitoring urban land cover with the use of satellite remote sensing techniques as a means of flood risk assessment in Cyprus. Sens Syst Next-Gener Satell XV, vol 8176. https://doi.org/10.1117/12.898185

Araya YH (2009) Urban land use change analysis and modeling: a case study of Setúbal and Sesimbra, Portugal. Master thesis, Institute for Geinformatics, University of Münster

Bailey KD (2009) Entropy systems theory. Systems Science and Cybernetics, Eolss Publishers, Oxford, UK, pp 152–169. http://www.eolss.net/ebooks/Sample%20Chapters/C02/E6-46-01-04.pdf. Accessed 1 Dec 2008

Barnes KB, Morgan JM III, Roberge MC, Lowe S (2001) Sprawl development: its patterns, consequences, and measurement. Towson University, Towson. http://www.chesapeake.towson.edu/landscape/urbansprawl/download/Sprawl.whitepaper.pdf

Batty M, Howes D (2001) Predicting temporal patterns in urban development from remote imagery. In: Donnay JP, Barnsley MJ, Longley PA (eds) Remote sensing and urban analysis. Taylor and Francis, London, pp 185 204

Batty M, Xie Y, Sun Z (1999) The dynamics of urban sprawl. Working paper series, paper 15, centre for advanced spatial analysis, University College, London

Cabral P, Geroyannis H, Gilg JP, Painho M (2005) Analysis and modelling of land-use and land-cover change in Sintra-Cascais area In: Proceedings of 8th agile conference, Estoril, Portugal

Census of India (2001) http://www.censusindia.net

Census of India (2011) http://www.censusindia.net

Cheng J, Masser I (2003) Urban growth pattern modelling: a case study of Wuhan City, PR China. Landsc Urban Plan 62:199–217

Civco DL, Hurd JD, Wilson EH, Arnold CL, Prisloe M (2002) Quantifying and describing urbanizing landscapes in the Northeast United States. Photogramm Eng Remote Sens 68(10):1083–1090

Clarke KC, Parks BO, Crane MP (2002) Geographic information systems and environmental modeling. Prentice Hall, New Jersey

Clarke KC, Hoppen S, Gaydos L (1996) Methods and techniques for rigorous calibration of a cellular automation model of urban growth. In: Proceedings of the third international conference/workshop on integrating GIS and environmental modelling, Santa Fe, New Mexico, 21–25 Jan 1996

Donnay JP, Barnsley MJ, Longley PA (2001) Remote sensing and urban analysis. In: Donnay JP, Barnsley MJ, Longley PA (eds) Remote sensing and urban analysis. Taylor and Francis, London, pp 3–18

Eiden G, Kayadjanian M, Vidal C (2012) Capturing landscape structures: tools. The European Commission. http://ec.europa.eu/agriculture/publi/landscape/ch1.htm. Accessed 8 June 2015

Epstein J, Payne K, Kramer E (2002) Techniques for mapping suburban sprawl. Photogramm Eng Remote Sens 63(9):913–918

Gomarasca MA, Brivio PA, Pagnoni F, Galli A (1993) One century of land use changes in the metropolitan area of Milan (Italy). Int J Remote Sens 14(2):211–223

Green K, Kempka D, Lackey L (1994) Using remote sensing to detect and monitor land/cover and land/use change. Photogramm Eng Remote Sens 60:331–337

Haack BN, Rafter A (2006) Urban growth analysis and modelling in the Kathmandu valley, Nepal. Habitat Int 30(4):1056–1065

Herold M, Clarke KC, Scepan J (2002) Remote sensing and landscape metrics to describe structures and changes in urban land use. Environ Plan A 34:1443–1458

Herold M, Menz G, Clarke KC (2001) Remote sensing and urban growth models demands and perspectives. In: Symposium on remote sensing of urban areas, Regensburg, Germany, vol 35

Hurd JD, Wilson EH, Lammey SG, Civco DL (2001) Characterisation of forest fragmentation and urban sprawl using time sequential Landsat Imagery. In: Proceedings of the ASPRS annual convention, St. Louis, MO, 23–27 Apr

ICIMOD (1999) Integration of GIS, remote sensing and ecological methods for biodiversity inventory and assessment. In: Issues in mountain development. http://www.icimodgis.net/web/publications/IssuesMountain61999.pdf

Jantz CA, Goetz, Scott J (2005) Analysis of scale dependencies in an urban land/use- change model. Int J Geogr Inf Sci 19(2):217–241

Jat MK, Garg PK, Khare D (2008) Modeling of urban growth using spatial analysis techniques: a case study of Ajmer city (India). Int J Remote Sens 29(2):543–567

Jat MK, Garg PK, Khare D (2007) Monitoring and modeling urban sprawl using remote sensing and GIS techniques. Int J Appl Earth Obs Geoinf 10(1):26– 43

Jensen JR, Cowen DC (1999) Remote sensing of urban/suburban infrastructure and socio economic attributes. Photogramm Eng Remote Sens 65(5):611–622

Jha R, Singh VP, Vatsa V (2008) Analysis of urban development of Haridwar, India, using entropy approach. J Civil Eng 12(4):281–288

Joshi PK, Lele N, Agarwal SP (2006) Entropy as an indicator of fragmented landscape. Curr Sci 91(3):276–278

Jothimani P (1997) Operational urban sprawl monitoring using satellite remote sensing: excerpts from the studies of Ahmedabad, Vadodara and Surat, India. Paper presented at 18th Asian conference on remote sensing held during 20–24 Oct 1997, Malaysia

Kumar JAV, Pathan SK, Bhanderi RJ (2007) Spatio-temporal analysis for monitoring urban growth-a case study of Indore city. J Indian Soc Remote Sens 35(1):11–20

Lata KM, Sankar Rao CH, Krishna Prasad V, Badrinath KVS, Raghavaswamy (2001). Measuring urban sprawl: a case study of Hyderabad. GIS Dev 5(12):22–25

Lo CP, Yang X (2002) Drivers of land-use/land-cover changes and dynamic modelling for the Atlanta, Georgia metropolitan area. Photogramm Eng Remote Sens 68(10):1062–1073

Pontius RG Jr, Claessens L, Hopkinson C Jr, Marzouk A, Rastetter EB, Schneider LC, Vallino J (2000) Scenarios of land-use change and nitrogen release in the Ipswich watershed, Massachusetts, USA. Paper presented at 4th international conference on integrating GIS and environmental modelling (GIS/EM4): problems, prospects and research needs. Banff, Alberta, Canada, 2–8 Sept 2000. http://www.colorado.edu/research/cires/banff/pubpapers/165/

Punia M, Singh L (2012) Entropy approach for assessment of urban growth: a case study of Jaipur, India. J Indian Soc Remote Sens 40(2):231–244

Punia, M., and Singh, L., (2011), Entropy Approach for Assessment of Urban Growth: A Case Study of Jaipur, India. J Indian Soc Remote Sens 1–14. https://doi.org/10.1007/s12524-011-0141-z

Rawat JS, Kumar M (2015) Monitoring land use/cover change using remote sensing and GIS techniques: a case study of Hawalbagh block, district Almora, Uttarakhand, India. Egypt J Remote Sens Space Sci http://dx.doi.org/10.1016/j.ejrs.2015.02.002

Sierra Club (1998) The dark side of the American dream: the costs and consequences of suburban sprawl. (http://www.sierraclub.org/sprawl/report98/)

Sudhira HS, Ramachandra TV, Jagadish KS (2004) Urban sprawl: metrics, dynamics and modelling using GIS. Int J Appl Earth Obs 5:29–39

Sudhira HS, Ramachandra TV, Jagadish KS (2003) Urban sprawl pattern recognition and modelling using GIS. Paper presented at Map India, 2003, New Delhi, 28–31 Jan 2003

The Regionalist (1997) Debate on theories of david rusk, vol 2, no 3. http://scpp.ubalt.edu/public/regional/detail.htm. issue3v2

Torrens PM, Alberti M (2000) Measuring sprawl. Working paper no. 27, Centre for advanced spatial analysis, University College, London. http://www.casa.ac.uk/workingpapers/

Torrens PM, O' Sullivan D (2001) Cellular automata and urban simulation: where do we go from here? Environ Plann B 28:163–168

Trani MK, Giles RH (1999) An analysis of deforestation: metrics used to describe pattern change. For Ecol Manage 114(2):459–470

Waddell P (2002) Urban Sim: modelling urban development for land use, transportation and environmental planning. J Am Plann Assoc 68(3):297–314

Weijers D (2012) The suitability of GIS methods for analyzing urban sprawl, and the influence of scale. Master degree thesis geographical information management & applications, Utrecht University

Weng Q (2001) Modelling urban growth effects on surface runoff with the integration of remote sensing and GIS. Environ Manage 28(6):737–748

Xi F, He HS, Hu Y, Wu X, Bu R, Chang Y, Liu M, Yu J (2009) Simulate urban growth based on RS, GIS, and SLEUTH model in Shenyang-Fushun metropolitan area north-eastern China. Urban Remote Sens Jt Event 1(3):959–968

Yang X, Liu Z (2005) Use of satellite derived landscape imperviousness index to characterize urban spatial growth. Comput Environ Urban Syst 29:524–540

Yang X, Lo CP (2003) Modelling urban growth and landscape changes in the Atlanta metropolitan area. Int J Geogr Inf Sci 17(5):463–488

Yeh AGO, Li X (2001) Measurement and monitoring of urban sprawl in a rapidly growing region using entropy. Photogramm Eng Remote Sens 67(1):83–90

Yu XJ, Ng CN (2007) Spatial and temporal dynamics of urban sprawl along two urban-rural transects: a case study of Guangzhou, China. Landsc Urban Plan 79:96–109

Chapter 18
Impact of Corridor Development on Rural Settlements Along Delhi–Jaipur Axis

Awadh Narayan Choubey and Taruna Bansal

Abstract Over the years several scholars and philosophers have done large amount of work on the rural—urban linkages as well as on the spatial pattern of non-agricultural employment. This has been attempted to understand the changes that both rural and urban settlements have witnessed over the years and to analyze the factors responsible for these changes. The result of these modifications has been that the line between urban and rural has not only thinned but has even blurred in some cases. In developing countries like India, though there may be some exceptions. This is because even today the distinction is clear when these settlements are examined on the dimensions of economy, occupational structure, levels of education, accessibility to services, demography, political set up and levels of migration. Here urban population is still defined as those contained within urban administrative regions. Although the peripheral regions are under the urban influence the degree varies from place to place in the context of concentration and accessibility. The result is rural transformation. One needs to look into this transformation and examine the pattern as this has lead the transition from the rural dominated to an urban dominated world. In this context the present paper attempts to study the process of rural transformation along the Delhi–Jaipur axis using a geospatial approach. The purpose is to understand the impact of growth of urban centers on rural settlements along this axis from 1991 to 2017.

Keywords Rural transformation · Urban growth · Delhi · Jaipur · Geospatial technology

---

A. N. Choubey · T. Bansal (✉)
Department of Geography, Jamia Millia Islamia, New Delhi, India
e-mail: tbansal@jmi.ac.in

A. N. Choubey
e-mail: awadhonline@gmail.com

S. Sahdev et al. (eds.), *Geoecology of Landscape Dynamics*,
Advances in Geographical and Environmental Sciences,
https://doi.org/10.1007/978-981-15-2097-6_18

18.1 Introduction

In the 1940s, the United Nations in their statistical reports adopted the categorization of areas into either rural or urban. This approach was based on the premise that to understand the social, economic, political, cultural and demographic processes one has to consider the place where human beings live and work. The most basic categorization put forward was that of rural and urban areas within the settlement system of human beings. In recent decades, this categorization has lost its importance as massive changes have occurred in the size, spatial extent, nature and morphology of the settlements. For some while, these changes have been deliberated upon in detail as because of these modifications the line between urban and rural has not only thinned but has even blurred in some cases. In fact in present times there is no clear dividing line between the rural and urban settlement and their dwellers. These two have intertwined in such a manner that instead of being isolated settlements they have become dependent on each other. Gottmann was among the first to recognize this phenomenal change when he gave the concept of 'Megalopolis' in 1961. Today there exist vast array of alternative approaches which analyze these changes to monitor, plan and provide basic services in these settlements.

In spite of these transformations, in developing countries like India, the spatial context is a matter of concern and importance in examining the demographic characteristics of the population and their behavior. This is because even today the distinction is clear when observed through this dichotomous approach where urban and rural settlements are examined on the dimensions of economy, occupational structure, levels of education, accessibility to services, accessibility to information, demography, political set up, ethnicity and levels of migration. Here urban population is defined as those contained within urban administrative regions. Although the peripheral regions are under the urban influence the degree varies from place to place in the context of concentration and accessibility. The result is rural transformation. One needs to look into this transformation and examine the pattern as this has lead the transition from the rural dominated to an urban-dominated world.

Rural transformation should not be analyzed only as a demographic change rather it is a cultural phenomenon. This is because it is a form of cultural convergence and thus has impact on both types of settlements.

18.2 Objectives

Conventional regional growth theory suggests that industrial growth in major cities result in instrumental changes in its hinterland. The development and economic advancements must 'trickledown' to rural areas which should bring dynamism in the economic structure of the 'traditional peripheries over time. But, this has not occurred in India. Rather, here, cities have prospered by exploiting the resources of their peripheries leaving them with economy characterized with backward and subsistence activities. In this framework, this paper has the following objectives:

(a) To understand the settlement structure of the rural settlements in the Delhi–Jaipur axis.
(b) To comprehend the process of rural transformation along the Delhi–Jaipur axis and the reasons for the same.

18.3 Study Area

In this paper, Delhi–Jaipur axis has been analyzed from geographical point of view considering the following factors—distribution of urban places of various orders, levels of development and land cover changes. The details of the study area are given below

National Highway—8
Distance (in Kms)—265
Districts—4
Villages—219 (in 2011)
Urban centers (Class I–VI)—59 (in 2011).

18.4 Database and Methodology

This paper examines the settlement structure of the rural settlements in the Delhi–Jaipur axis and analyses how over time the process of urbanization has transformed the nature of rural settlements in the axis. In this analysis the rural transformation has been examined for three census years—1991, 2001 and 2011. This has been done with the help of composite indices computed for demographic parameters, workforce participation and infrastructural facilities. The following indicators have been used for the above calculations.

18.4.1 Demography

(a) Density of population,
(b) Sex-ratio,
(c) Sex-ratio (0–6),
(d) Percentage of Scheduled Population to the total population,
(e) Female literacy Rate.

18.4.2 Economy

(f) Crude workforce participation rate (percentage of workers to the total population),
(g) Crude female workforce participation rate (percentage of female workers to the total female population),
(h) Proportion of workers in agricultural activities (percentage of workers as cultivators and agricultural labourers to the total workers),
(i) Proportion of workers in household industry (percentage of workers in the household industries to the total workers),
(j) Proportion of workers in non-agricultural activities {percentage of workers in the other activities (apart from cultivators, agricultural labourers and household industries) to the total workers}.

18.4.3 Infrastructure

(k) Availability of medical facilities in the villages (Yes/No),
(l) Presence of educational institutions (Yes/No),
(m) Accessibility to drinking water facility (Yes/No),
(n) Availability of post office facility (Yes/No).

The purpose of this exercise is to analyze the characteristics and changes that have occurred in the rural settlements of this Axis in the last three decades. Apart from these three indices, an index has been computed to analyse the levels of development (taking all the fourteen indicators); to comprehend the changes that the villages have undergone over the last three decades. The development index also has been computed for the axis.

18.5 Analysis and Discussion

Over the past half century India's urban system has undergone tremendous change which has resulted in alterations within the regional urban dynamics (Denis and Marius-Gnanout 2011). These transformations have been led by the megacities of Mumbai, Delhi and Kolkata. In the present study also the transformations have been influenced by the dynamic character of the Delhi metropolis. But when analyzed at the meso level that is at the level of each individual axes it becomes crystal clear that in the present era of decentralization and economic liberalization the role of hierarchies in the regional urban systems has diluted. Though there may be some exceptions, especially where regional hierarchies have been perturbed by some mega metropolises which play dominant role in the economic setting of the country as a whole.

Whatever the case may be the process of agglomeration associated with urbanization extends well beyond the mega and metro cities. It has spilled to the small and medium towns as well as to the village agglomerates located in their hinterland. This process is changing the organization of the space as rural settlements are getting access to opportunities and services due to their proximity to the urban areas. Similarly, this axis has its own regional dynamics which have been discussed here.

18.5.1 Rural Transformation Along the Delhi–Jaipur Axis

This axis is of vital importance in present times as it is part of the ambition Delhi–Mumbai Industrial Corridor. It is connected with National Highway 8 through four districts—Gurgaon, Rewari, Alwar and Jaipur. The two cities of Delhi and Jaipur are at the distance of 265 kms. The axis consists of 59 urban centres and 220 rural settlements in 2011 (Table 18.1). The most interesting part of this axis is that it has two primate cities at the two ends. Over time Delhi has emerged as the Primate City in the northern part of the country. It is not surprising that the city constitutes nearly 66.2% of the population residing in the National Capital Region. The mushrooming growth of Delhi over the years seems to be in conflict with heritage; this has in some parts restricted the process of development in terms of land use as well as built space. The result has been that population as well as economic activities spilled in the nearby towns. Another important aspect is that there was government intervention in reducing the pressure from Delhi and various towns were selected to bear this burden. Initially these were called Satellite towns of Delhi which later were recognized as DMA towns.

Gurgaon is one such town which is located on this axis. Over the period of time the town has attained the position of industrial hub which holds a significant position in the National Capital Region (NCR). This has happened basically due to its close proximity to Delhi. Its potential gave into notice in the decade of 1980s when the Haryana Government initiated the process of industrial development here. Moreover, the economic liberalization brought involvement of private builders; this transformed

Table 18.1 Distribution of rural settlements in the Delhi–Jaipur axis (1991–2011)

Axis	1991	2001	2011	Number of villages transformed into urban centers from 1991 to 2011
Delhi–Jaipur axis (NH 8)	225	225	219 (6)	6 (2.7%)*

Source Data computed from Census of India 1991, 2001 and 2011
Note The data in parenthesis denotes the number of villages that have transformed into urban centers over the next census year and the data in the parenthesis* denotes the percentage of villages transformed

the entire landscape of the town as now it was dotted with posh residential and commercial complexes. The result was high rate of migration especially of highly skilled professionals. In spite of these developments somehow its growth has been in the shadow of giant metropolis, Delhi, which has witnessed dichotomous development. On one side there is ultra modern residential and industrial development and on the other side—especially the core areas, is plagued with the problems as these areas could not integrate themselves with the overall process of development that Delhi has witnessed (Denis and Marius-Gnanout 2011). As a result it has not spread growth impulses in its neighbouring hinterland.

Jaipur also known as the 'Pink City' has the locational advantage as it lies approximately 250 km away from both New Delhi and Agra and forms the 'Golden Triangle' tourist circuit. It also has a growing IT sector with approximately 50% of its working population engaged in the tertiary sector. Surprisingly, in spite of these advantages the city has grown as a primate city and thus has failed to spread growth impulses in its rural hinterland. This is evident from the fact that from 1991 to 2011 not a single village in the vicinity of Jaipur city was transformed into an urban centre.

There have been some developments at two different stretches on this axis. One such development has occurred near Neemrana, where Japanese Industrial Zone was set up in 2008 under the Special Economic Zones (SEZ) scheme of the Government of India with Japanese collaboration. Twenty seven Japanese companies specially related to automobiles and auto ancillaries set up their industrial units and have a future plan of developing his industrial zone by providing residential and other services in its vicinity. This has resulted in the growth and development of some villages in its vicinity.

The second such development has occurred near Manesar from where the ***Delhi Western Peripheral Expressway*** *or* ***Kundli–Manesar–Palwal Expressway*** *pas*ses. This expressway was first proposed in 2003 as of now the Palwal–Manesar stretch is operational due to which some villages near this expressway have shown signs of development. These include Ghatal Mahaniwas, Dharuhera and Maheshwari. One can say that this stretch is also coming up as a small industrial hub as the Haryana Government has envisaged the process of development here through various land uses which can be aptly described as Fashion City, Knowledge City, Medi City, Leisure and Entertainment Cities, Cyber City and Dry-Port facility. These have been put forward and brought simultaneously with other developments in this region (Fig. 18.1).

The above account clearly depicts that the development of rural settlements in this axis is restricted to some stretches mainly because both Delhi and Jaipur have emerged as Primate Cities over time. Whatever changes have occurred is the result of special incentives taken by the government.

18.5.2 Demographic Index Along the Delhi–Jaipur Axis

In this section a link has been tried to find by analyzing the basic demographic indicators of the rural settlements and the changes they have witnessed in last twenty

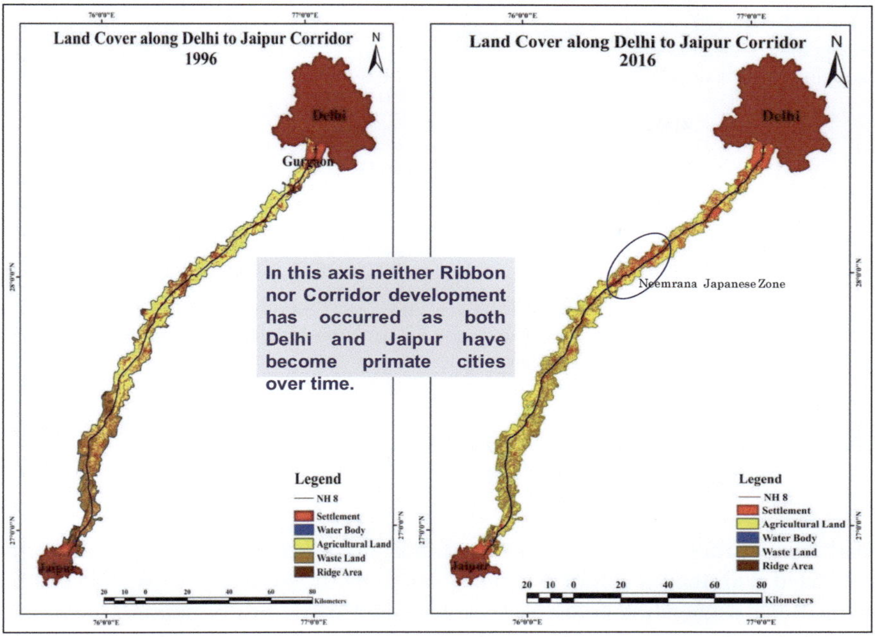

Fig. 18.1 Land cover changes along the Delhi–Jaipur axis (1996–2016). *Source* Landsat 1-5 MSS C1 Level 1 1996 and Landsat 8 OLI/TIRS C1 Level 1 2016

years. Along the NH 8 the growth pattern of urban centers as well as transformation process between Dharuheda and Amber is negligible. The development has occurred either near Gurgaon or Manesar or Neemrana. And interestingly in all these three cities there have been government interventions to bring growth and splurge their spread effects in the neighbouring hinterland. Gurgaon in recent years has been developed as DMA town; Manesar has been linked with the Kundli–Manesar–Palwal expressway and Neemrana has developed due to the establishment of the Neemrana Japanese Industrial Zone.

In this case the argument is rooted in the premise given by Batty (2001) who states '*history matters*'. Most of the urban centers on this axis are still dependent on traditional industry which leads to some kind of persistence in urban development. The data in Table 18.2 depict similar scenario as the villages show stagnation as a large proportion of villages (between 50 and 60%) are still in the low category of 2.51–5.00.

Table 18.2 Index of demographic indicators for the Delhi–Jaipur axis (NH8)

Demographic index	1991	2001	2011
More than 10	0.00 (0)	0.44 (1)	0 (0)
7.51–10.0	1.78 (4)	1.78 (4)	2.28 (5)
5.01–7.50	44.00 (99)	36.00 (81)	36.53 (80)
2.51–5.00	53.33 (120)	61.33 (138)	61.19 (134)
Less than 2.5	0.89 (2)	0.44 (1)	0 (0)
Total	100.0 (225)	100.0 (225)	100.0 (219)

Source Data computed from Census of India 1991, 2001 and 2011
Note The data in the parenthesis denote the number of villages in each category

18.5.3 Workforce Participation Index Along the Delhi–Jaipur Axis

Workforce participation is an important indicator of economy, which is directly associated with the growth of a region's economy. In the words of Koppel (1992) occupational segregation is an essential element of rural transformation. One should know that rural transformation is much more than agrarian transformation. It should be looked upon as a social, cultural and political mobilization which has cultural expression.

Some exceptional change has occurred in the Delhi–Jaipur axis from 2001 to 2011 where some villages have moved to the higher category. This can be the result of the growth of Gurgaon as the Delhi Metropolitan Town and establishment of the Neemrana Japanese Industrial Zone. These two initiatives have attracted a huge fraction of workforce from the neighbouring hinterland. The growth of service sector in Gurgaon can be another reason for this shift (Table 18.3).

18.5.4 Infrastructural Facilities Index Along the Delhi–Jaipur Axis

In a Socialist country like India the distribution, accessibility and availability of social and economic overheads like medical, educational, postal, basic amenities, transport and communication facilities is of paramount importance. The relationship between infrastructural facilities and urbanization are a mirror of a region's urban environment. The development process and the growth of economy are directly linked with these amenities. The positive relation between these two is crystal clear in the study area. The data in Table 18.4 makes it apparent. The rural settlements in along the axis have shown some modifications. A large number of villages in all these axes have moved into the higher category.

Table 18.3 Index of workforce participation for the Delhi–Jaipur axis (NH8)

Workforce participation index	1991	2001	2011
More than 10	2.22 (5)	2.11 (7)	1.83 (4)
7.51–10.0	8.00 (18)	1.78 (4)	4.11 (9)
5.01–7.50	33.78 (76)	18.67 (42)	28.77 (63)
2.51–5.00	48.89 (110)	76.00 (171)	65.30 (143)
Less than 2.5	7.11 (16)	0.44 (8)	0.00 (0)
Total	100.0 (225)	100.0 (225)	100.0 (219)

Source Data computed from Census of India 1991, 2001 and 2011
Note The data in the parenthesis denote the number of villages in each category

Table 18.4 Index of infrastructural facilities for the Delhi–Jaipur axis (NH8)

Infrastructure facilities index	1991	2001	2011
More than 10	0.00 (0)	0.00 (0)	4.57 (10)
7.51–10.0	0.00 (0)	8.89 (20)	9.14 (20)
5.01–7.50	12.00 (27)	13.33 (30)	18.72 (41)
2.51–5.00	21.78 (49)	24.44 (55)	26.48 (58)
Less than 2.5	66.22 (149)	53.34 (120)	41.10 (90)
Total	100.0 (225)	100.0 (225)	100.0 (219)

Source Data computed from Census of India 1991, 2001 and 2011
Note The data in the parenthesis denote the number of villages in each category

Table 18.5 Index of development for the Delhi–Jaipur axis (NH8)

Levels of development	1991	2001	2011
Above 25	0.00 (0)	0.44 (1)	3.18 (7)
20.1–25.0	0.89 (2)	0.89 (2)	1.36 (3)
15.1–20.0	8.44 (19)	11.56 (26)	16.82 (37)
10.1–15.0	15.56 (35)	69.77 (157)	72.27 (159)
Below 10.0	75.11 (169)	17.78 (40)	5.45 (12)
Total	100.0 (225)	100.0 (225)	100.0 (220)

Source Data computed from Census of India 1991, 2001 and 2011
Note The data in the parenthesis denote the number of villages in each category. Jaipur has emerged as a Primate city which has hampered the development process in this axis. But still there is some development as nearly 70% of villages have moved from lowest level to the next level (Table 18.5)

The above analysis on all the indicators through composite indices had a purpose. The aim was to bring out a pattern of development as well as to understand the performance of basic demographic indicators, workforce participation and availability of the infrastructural facilities at the village level in this axis. The analysis of these indicators therefore shows that the process of development in this region is not dependent on one particular factor. It is a conglomeration of all the factors which even include their location and proximity to urban centers. Moreover, these developmental indicators have a high degree of mutual inter-dependence. The development index computed using all the 14 indicators is a step towards this understanding.

18.6 Conclusions

In older days it was easy to define city as it had a defensive wall around it. In due course of time, the wall was diminished and the city expanded its physical limits. In more recent times, the settlement system has evolved in such a manner that the distinctive line between urban and rural has not only thinned rather number of rural settlements have either merged with urban areas or have been engulfed by them. These transformations have made distinct influence on the socio-cultural and economic fabric of the urban areas as well as its rural hinterland. Similar transformations have been experienced by all the villages along the Delhi–Jaipur Axis from 1991 to 2011.

Jaipur has evolved as primate city and primacy has a nonlinear relationship with growth and development of a region. In this axes the development has only occurred in two stretches: one near Gurgaon and Manesar (development plans initiated by government) and another near Neemrana where Neemrana Japanese Industrial Zone has been established. The performance of demographic and workforce participation indicators is synchronized with these findings. In the case of infrastructural facilities the results are varied as infrastructure is directly linked with the process of urbanization. Since all the villages are located on National Highways they show some sort of growth. Moreover there have been number of schemes of the government to provide these basic amenities in the villages. Proximity to Delhi has somehow favoured these rural settlements in gaining better infrastructural facilities. The data clearly shows that in 2011 most of the villages had educational and medical institutions as well as drinking water. Availability of post office was not common.

The entire analysis clearly brings out the relationship between urbanization and rural transformation. It would be right to say that rural transformation is much more than a demographic phenomenon. The changes are even socially and culturally expressed through occupational differentiation and participation of females in the labour market.

Larger cities like Delhi and Jaipur should not be ignored as they have larger economies and are likely to produce innovations. Smaller cities must depend more on adopting these innovations rather than diffuse from others. These innovations may lead to urban corridors only if a city does not become a primate city.

References

Batty M (2001) Polynucleated urban landscapes. Urban Stud 38(4):635–655

Denis E, Marius-Gnanout K (2011) Towards a better apprasila of urbanization in India: a fresh look at the landscape of morphological agglomerates. Working Paper Series 2, USR 3330, Savoirs et Mondes Indiens, Institut Francias de Pondicherry, Pondicherry and Centre de Sciences Humaines, New Delhi

General R (1991) Census of India 1991. Provisional population totals. New Delhi: Government of India

General R (2001) Census of India 2001. Provisional population totals. New Delhi: Government of India

General R (2011) Census of India 2011. Provisional population totals. New Delhi: Government of India

Koppel B (1992) The rural–urban dichotomy reexamined: beyond the ersatz debate. In: Ginsburg N et al (eds) The extended metropolis: settlement transition in Asia. University of Hawaii Press, Honolulu, pp 47–67

Priemus H, Zonneveld W (2003) What are corridors and what are the issues? Introduction to special issue: the governance of corridors. J Transp Geogr 11:167–177

Chapter 19
Analysis of Land Use/Land Cover Change Detection Using Remote Sensing and GIS of Fatehgarh Nau Abad Village, Bathinda, Punjab

Balwinder Singh and Chander Gagandeep Singh

Abstract Most of the Indian population is residing in villages to meet their basic necessities by cultivating agriculture as their main occupation. Land witnesses changes due to population increase and socio-economic activities, along with showing the sphere of influence of an urban centre on the nearby areas with the passage of time. As land use/land cover is dynamic in nature so, it becomes very important to understand how villages have changed and are changing its land use/land cover pattern at the micro-level. This research paper applied the potentiality of Remote Sensing and GIS to map, detect and quantify spatio-temporal land use/land cover changes by using the satellite images of Land sat (TM) of 1988, 1998 and high resolution images of 2008 and 2018 downloaded from Google Earth. The study was carried out in the village namely Fatehgarh Nau Abad of Bathinda District in Punjab. The different land use and land cover maps were prepared to analyze change detection of land use and land cover for the 4 distinct years covering the time period of 30 years, i.e. 3 decades. The result shows the decadal changes of land cover that the area under built up has increased from 16 (2.78%) to 39.51 (6.8%) in ha on non-built up land but there were also interchange of land between different land use and land cover categories in the study area. The developed spatial data base at the village level can be useful for rural planning, agriculturalists, and natural resource management.

Keywords Remote sensing · GIS · Land use · Land cover · Change detection

B. Singh (✉)
Department of Geography, Punjabi University Guru Kashi College, Talwandi Sabo, Bathinda, Punjab, India
e-mail: balwindersidhu67@gmail.com

C. G. Singh
Maharaja Ranjit Singh P.T.U., Bathinda, Punjab, India
e-mail: cgs.patiala87@gmail.com

S. Sahdev et al. (eds.), *Geoecology of Landscape Dynamics*,
Advances in Geographical and Environmental Sciences,
https://doi.org/10.1007/978-981-15-2097-6_19

19.1 Introduction

Agriculture is the main profession of the rural population residing in India to meet their basic needs. Increase in population and socio-economic activities are putting a burden on the land with the passage of time. It has become very important for properly planning, rational assessment and scientific utilization of every parcel of land. It becomes possible only if the whole complex of land use is studied at the district, tehsil or even village level by considering the conditions of local physical and socio-economic (Mohammed 1978). "The term *land cover* relates to the type of features present on the surface of the earth. The term land use relates to the human activity or economic function associated with a specific piece of land" (Lillesand et al. 2008). In other words, land use is the inclusive of natural environment for socio-economic purposes is changed, managed and modified into built up; while land cover is defined as the physical material (grass, tree, bare ground, etc.) at the earth surface. The interactions between land use and land cover attract the geographers to know it for longstanding time (Meyer and Turner 1996). The primary concern relevant to the geographer is to have insight of the prevalent relationship between the natural environment and the human being (Kumar 1986). The importance of land use mapping is felt in concern with evolution, better management, preservation, and conservation of natural resources of any area (Gangaraju et al. 2017). So, it becomes essential to know, to show and monitor the occurring trends in land use/land cover at the village level. For this, the modern technology of Remote Sensing and GIS is very effective, useful tool to analyze, detect the changes in land use/land cover by taking spatio-temporal data for the future planning of sustainable and management of natural resources. Application of remotely sensed data taken at different times made possible to study the changes in land use and land cover in less time, at low cost and with better accuracy (Kachhwala 1985) along with GIS software that provides the facility to analysis, update and retrieve data (Chilar 2000). This is better assessed as change detection is the process to identify differences in the state of any object or phenomenon by observing it from time to time (Singh 1989). It requires the comparison of two images of an area taken at different times for drawing land use and land cover changes. Prasoon and Pushpraj (2015) have analysed spatial and temporal distribution of land utilisation in Mopka village of Bilaspur city for studying changes in land use. Gangaraju et al. (2017) studied land use/land cover change detection in Pindrangi village. Rao et al. (2003) identified aquaculture hotspots at the village level. Mitra and Singh (2014) did land use/land cover mapping and change detection for a village. Wang et al. (2017) also used Remote Sensing and GIS to quantify changes of villages in the urbanizing Beijing metropolitan region.

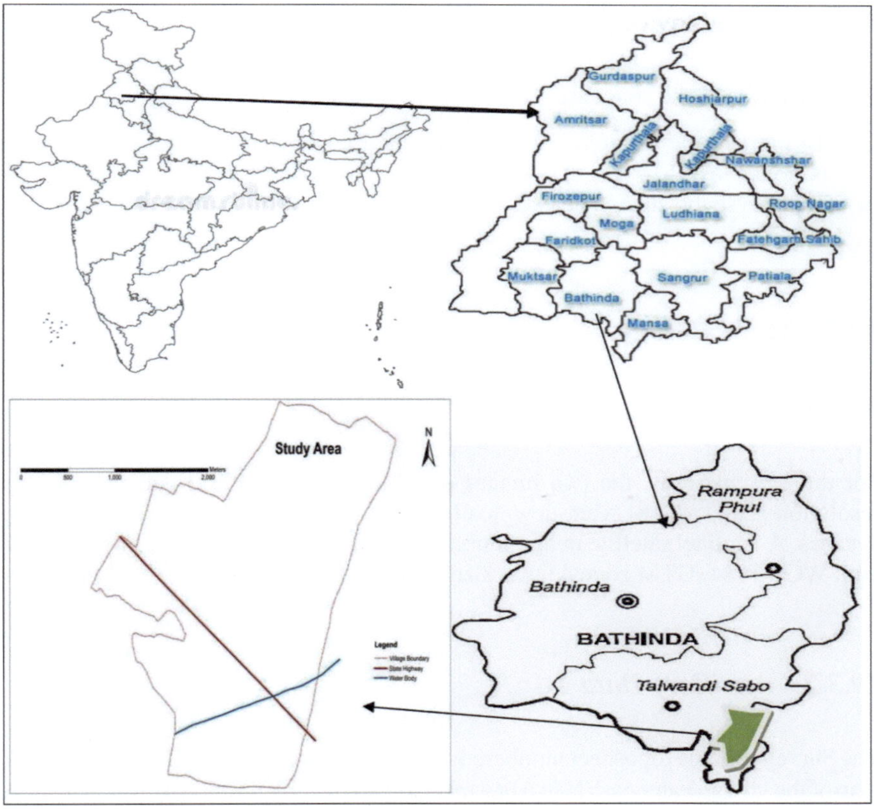

Fig. 19.1 The study area. *Source* India Map (www.dreamstime.com), Punjab Map (webindia123.com), Bathinda Map (onefivenine.com), and Study Area Map (by the authors)

19.2 Study Area

This present study was carried out in Fatehgarh Nau Abad village (Fig. 19.1), which falls under Bathinda District of Punjab state. It is located between 29° 58′32″ N Latitude and 75° 07′19″ E Longitude. It has total geographical area of 577.0 in ha. The population in the village is 1748 according to 2011 census. The village revenue boundary is surrounded by town Talwandi Sabo in the north, and by villages namely, Seikhpura in the east, Jaga Ram Tirth in the south, and Gurusar Jaga in the west. The main occupation of the village is agriculture. It is located four kilometres away south of Talwandi Sabo tehsil along the state highway. The town is a religious place and emerging as an educational hub having one private university and another in the study area with influencing socio-economic activities on the surrounding village land.

19.2.1 Objectives of the Study Area

The objectives of the study are

(i) To prepare the different land use and land cover maps at the different time periods.
(ii) To analyse the change detection.

19.3 Methodology

19.3.1 Data Base

For the present study, the two images of Landsat (1988, 1998) and two of high resolution (2008, 2018) were downloaded from Google Earth in distinct years on decades. A Sentinel satellite image georeferenced (10 m spatial resolution) geocoded with WGS 1984, UTM zone 43 was also acquired from USGS earth explorer.

19.3.2 Ancillary Data

The Survey of India toposheet numbered 44 o/1 on the scale of 1:50 000 and cadastral map of the village Fatehgarh Nau Abad taken from concerned authority was also used.

19.3.3 Methodology

GIS open free software QGIS version 2.18.12 with GRASS 7.2.1 was used to georeference the Google earth data and cadastral map with the help of Sentinel georeferenced image and to extract village revenue boundary. Digitization-based on-screen and visual interpretation was performed to map land use/land cover in the form of polygons based on the geometric boundaries of each features and followed by field observation. Linear features such as road, branch of canal were shown as a buffer. As the spatial resolution of the Landsat images of 1988 and 1998 is poor and coarse. So, for the preparation of land use and land cover maps for 1988 and 1998, the information related to some features was collected by asking the information from the people of this village such as where and when was the area under sand dunes and orchards, when did they built their (isolated) houses and in which year religious, educational institute, water works, dharmshala (community hall) and marriage palace came into existence by showing them high-resolution images of 2008 and 2018 to mark polygons on the map during the field verification. The two images of different time periods were superimposed to show the changes in land use/land cover. The digitised data was quantified to get the area of each polygon. To show change detection during 1988–2018, the change matrix was also prepared for the study area.

19.3.4 Land Use and Land Cover Classification

The scheme for land use and land cover classification was devised to classify land use and land cover (Anderson et al. 1976; NRSA 2006). The adopted land use and land cover classification scheme for this study identified the 20 categories of land use/land cover namely crop land, orchard, vegetation, sand dunes, vacant land, branch of canal, pond, residential, educational institution, religious place, dharamshala (community hall), cremation/buried ground, water works, marriage palace, factory, shops, petrol pump, cart track (locally known as *pahi* leading to agricultural field), main road and link road.

19.4 Results and Discussion

19.4.1 Land Use and Land Cover

Land use and land cover for the different years was mapped to know the existing features related to 1988, 1998, 2008 and 2018 (Figs. 19.2, 19.3, 19.4 and 19.5).

19.4.2 Change Detection for Land Use and Land Cover

Change detection simply means to demarcate and determine an area of what is changing actually to what extent, by which one class of land use is changing into another over time period. The information extracted from the different four land use and land cover maps was analysed to show land use changes in the study area. There were no changes detected for the categories such as branch of canal 0.82 (0.14%), followed by pond 1.0 (0.17%), religious place 0.2 (0.03%), water works 1.6 (0.3%), cremation ground 0.69 (0.12%), main road 2.5 (0.43%), link road 4.15 (0.72%), and cart track 4.35 (0.75%) in ha.

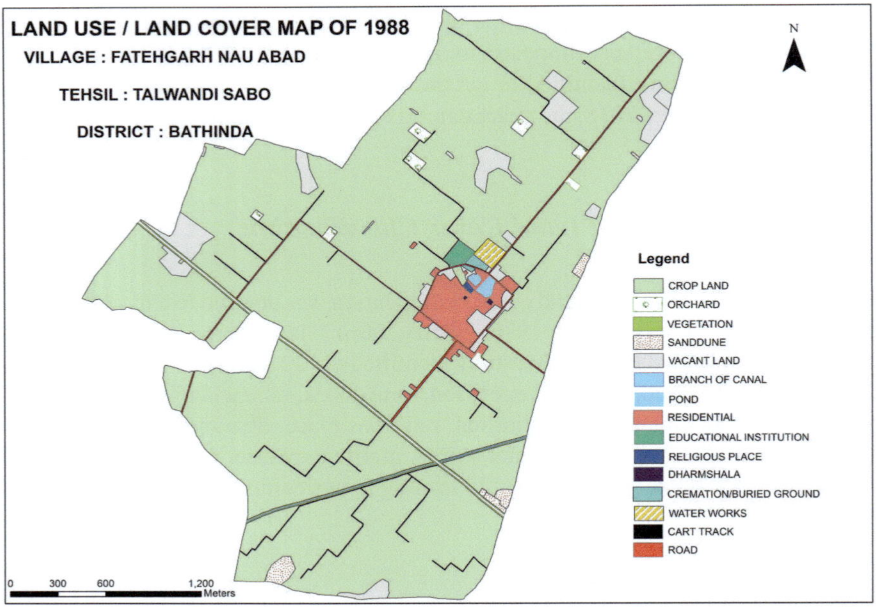

Fig. 19.2 Land use and Land cover map of 1988. *Source* prepared by the authors from Landsat (TM) Google Earth

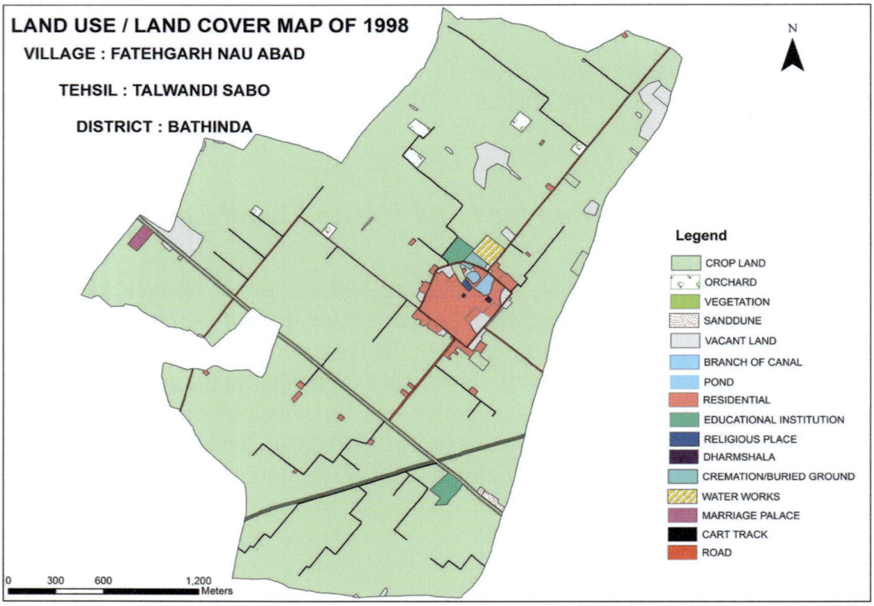

Fig. 19.3 Land use and Land cover map of 1998. *Source* prepared by the authors from Landsat (TM) Google Earth

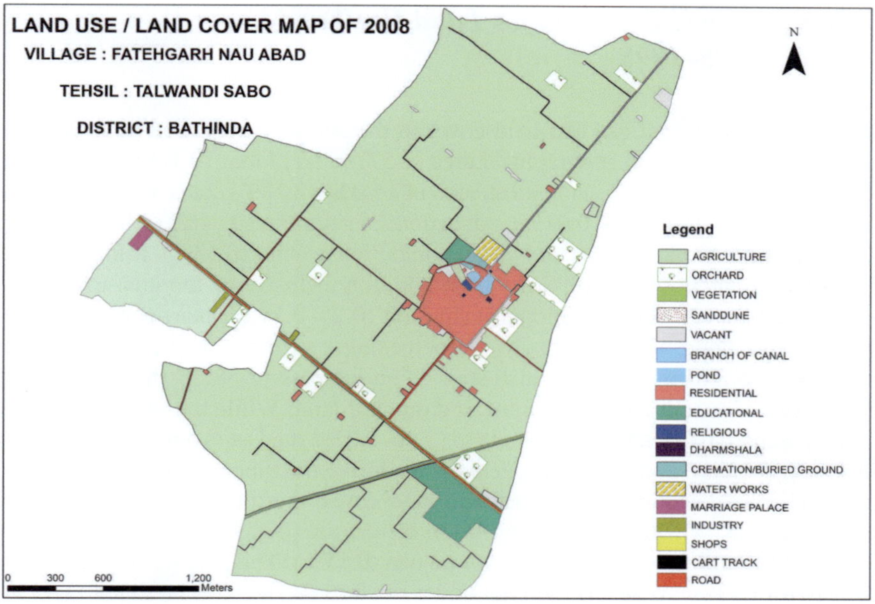

Fig. 19.4 Land use and Land cover map of 2008. *Source* prepared by the authors from Google Earth (High Resolution)

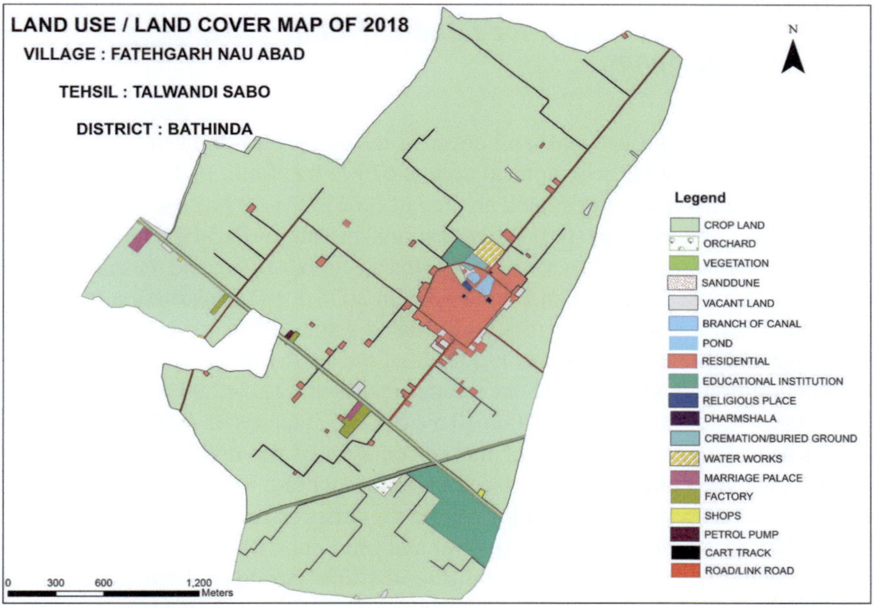

Fig. 19.5 Land use and Land cover map of 2018. *Source* prepared by the authors from Google Earth (High Resolution)

19.4.3 Change Detection for Land Use/Land Cover (1988–1998)

The categories of land use and land cover in the study area showed both increase and decrease in the total geographical area of 577 in ha (Table 19.1, Fig. 19.6). The total built up area in 1988 occupied an area of 16.0 ha (2.78%) consisting of residential (2.07%), dharamshala (0.01%), educational institution (0.24%), religious place (0.03%), water works (0.3%) and cremation/buried ground (0.12%). It increased to 21.09 ha around3.66% in 1998. It is the addition/inclusive of the area under residential (2.07%–2.5%), educational institution (0.25–0.54%), marriage palace (0–0.15%), and along with the plus of the percentages of the rest categories in built up 1998. The area of crop land also showed an increase of 88.86–90.4% from the conversion of vacant land, sand dune and orchard land. While the other categories such as orchard (0.7–0.53%), vacant land (3.96–2.23%) and sand dune (0.63–0.12%) showed a decrease under their respective area during the time period of 1988–1998. In the change detection of land use and land cover, the total built up showed an increase of 0.87% with the addition/plus of residential (0.43%), educational institution (0.29%) and followed by marriage palace (0.15%). The area of crop land also increased (1.52%). Whereas the area under orchard (0.16%), vacant land (1.72%) and sand dune (0.51%) decreased.

19.4.4 Change Detection for Land Use/Land Cover (1998–2008)

The categories of land use and land cover in the study area showed both increase and decrease in the total geographical area of 577 in ha (Table 19.1, Fig. 19.7). The total built up area in 1998 occupied an area of 3.66% (21.09 ha) consisting of residential (2.5%), dharamshala (0.01%), educational institution (0.54%), religious place (0.03%), water works (0.3%), cremation/buried ground (0.12%), and marriage palace (0.15%). It increased to 33.04 ha around 5.7% in 2008 due to construction for educational purposes, economic activities and isolated settlements showing expansion of the village and the socio-economic influence of the nearby town on the village land. It is the addition/inclusive of the area under residential (2.5–2.93%), dharamshala (0.01–0.02%), educational institution (0.54–2.08%), factory (0–0.09%), shops (0–0.01%) and along with the plus of the percentages of the rest categories in built up 2008. The area under orchard also showed an increase of 0.53–2.74% from the conversion of crop land into orchard. While the other categories such as crop land (90.4–87.73%), vacant land (2.23–0.72%) and sand dune (0.12–0.006%) showed a decrease under their respective areas during the time period of 1998–2008. In the change detection of land use and land cover, the total built up showed an increase of 2.1% with the addition/plus of residential (0.43%), dharamshala (0.003%), educational institution (1.54%), shops (0.01%) and followed by factory (0.09%). The area

Table 19.1 Classification of land use/land cover and change detection

LU/LC categories		1988		1998		2008		2018		Change Detection					
										1988–1998		1998–2008		2008–2018	
		Hectares	%	Hectares	%	Hectares	%	Hectares	%	Hectares	%	Hectares	%	Hectares	%
BUILT UP(1–10)		16.0	2.78	21.09	3.66	33.04	5.73	39.51	6.8	5.09	0.88	11.96	2.1	6.54	1.1
1	Residential	11.96	2.07	14.44	2.5	16.92	2.93	19.12	3.31	2.48	0.43	2.48	0.43	2.2	0.38
2	Dharamshala	0.11	0.01	0.11	0.01	0.13	0.02	0.13	0.02			0.02	0.003		
3	Educational inst.	1.44	0.25	3.14	0.54	12.03	2.08	14.4	2.5	1.7	0.29	8.9	1.54	2.4	0.4
4	Religious place	0.2	0.03	0.2	0.03	0.2	0.03	0.2	0.03						
5	Water works	1.6	0.3	1.6	0.3	1.6	0.3	1.6	0.3						
6	Cremation/Buried ground	0.69	0.12	0.69	0.12	0.69	0.12	0.69	0.12						
7	Marriage palace			0.91	0.15	0.91	0.15	1.3	0.23	0.91	0.15			0.4	0.07
8	Factory					0.5	0.09	1.7	0.3			0.5	0.09	1.2	0.21
9	Shops					0.06	0.01	0.23	0.04			0.06	0.01	0.2	0.03
10	Petrol pump							0.14	0.02					0.14	0.02
11	Crop land	512.74	88.86	521.49	90.4	506.18	87.73	516.4	89.5	8.75	1.52	−15.31	−2.65	10.22	1.8
12	Orchard	4.0	0.7	3.08	0.53	15.8	2.74	0.56	0.09	−0.92	−0.16	12.72	2.2	−15.24	−2.6
13	Vacant land	22.84	3.96	12.9	2.23	4.14	0.72	2.7	0.47	−9.94	−1.72	−8.8	−1.52	−1.44	−0.25
14	Sanddune	3.64	0.63	0.7	0.12	0.04	0.006	0.04	0.006	−2.94	−0.51	−0.66	−0.11		
15	Vegetation	4.98	0.86	4.98	0.86	4.98	0.86	4.98	0.86						
16	Branch of canal	0.82	0.14	0.82	0.14	0.82	0.14	0.82	0.14						
17	Pond	1.0	0.17	1.0	0.17	1.0	0.17	1.0	0.17						

(continued)

Table 19.1 (continued)

LU/LC categories		1988		1998		2008		2018		Change Detection					
										1988–1998		1998–2008		2008–2018	
		Hectares	%	Hectares	%	Hectares	%	Hectares	%	Hectares	%	Hectares	%	Hectares	%
18	Main road	2.5	0.43	2.5	0.43	2.5	0.43	2.5	0.43						
19	Link road	4.15	0.72	4.15	0.72	4.15	0.72	4.15	0.72						
20	Cart track	4.35	0.75	4.35	0.75	4.35	0.75	4.35	0.75						
	Total	577.0	100	577.0	100	577.0	100	577.0	100						

Source Prepared by the authors

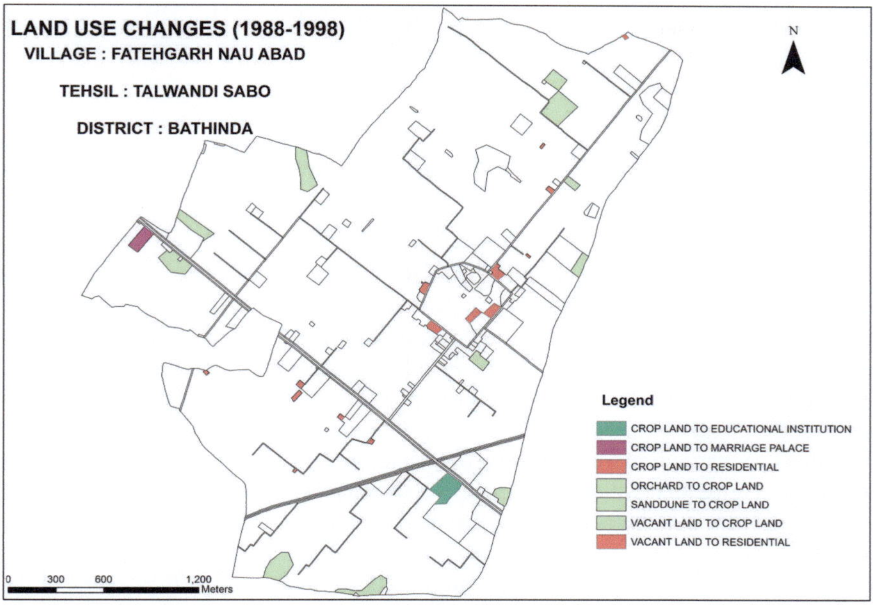

Fig. 19.6 Land use changes (1988–1998) in the study area. *Source* prepared by the authors from Google Earth (Landsat TM)

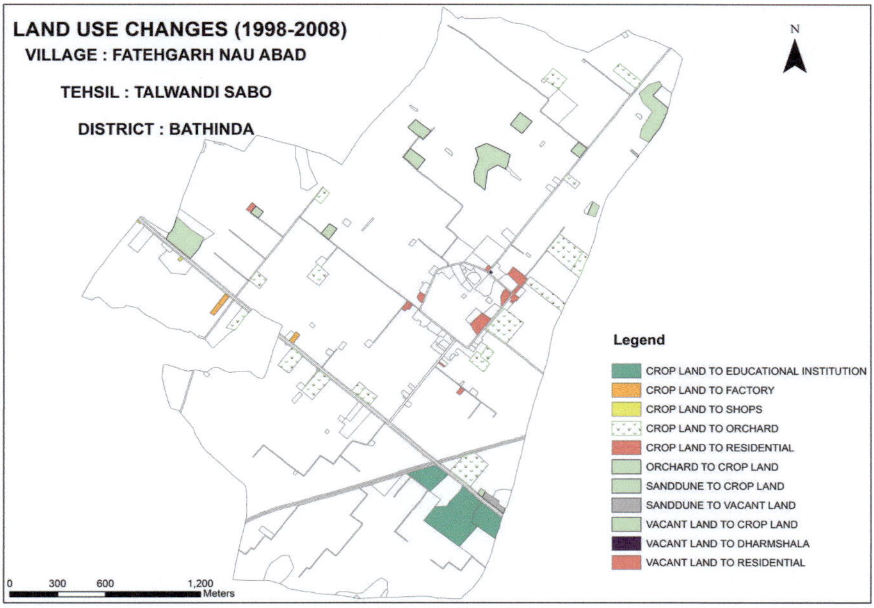

Fig. 19.7 Land use changes (1998–2008) in the study area. *Source* prepared by the authors from Google Earth (High Resolution)

of orchard also increased (2.2%). Whereas the area under crop land (2.65%), vacant land (1.52%) and sand dune (0.11%) decreased.

19.4.5 Change Detection for Land Use/Land Cover (2008–2018)

The categories of land use and land cover in the study area showed both increase and decrease in the total geographical area of 577 in ha (Table 19.1, Fig. 19.8). The total built up area in 2008 occupied an area of 5.73% (33.04 ha) consisting of residential (2.93%), dharamshala (0.02%), educational institution (2.08%), religious place (0.03%), water works (0.3%), marriage palace (0.15%), factory (0.09%), shops (0.01) and cremation/buried ground (0.12%). It increased to 39.51 ha around 6.81% in 2018 due to construction for expansion of educational institution, economic activities and isolated settlements showing the socio-economic influence of the nearby town on the village land. It is the addition/inclusive of area under residential (2.93–3.31%), educational institution (2.08–2.5%), marriage palace (0.15–0.23%), factory (0.09–0.3%), shops (0.01–0.04%), petrol pump (0–0.02%) and along with the plus of the percentages of the rest categories in built up 2018. The area under crop land also showed an increase of 87.73–89.5% from the conversion of orchard and vacant land.

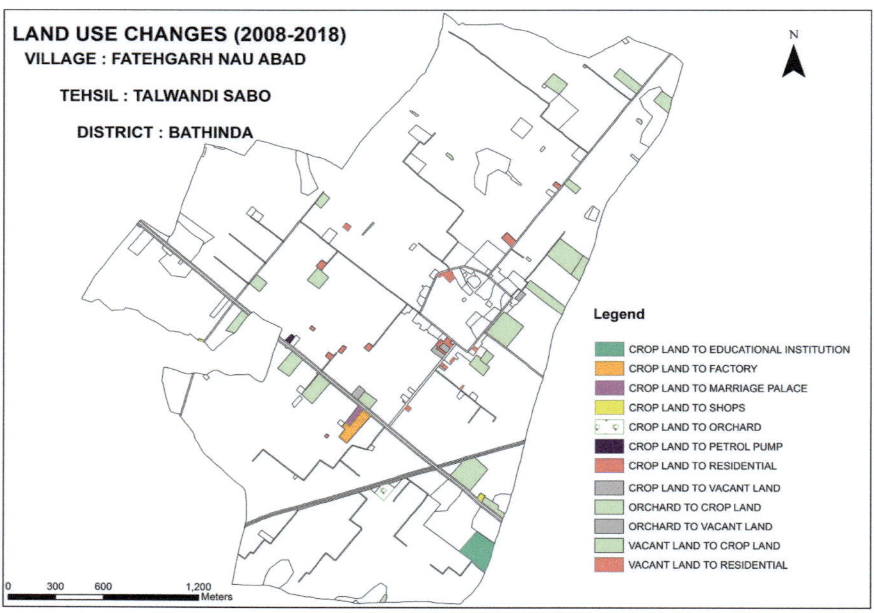

Fig. 19.8 Land use changes (2008–2018) in the study area. *Source* prepared by the authors from Google Earth (High Resolution)

Some of the vacant land was utilized for residential areas. While the other categories such as orchard (2.74%–0.09%), vacant land (0.72%–0.47%) showed a decrease under their respective areas during the time period of 2008–2018. In the change detection of land use and land cover, the total built up showed an increase of 1.1% with the addition/plus of residential (0.38%), educational institutions (0.4%), marriage palace (0.07%) and followed by factory (0.21%). The area of crop land (1.8%) also increased. Whereas the area under orchard (2.6%) and vacant land (0.25%) decreased.

19.5 Conclusion

This study carried out for analysis, monitoring of land use/land cover change detection in Fatehgarh Nau Abad village using Landsat and high resolution images of the different years like 1988, 1998, 2008 and 2018 download from Google earth reveals the result of important land use and land cover changes from 1988 to 2018 in the study area. The generated statistical data shows the changes of sand dunes, vacant land into crop land and crop land into orchard, vacant land and built up area. The total built up has increased from 16 ha (2.78%) in 1988 to 39.51 (6.8%) in 2018 for meeting the socio-economic needs of the area and is responsible for major land use/cover changes leading to encroachment on the village land toward or along the state highway (main road) due to expansion of the village and Talwandi Sabo town. This comprehensive spatio-temporal data can be useful for rural-planning, natural resources management at the micro-level.

References

Anderson JR, Hardy EE, Roach JT, Witmer RE (1976) A land use and land cover classification system for use with remote sensing data. Geological Survey Professional Paper 964

Chilar J (2000) Land cover mapping of large areas from satellites: status and research priorities. Int J Remote Sens 21(6–7):1093–1114

Gangaraju M, Anitha P, Krishna TV, Sravan KA (2017) Spatio-temporal changes of land use/land cover of Pindrangi village using high resolution satellite imagery. J Remote Sens GIS 6(2):1–4

Kachhwala TS (1985) Temporal monitoring of forest land for change detection and forest cover mapping through satellite remote sensing. In: Proceedings of the 6th Asian conference on remote sensing. National Remote Sensing Agency, Hyderabad, pp 77–83

Kumar J (1986) Land use analysis: a case study of Nalanda district. Bihar Inter-India Publications, New Delhi, India

Lillesand TM, Kiefer RW, Chipman JW (2008) Remote sensing and image interpretation. Willey, New York

Manual National Land Use Cover Mapping using Multi–temporal Satellite Data, NRSA, Hyderabad, May 2006

Meyer WB, Turner BL (1996) Land use land cover change: challenges for geographers. GeoJournal 39(3):237–240

Mohammed A (1978) Studies in agricultural geography. Rajesh Publications, New Delhi, India, pp 1–6

Mitra M, Singh S (2014) Land use land cover mapping, change detection and estimating at cadastial level through Remote Sensing and GIS application: a case study of Dhobi village, Hisar, India. Int J Geomat Geosci 5(2):357–361

Prasoon S, Pushpraj S (2015) A study of spatial and temporal distribution of land utilisation pattern in Mopka village using RS & GIS. Int J Agril Res Innov Tech 5(1):1–9

Rao KN, Krishna GM, Naik DR, Malini BH (2003) Remote Sensing and GIS applications in the identification of aquaculture hotspots at village level. J Indian Soc Remote Sens 31(2):71–80

Singh A (1989) Digital change detection techniques using remotely sensed data. Int J Remote Sens 10(6):989–1003

Wang K, Zhou W, Xu K, Liang H, Yu W, Li W (2017) Quantifying changes of villages in the urbanizing Beijing Metropolitan Region: integrating remote Sensing and GIS analysis. Remote Sens 9(448):2–16

Chapter 20
Sustainable Energy Development and Participatory Management Scenario in the Sundarban: A Case Study in the Sagar Island, West Bengal, India

Anwesha Haldar and Ajanta Bhattacharya

Abstract The Sundarban is an ecologically sensitive region with the inhabitants lacking in proper basic amenities, purchasing power, technological awareness and capacity generation for sustainable development of the community. The Sagar Island has been facing the constant wrath of environmental disasters and damage due to their poor adaptive technologies offered by the planners and policy makers. Even though, the introduction of renewable technologies as an alternative source of power has not been equally welcomed everywhere nevertheless the dependence on forest wood as the prime fuel is no longer prevalent and people have voluntarily involved themselves in forest conservation. This paper tries to make a contribution by upholding the dualism in the society between direct gains and selfless sustainability; whereby an integrated framework can be formulated for a comprehensive developmental goal. To understand the scenario, the awareness and perception study of the inhabitants towards social forestry and renewable energy sources have been carried out. From the results, it is found that even though Indian energy policy encourages renewable energy but it is mostly controlled by the supply channel and public demand leading to more reliance on conventional thermal gridded electricity. Thus, public awareness and participatory management of locals can ultimately provide a sustainable way to in situ resource management in the Sagar Island.

Keywords Capacity building · Solar power · Social forestry · Sustainability · Participatory management

---

A. Haldar (✉) · A. Bhattacharya
Department of Geography, University of Calcutta, Calcutta, India
e-mail: ahgeog_rs@caluniv.ac.in

A. Bhattacharya
e-mail: bhattacharya.ajanta@gmail.com

S. Sahdev et al. (eds.), *Geoecology of Landscape Dynamics*,
Advances in Geographical and Environmental Sciences,
https://doi.org/10.1007/978-981-15-2097-6_20

20.1 Introduction

The increase in awareness about the state of environment caused by over-exploitation of natural resources for economic development has evoked major discussions and debate worldwide. This has led to the concept of Sustainable Development that cut across ecological, economic, social and cultural dimensions (Ramakrishnan 2008). Environmental and societal degradation has been further exacerbated by human-induced 'global change' which is a collective term for climate change, land use changes, impacts of globalization, population change, economic and resource alterations, land degradation and desertification, biological invasion by exotics and biodiversity depletion. The dynamics consequently, interact among themselves to leave a lasting impact on the affected communities and in turn affects the entire nation in terms of economic losses, migrations, social conflicts, risks to human health, compromise on energy, food, and water security, loss of ecosystem services, reduced resilience of human societies and a decline of the environmental health which is basis of our well-being (Duarte 2012). This paper deals with the two major thrust areas of Sustainable Development Goals (SDG, 2015) namely affordable and clean energy (7th SDG), and life on land (15th SDG), i.e. protecting, restoring and promoting sustainable management of forests to reverse land degradation.

The estuarine system of the Ganga-Brahmaputra deltaic region forms the largest mangrove eco-system in the world shared between India and Bangladesh and the extension of the region within India is roughly between 21° 30′ N to 22° 30′ N and 88° 05′ E to 89° 05′ E (Fig. 20.1). India being the largest democracy and with more than 1.3 billion residents, is faced with numerous challenges to provide even the

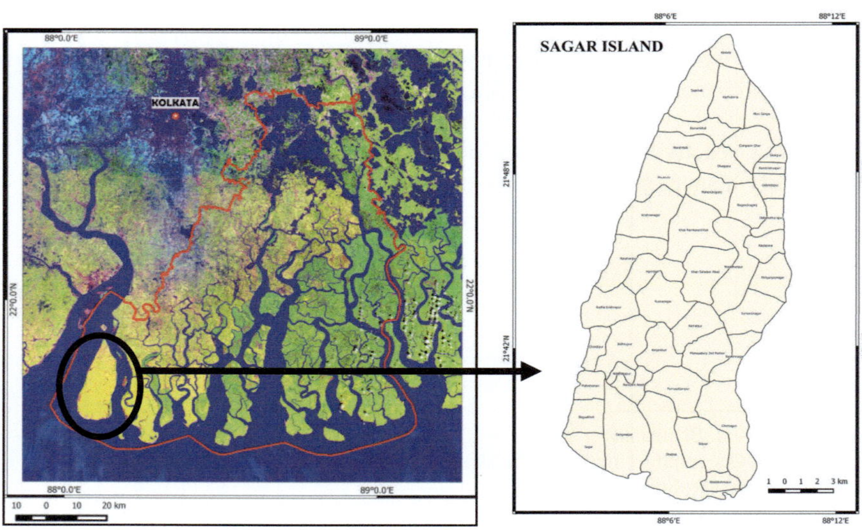

Fig. 20.1 Location of the Sagar Island

bare necessities of life to its people. At this conjecture it is certainly a sensitive issue to introduce experimental technologies in renewable resources when people are deprived from basic food, clothing and shelter. At present the Indian Sundarban covers an area of about 4,266.6 km^2 and over 4.4 million of the most impoverished and vulnerable people in India. About half of this population lives below the poverty line (BPL), with poverty incidence highest in the Administrative Blocks close to the vast mangrove forest (Planning Commission of India 2013; World Bank 2014). Over and above the region is frequently plagued with climatic hazards such as storm surge, cyclones, erosion, embankment breaching and saline water ingression. Hunter (1875) and Ghosh et al. (2018), presents a case study of South 24 Parganas district that focuses on the traditional adaptation practices used by the susceptible communities in this vulnerable area. It was also as a potential decision-making ground to formulate any policy and financial intervention.

Geo-hydrologically, it is separated from the mainland of West Bengal by a 3.5 km stretch of the "Muri Ganga" river. The total area of the Sagar Island is around 286.03 km^2. Even though the physical and social characteristics of the Sagar Island are almost similar to that of the other islands of Indian Sundarban region, it is rapidly emerging as the most densely populated island. Thus with decreasing carrying capacities there is a marked increase in its natural and social vulnerabilities. In this study, the Sagar island (Fig. 20.1) in the south-western most part of the Sundarban is one such place were majority of the population face scarcity from safe drinking water, healthcare facilities, proper sanitation, electricity connection, educational institutions and job opportunities (Haldar 2014). Thus, resulting in poor resilience and low capacity generation of the locals to attain sustainable development.

20.2 Objectives of the Study

The present era witnesses the importance of participatory planning all over the world. Participatory approach is now an integral part of conservational planning and management of the different sectors of the society. In energy sector, participatory planning enhances conservation of energy and community renewable energy use has been observed worldwide. In India, the present scenario encourages the rapid and cost-effective developments in the field of renewable technologies. As we all know India has made some progressive technological innovations and administrative policies towards promoting renewable power systems especially solar power; but unfortunately some pockets still exists and majorly in the Sagar Island, it is observed that renewables have failed miserably. The mal practices in participatory management for not being able to conserve the solar plants, standalone batteries being one of the direct problems. However, it is observed that people have readily come forward to conserve the mangroves in Sagar and here participatory management have been hugely successful. This dichotomy has raised the question on human behaviour and responses towards non-conventional infrastructure and environment. Thus a proposition can be assumed that low awareness regarding energy technologies has prevented mass propagation of solar sources while high awareness regarding forestry and biodiversity improves participatory management results. The paper aims to critically address the proposition, attend to the possible issues and decide whether the renewable power source is supposed to bring in more development or development harbours in acceptance of a new technology. In this discussion, the changes of perspective of participatory management planning in different fields of the environment management have been focused. The top-down management system has failed to realise the true sustainable indicators. Thus, the planning of this area needs reformulation of present management policy by introducing the local know-how in development of the area. The participatory management has become successful in different countries as a new paradigm for sustainable development and regional management planning, and this participation of locals mainly helps in recognising and combining of social, economic and ecological values. This study aspires to discuss the present management strategies in utilisation of energy and forest resources and implementation of participatory management with the involvement of local people.

20.3 Methodology

The work has been done through qualitative analysis of information from available literatures to identify the problem areas. Field surveys by purposive random sampling in questionnaire surveys in households and target group discussions in the villages of the Sagar Island have been carried out to understand and record their experiences and perceptions. The respondents all had given verbal consent in participating in the questionnaires surveyed and confirmed that they answered to the best of their knowledge. An ethical committee supervised the authenticity of data and information in University of Calcutta. Visit to solar plants in Sagar and several plots where social forestry was done and discussions with the erstwhile beneficiaries were carried out. Census of India (2011) data has been analysed. Livelihood susceptibility index is computed to determine by various factors such as physical, social, economic and environmental processes, which increase the susceptibility of a community to the impact of hazards. This was also computed with the help of census data by compound score method following UNDP (1990, 1992). This was followed by data processing in Microsoft-excel, maps were created on the remote sensing and GIS based software Q GIS 2.8.7 and Geomatica 15. Finally the information collected was tabulated for interpretation and reporting.

20.4 Livelihood Susceptibility of the Study Area

Poverty, occupation, caste, ethnicity, exclusion, marginalization and inequalities in material consumption of a society or community affect on the livelihood susceptibility. Livelihood susceptibility study refers to the resilience of communities when confronted by external stresses on our lives. These stressors in the form of both natural or anthropogenic disasters, or disease outbreaks create livelihood maladjustments in the society. The study accounts for deprivation in income generation, physical habitat and infrastructure as well as access to state provided resources, political power and governance and provide a better understanding of household conditions as well as local planning strategies for development of an area (McGee and Broke 2001). For this reason, all the parameters are taken for analysing the quality of life. These parameters are ownership and permanency of household, quality and distance from source of drinking water, use of electricity, latrine facility, presence of waste water outlet, use of banking service and other different assets derived from Census of India (2011) data. For the evaluation of the index, the relative approach of UNDP for developing HDI (UNDP 1990, 1992) is applied. The analytical approach essential for the calculation of the index and this index is calculated with site specific data of the study area (Fig. 20.2).

For this study the livelihood study is calculated of the above said indicators in the following manner

$$\mathrm{VI}_{ijk} = \left(\mathrm{X}_{ijk} - \mathrm{MIN}_{ijk}\right)/\left(\mathrm{MAX}_{ijk} - \mathrm{MIN}_{ijk}\right)$$

where, i = Variables (1, 2, 3, … I), j = Components (1, 2, 3, … J), k = Blocks (1, 2, 3, … K).

In the study the score varies from 2.20 to 5.40. This score is proportionate with the development in services and other infrastructure. In this case, the livelihood index is found to be more in developed mouzas which is near the central part of the island. Rudranagar being the most developed town here has the highest developmental scores than others. Rudranagar, Radha Krishnapur, Kamalpur, Khan Saherber Abad, etc. shows high scores while Bakimnagar, Naraharipur shows lower value, which represents that livelihood services are yet to be delivered in the lower value areas. Comparing with the density of population it is found that the denser areas of Sagar are enjoying more services than others (Fig. 20.3) or it may be interpreted that amenities have attracted people from the surrounding areas to cluster in these villages.

20.5 Participatory Management in the Renewable Energy Sector in Sagar Island

Electricity is a basic service now for sustaining life. It is of great responsibility for the local governing bodies to provide 100% rural electrification as it not only raises the well-being of the people but also opens up to a wide range of opportunities to improve their social, economic and environmental livelihood. The Sundarban are an energy deficient region. Most reclaimed villages on the islands and at the fringes of the densely populated mainland are yet to be connected to the conventional power grid. According to the Census of India (2011) the power scenario (Fig. 20.4, 20.5 and 20.6)

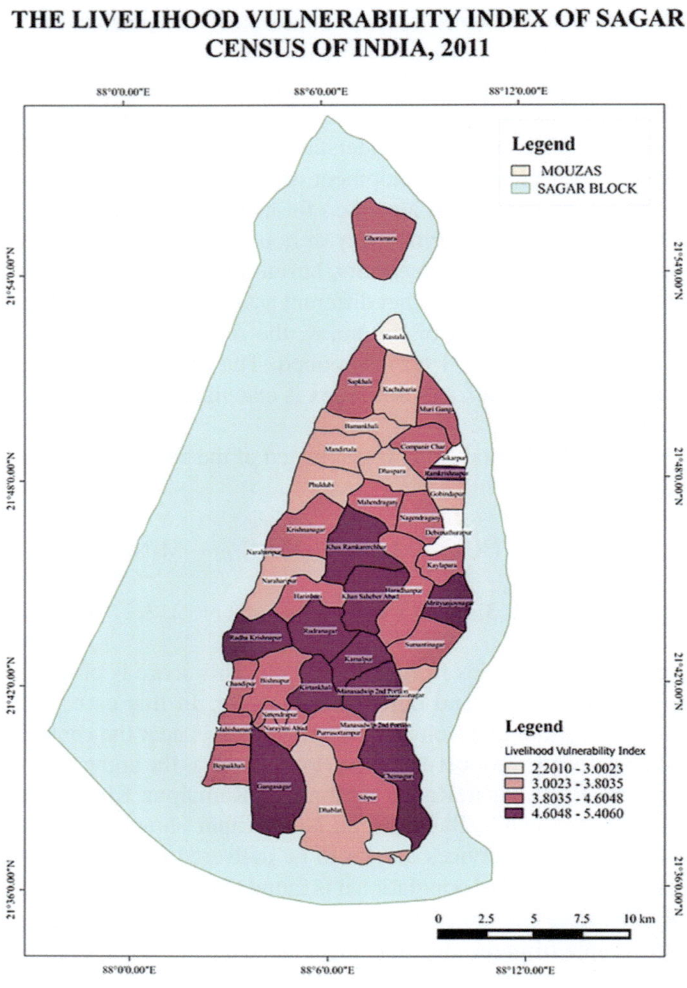

Fig. 20.2 The livelihood vulnerability index of Sagar block (Census of India 2011)

of Sagar island has been mapped. It is generally noticed that for lighting, kerosene dominated the sector with high solar power prevalence in few mouzas of Muriganga, Gobindapur, Dhablat and Kirtankhali. The total energy expenditure surveys in the Sundarban in 2014 showed the dominance of biomass (73%), followed by kerosene (14%), electricity 4%, and diesel 4% (World Bank 2014).

Regardless of the gradual fall in capital costs behind solar energy technologies and steady rise in fossil fuel prices and depleting reserves, solar powered electricity cannot compete with the strong conventional electricity market. Besides the economic disadvantage, solar energy technologies face a number of technological, financial and institutional barriers that further constrain their large-scale deployment. It is also to

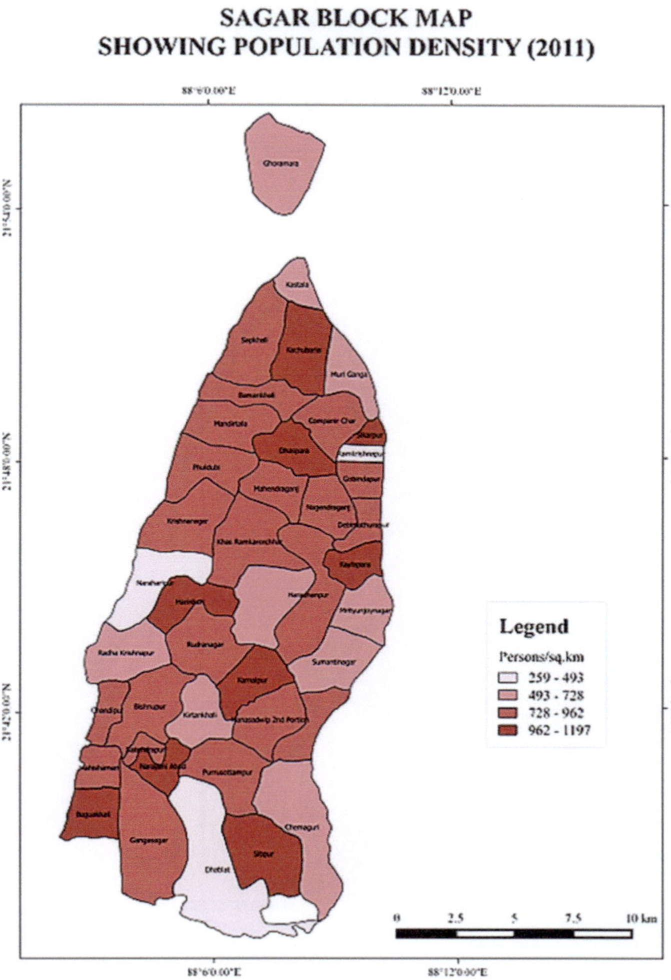

Fig. 20.3 Block Map of Sagar showing the Mouza-wise population density (Census of India 2011)

be noted that even though large programs and projects have been sanctioned by the Union and State Governments, almost negligible amount of subsidy or awareness have been reached to the remote islands (Haldar 2015).

Almost 20 years ago, the Sagar Island in the Bay of Bengal became a prominent example of India's efforts to power the rural, remote, impoverished areas with solar energy. Between 2000 and 2008, renewable power technologies, including household photovoltaic panels, solar home and street lighting, wind power generating stations and mini grids, were installed by WBREDA on Sagar with large state subsidies. The government helped cover generation and distribution expenses at a cost of around 15 crores rupees. The solar and wind—generation plants were generating close to 1

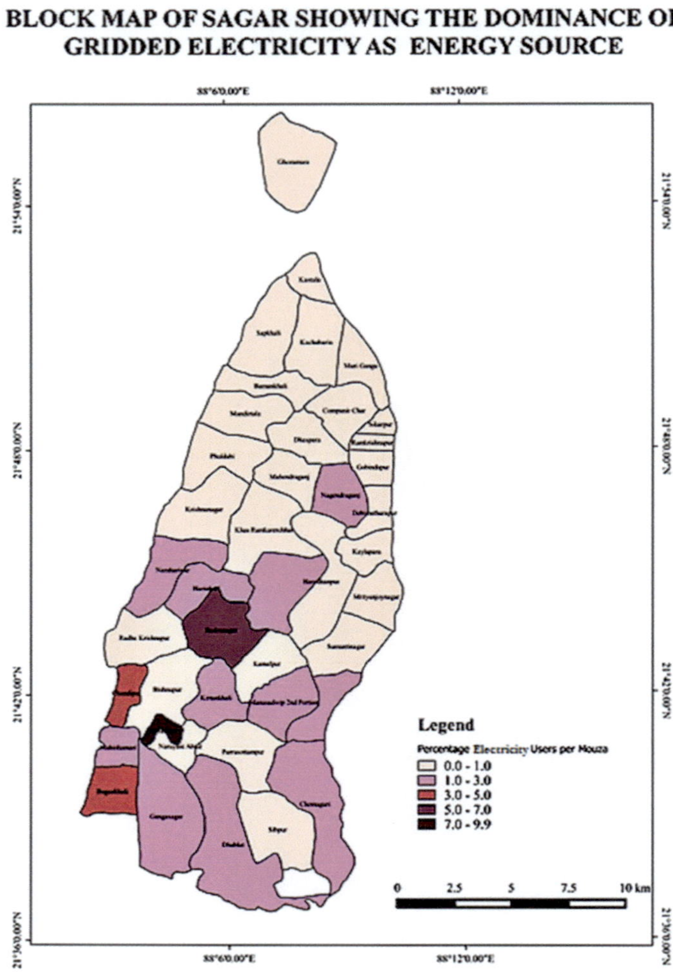

Fig. 20.4 Dominance of grid electricity as the main energy source. *Source* Census of India (2011) data

megawatt (MW) of electricity, distributed through mini grids to about 2,000 households. But a government decision to pull off all renewable energy subsidies and to hook up the island with the electric grid has left the solar schemes in tatters.

Sagar is now connected with the grid under the erstwhile ambitious national scheme of Rajiv Gandhi Gramin Vidyutikaran Yojna (RGGVY) and later the Deen Dayal Upadhyaya Gram Jyoti Yojana since 2015, both of which aimed to bring electric power to all villages by 2020. People below the poverty line to get free electricity and pay rates as low as 3.5 rupees per unit. As broadly estimated the present average domestic load in Sagar Block is 140 kW per consumer and average

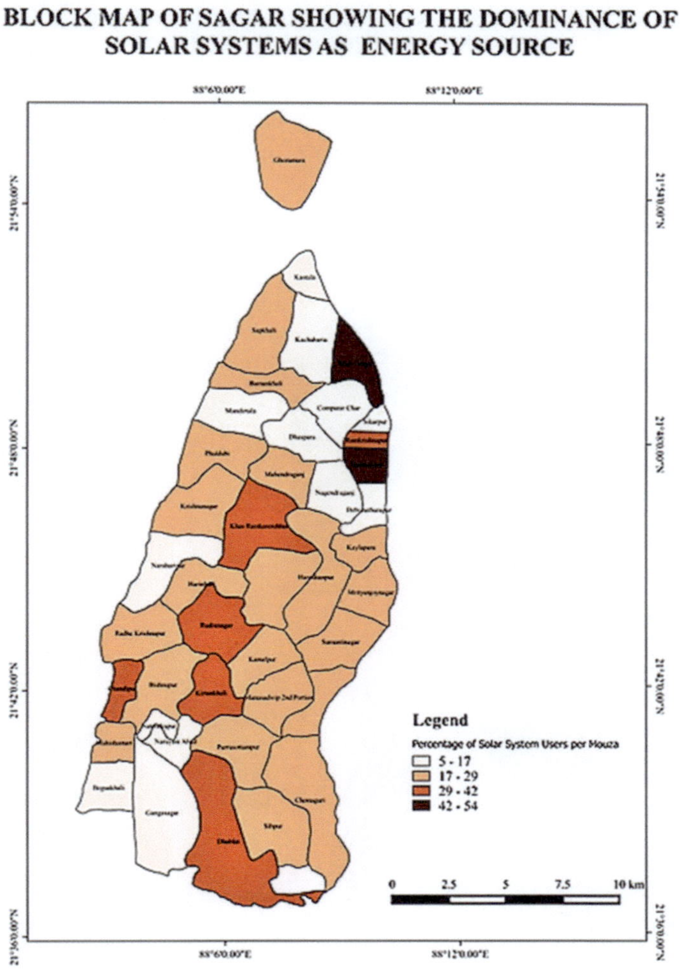

Fig. 20.5 Dominance of solar system as the main energy source. *Source* Census of India (2011) data

commercial load is 620 kW per consumer, and accounting with the ever increasing demand of power in the island, the requirement of power is estimated to rise to about 9.5 MW by 2020. This is very difficult for the solar plants to generate continuous good quality power of this amount for the entire island.

Initially, Sagar had been excluded from the national grid scheme due to technical unfeasibility but was reversed largely for business, political and also for religious reasons electricity development even at the cost of environmental conservation. The Sterling and Wilson company, a Shapoorji Pallonji Group owned engineering firm, was able to complete the commissioning of the entire grid-connected electrification

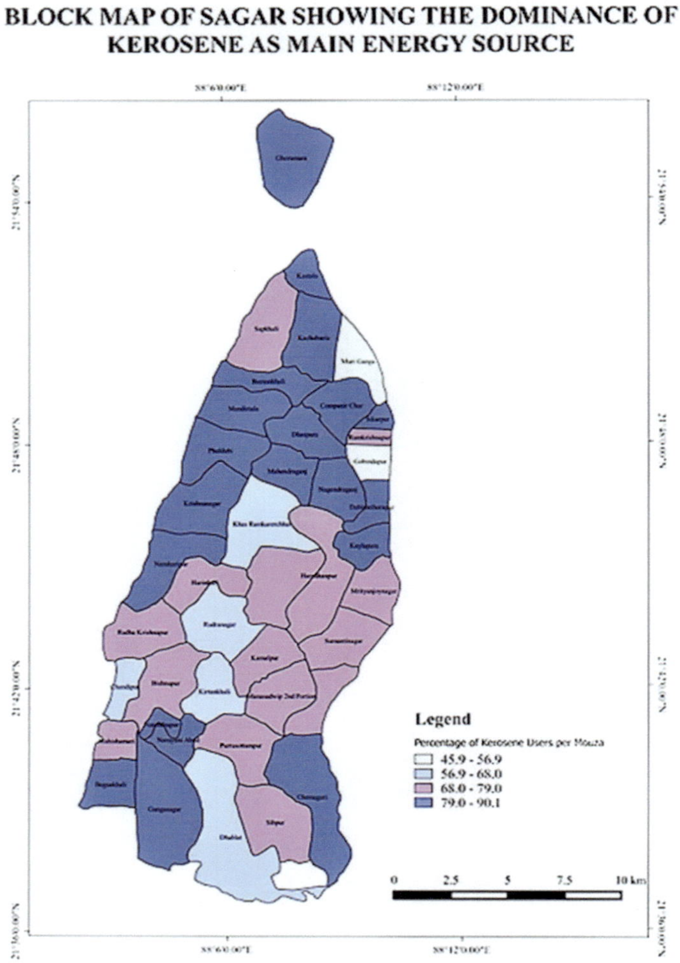

Fig. 20.6 Dominance of kerosene as the main energy source. *Source* Census of India (2011) data

project by January 2016 in Sagar Island which proposes to benefit 2,26,000 people across 43,000 households in the area (Business Standard 2016). Now the same inhabitants, who had previously whole heartedly welcomed solar systems, are leaving the connection and moving away to grid connections.

Since the time Sagar Island started to access power from the grid, most islanders have gradually moved away from using solar energy, and the solar plants are left at the mercy of the nature. Thus the generation solar capacity in the District of South 24 Parganas is rapidly declining and the state of the generation equipments are deteriorating, a trend that will continue if demand for solar power declines further, experts have warned (Haldar 2014). Local consumers in Sagar may be pleased with

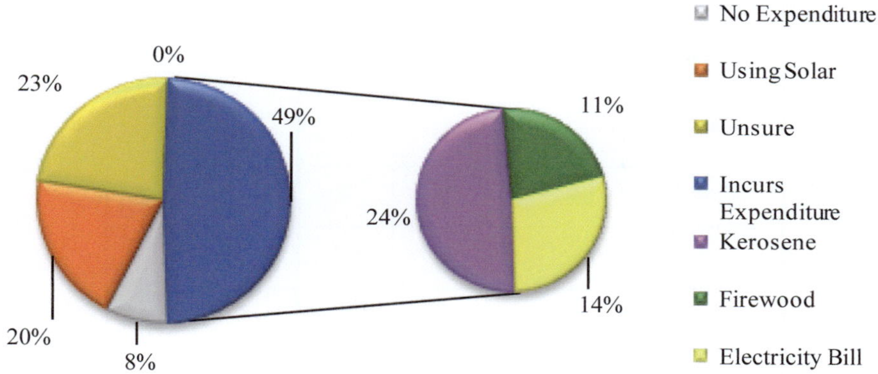

Fig. 20.7 Monthly expenditure on power. *Source* Primary survey by authors

the policy U-turn, as they will get somewhat cheaper compared to the solar-diesel hybrid systems and perhaps more reliable electricity, but the side-effects are many. Sagar's solar energy facilities are being neglected, and are already defunct, hindering the India government's target of installing 20,000 MW of solar power capacity by 2020. Electricity poles and overhead lines are at a major treat during the monsoons and especially during cyclonic storms which are recurrent in Sagar for its geographic location at the head of the vast Bay of Bengal.

Presently more than 40% of the inhabitants are still without any type of power connections on the island therefore a large part of the population is solely dependent on kerosene for lighting, cooking and other domestic needs. Kerosene is presently being disseminated at 500 ml to 1 L per ration card in the public distribution system stores every fortnight at 25 rupees per litre. Most of the time this fuel is not sufficient for those who are also using kerosene for lighting along with cooking, hence they have to resort to the kerosene from 'black market' which is sold at prices exceeding INR. 50/- per litre. An average power expenditure break-up of the surveyed population has been shown (Fig. 20.7).

Electricity is mostly of use to women for comfort and entertainment and to children for education after the Sun sets, while men remain mostly out of the house thus enjoying the benefits less. Televisions are the only source of entertainment in these remote villages. Children can study for the extra few hours and fan has brought in much comfort in their lives for all (Fig. 20.8). The Government is trying to bring in development through rural electrification whereby people can engage in alternative occupations, run small cottage or agro-industries, diversify income opportunities, ensure higher educational standards, better standard of living and thus social equity and stability in the society.

From our recent field studies at the Sagar Island it is observed that electricity has only reached the major connecting points of the island and lighting is majorly tourism based i.e. only those places have electric connection where pilgrims, tourists or government officials do access. This leaves thousands of marginal

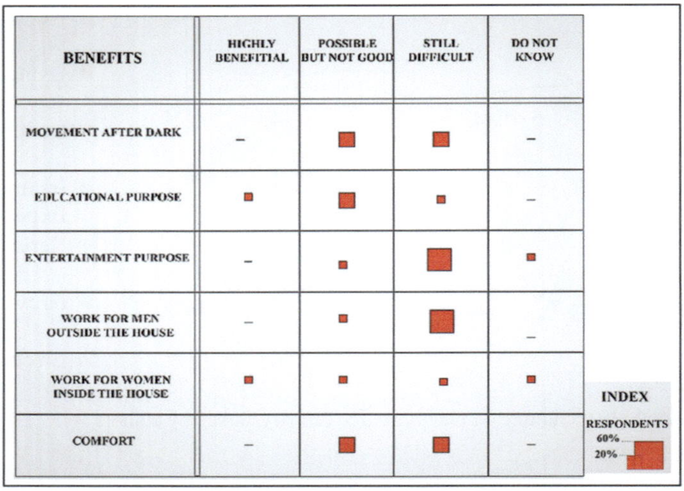

Fig. 20.8 Perceptional benefits of electricity. *Source* Primary survey by authors

dwellers in the interiors of the island without proper access to power. These regions are the Manasadwip-Jibantala colonies, Bisalakhipur-Botkhali area and Chandipur-Mahisamari area. When rough weather persists for more than 3 h, for regions that are remotely located or are separated by natural barriers, grid connectivity becomes a great problem.

The major factors of concern for lighting up Sagar C.D. Block are feasibility of electricity distribution and placing of poles across creeks and wide river networks of the Sundarban, environmentally stable connection lines that can withstand natural hazards and cost effective supply to individual homes so that it does not become an economic burden on the population. As surfaced from the surveys and interviews, almost no subsidy on solar connection is available from the government now but the local cooperative banks have certain schemes and in turn are making small business for themselves. The scheme is that a certain percent of the down payment is waivered and the equal monthly installment facility is provided to clients having their own bank account in that branch and buying more than 40 W solar panels. As most of the poor farmers do not have a bank account or are reluctant in using larger solar panels, they cannot avail this subsidy hence wide spread use of solar lags behind still today.

The other drawbacks are frequent mal practices and tampering with the solar batteries and funds in the community based solar plants has been reported by the locals. The ill maintenance of the batteries has led to poor performance and therefore needs frequent changes (Fig. 20.9). The gridded solar plants and diesel generator cooperatives in the favourable climatic season could only provide 6–8 h of electricity. Hooking, illegal connections, theft from solar plants was a common scenario, local caretakers were bribed and threatened not to take any legal step. The nonchalant attitudes of the panchayat and solar plant engineers posted there have led to the consecutive closure of the solar power plants.

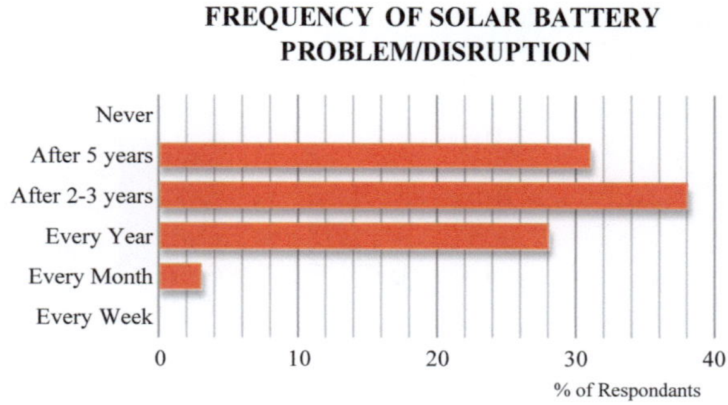

Fig. 20.9 Frequency of solar battery problem/disruption. *Source* Primary survey by authors

An off-grid/decentralized solar collection scheme is far more energy efficient than a grid-connected or centralized one as they conserve more than 30% of the electricity that is lost in transmission and distribution in the current system. An off-grid solar system is also far more resilient to natural disasters as there is no single point of failure that can bring down the whole grid, as is the case with grid-connected/centralized power generation and also saves a lot of space which is required to build stations. Although decentralized off grid power generation is the most feasible and only option in these locations (Haider 2012), locals have no urge to voluntary come together and revive the community renewable energy plants. Respondents here seem to have a good knowledge about the pollution caused from burning fossil fuels like kerosene, diesel, etc., but not all realise the importance of using an energy and non-polluting power supply (Fig. 20.10). In a similar way perceptions are clear in case of renewable being eco-friendly source but their tendency to continue and personal satisfaction level with solar is low compared to thermal grids (Fig. 20.11). Solar must be properly advertised to increase awareness and attract more users into the renewable sector. People's participation in developing and continuing with the solar systems failed. Even though community lighting services in panchayats, schools, clubs and market

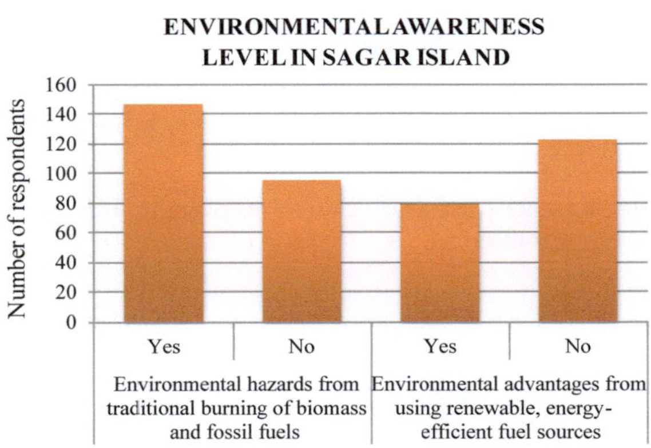

Fig. 20.10 Environmental awareness level in Sagar block. *Source* Primary survey by authors

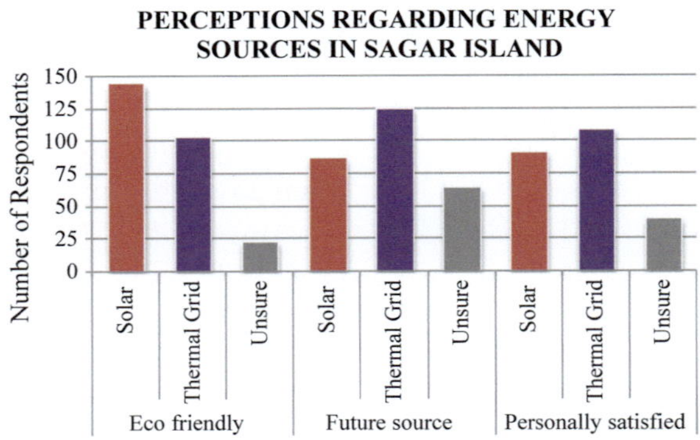

Fig. 20.11 Perceptions regarding energy sources in Sagar block. *Source* Primary survey by authors

places was successful but where vigilance and direct gain of individuals were negligible such as for street lighting and solar plants, renewable failed miserably. People was solely profit minded and calculated only personal benefits in the solar sector as can be seen with the overwhelming success of standalone solar panel home lighting systems (Fig. 20.12). Hence local awareness, collective responsibility and peoples' participation can only revive the renewable power sector on the island.

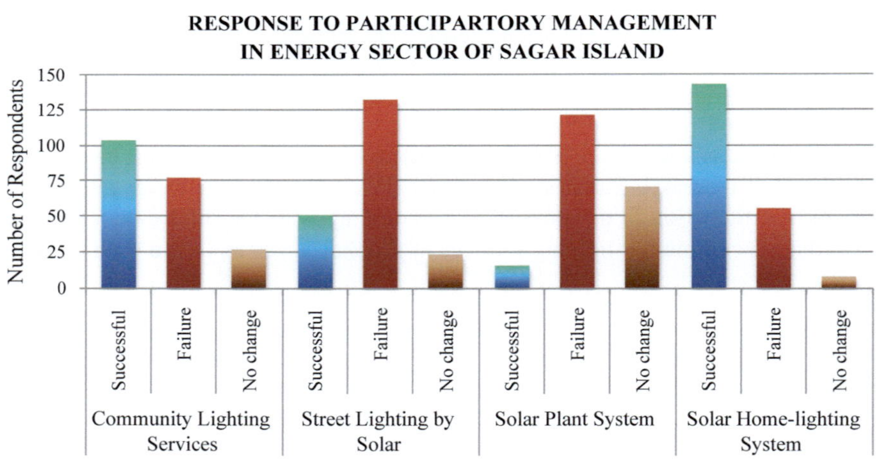

Fig. 20.12 Response to participatory management in energy sector of Sagar Island. *Source* Primary survey by authors

20.6 Social Forestry and Participatory Planning in Environmental Conservation

After 40 years of state owned conservative management of mangroves, India has emerged with the concepts of community based management of mangrove resources, better known as 'Joint Mangrove Management' (JMM), since the last decade. One of the aims of JMM is the involvement of Non-Governmental Organizations (NGOs) and Community Based Organizations (CBOs) in development, conservation and restoration of mangrove resources which has been largely described as an ameliorative management of these exclusive coastal resources. Under this backdrop, this study observed that with the generous participation of NGOs and CBOs in JMM, sustainable development of forestry especially of the mangrove species proved more successful than the energy sector. In case of Sagar Island, NGO's as well as local governance play an important role building a comprehensive relation with the local people whereby both are benefitted. The panchayats and NGOs can showcase these endeavours, funded by both governmental and non-governmental agencies and the locals are relieved from recurrent land erosions and get steady supply of forest-based resources. From the surveys, it was observed that 84% of the respondents in Sagar island were willing to continue with social forestry, among them 71% said that they conscious about the importance of planting saplings while the rest 29% responded that they first joined this interest group due to the wages but after attending meetings of NGO and local panchayats they are now aware about its importance (Fig. 20.13). Most depend on these forests for honey collection, dry timber gathering and for conservation of land from erosion and storm surges. Hence conserving these forests meant a sustainable addition to their livelihood. Apart from material gains, these

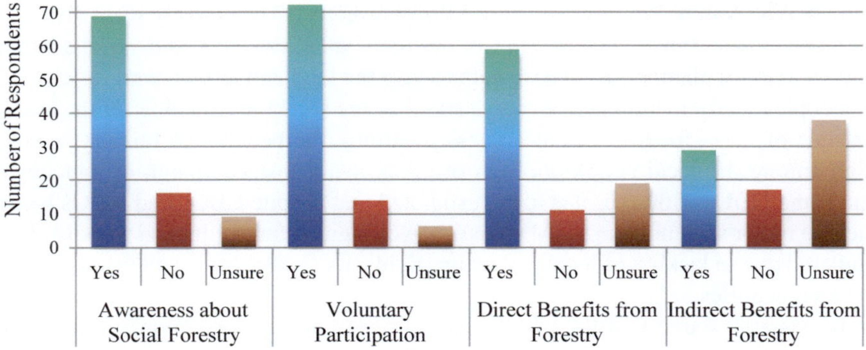

Fig. 20.13 Response to participatory management in social forestry sector of Sagar Island. *Source* Primary survey by authors

Fig. 20.14 *Changing Coastline of Sapkhali Mouza, after Mangrove Plantations. Map source* Google Earth Images © 2010–2018

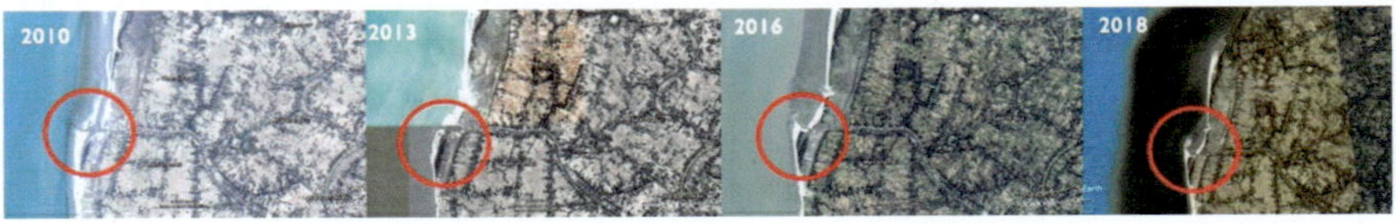

Fig. 20.15 *Changing Coastline of Chandipur-Mahisamari beach. Map source* Google Earth Images © 2010–2018

forests save the area from the cyclonic storms which has devastating effects on the entire region if the coastal fringes are left barren.

Significant interference was faced when people were unaware of the benefits of JMMs. But presently in the coastal edges of Sagar island, where the erosion is much more, NGO's and local governance started giving special importance to protect the soil and showed how social forestry can not only save the eroded areas but also reclaim the accreted and newly formed 'chars'. Since this time, larger number of people started showing interest in JMM and more vulnerable areas of Sagar could be brought under the purview of social forestry on this island. Coastal zones like Chandipur-Mahisamari beach area (Google Earth images showing the significant increase in forest cover in Fig. 20.14), Khastala-Sapkhali area (Google Earth images showing the beach conservation in Fig. 20.15), and coastal zones in the north of Beguakhali are the zones where people are engaged in the social forestry with the help of the panchayat and NEWS (Nature Environment and Wildlife Society). They mostly plant different varieties of Mangroves species and Coastal She-Oak (Casuarina equisetifolia better known as *Jhau* plantations). These help to grip the land and reduce land losses.

Peoples' participation here is active and they are solely responsible for the maintenance of these forests. The awareness regarding the benefits of forestry is very high among the local people through different awareness campaign of panchayat and some NGOs. However it is to be also noted that despite CBOs and NGOs being the two major stakeholders in JMM that showed significant results, the work is often hindered due to lacking of legal and tenure rights, local workforce and know-how of the technicalities.

It was also found that the effectiveness of the participatory approach with the inhabitants and local governance was more effectual from the vegetation cover figures which showed a sharply decline with rapid reclamation until 2011, after which it significantly increased when the JMM was implemented in the Gram Panchayats of Sagar C.D. Block. Participatory planning of the area witnessed a significant change in livelihood and land conservation due to community involvement. The Sundarban, being multi-dimensional in nature, has increasing pressure on land that needs to be rearranged for the planning policy at the bureaucratic levels. Local people are the prime beneficiaries of the resources and also the worst affected by the exploitation of the resources. Over the years of intense and regular observation of the environmental processes, they have gained immense expertise and experience to plan for land and forest conservation and other local resource managements that will eventually reduce land degradation, soil erosion and other social malpractices like poaching, illegal felling etc. in the targeted areas of the Sagar Island.

20.7 Social Dualism in Participatory Planning

The prime global focus of the 1970s was on the relationship between energy and economics, and at that time researchers and policy makers were not much concerned with the implications this would have on the environment. The global awareness and uproar over rising carbon and other fossil fuel emissions, environmental pollution and wasteful use of natural resources came about in the 1980s when sustainability in developmental strategies was first brought into focus. In this study area it was observed that the introduction of grid electricity reduces the use of solar energy and people are reluctant in welcoming the sustainable modern technologies inspite of the initial public participation to bring in electricity to the villages. Being one of the densely populated areas, the Sagar Island is the foremost of the Sundarban islands to start social forestry for environmental conservation (Mondal 2013). The top-to-down approach started this conservational practice but with the advent of participatory approach through Joint Forest Management, the progress of work have rapidly enhanced and such practices have become more realistic when done in an environment-friendly way (Dev Roy and Alam 2012). Participatory management plays a very important role in the enhancement of local people's awareness for the social forestry (Tmang 1994) and also the conservation of mangroves (Safa 2005). But this awareness building method is missing in the use of community solar power plants in the study area. There are some financial supports and subsides for the forestry development but the same is absent in the case of solar energy planning for this area. The Government of India, in their new agenda, have emphasised on the importance of renewable and afforestation programmes but here the use of solar power is reducing day by day while afforestation and social forestry for both species valuation and economic importance is on the rise. Thus the planning in more participatory ways and involving local expertise, skill and workforce in the planning avenues are needed to overcome these challenges.

20.8 Conclusion

As the concept of sustainable development took a realistic stance, it shifted focus on economic development rather than having only a welfare approach towards societal and environmental conservation. Sustainable Development Goals of 2015 addresses these changing concerns for nations to act conforming the new guidelines. The new planning interventions should aim at micro level problems and improving the basic social conditions together with education, health and nutrition, and focus on providing sufficient sustainable livelihood opportunities to increase individuals' adaptive capacity. This study on the awareness and perception of the inhabitants towards renewable energy sources and social forestry attempted to make a comprehensive analysis of the impact of the resources on local sustainability and its control by the supply channel and public demand.

With the constant rise in population density, sustainable energy and social forestry seems the most eco-friendly way of conserving our resources in the Sundarban region. But no technology, however effective, can be imposed on a community. It takes time, experience and level of development in accepting them. Energy consumption is intimately bound up with the natural environment and the environmental consequences of the growth in world energy demand would be catastrophic with far-reaching consequences. The resultant climate change with intensified cyclonic storms causes mass coastal damage and human destructions both physically and psychologically. The gradual transition from non-renewable to renewable resources should be done through encouragement and enhancing public awareness towards resource conservation. According to the planners and researchers, the use of non-renewable energy can be maximised by involving the local people in a small area. Likewise, for increasing the forested zone, location identification and sapling plantations can be best done with local help. Therefore, to achieve our sustainable development goals it is very much necessary to enhance community participation of the locals and incorporate their knowledge in micro-level policy building.

Acknowledgements This study was partially supported by University Grants Commission (UGC). We are grateful to the respondents of Sagar Island who have willingly participated in the surveys. We thank our co-scholars for helping with our field surveys and the ethics committee of University of Calcutta for supporting our work. We sincerely express our gratitude to Prof. Lakshminarayan Satpati and Prof. Sumona Bandyopadhyay, Dept. of Geography, University of Calcutta for supporting and providing valuable advice at every phase of our study.

References

Business Standard (2016) e-paper, January 19, 2016. http://www.business-standard.com/article/companies/sterling-and-wilson-commissions-power-grid-connectivity-in-west-bengal-s-sagar-islands-116011900867_1.html. Accessed 16 Dec 2016

Census of India (2011) District census handbook South 24 Parganas. Govt. of West Bengal

Dev Roy KA, Alam K (2012) Participatory forest management for the sustainable management of the Sundarbans Mangrove forest. Am J Environ Sci

Duarte C (2012). http://theconversation.com/what-is-global-change-6447

Ghosh U, Bose S, Bramhachari R (2018) Living on the edge: climate change and uncertainty in the Indian Sundarbans. STEPS working paper 101. STEPS Centre, Brighton

Haider SZ (2012) Tariff-Its importance for sustainability of power sector; in energy, environment and sustainable development. In: Uqaili MA, Harijan K (eds). Springer, Wien

Haldar A (2014) Energy future, vol 3, issue 1 October—December 2014. The Energy and Resources Institute (TERI), New Delhi

Haldar A (2015) Energy future, vol 4, issue 1 October—December 2015. The Energy and Resources Institute (TERI), New Delhi

Hunter WW (1875) A statistical account of Bengal, 1 (Districts of 24 Parganas and Sundarbans). Trubner and Co, London

McGee R, Brock K (2001) From poverty assessment to policy change: processes, actors and data. Institute of Development Studies, Brighton, England

Mondal D (2013) Assessing social vulnerability to coastal hazards: an examination on Sagar Island of Sundarban delta

Planning Commission of India (2013) Press note on poverty estimates, 2011–12. Government of India. http://planningcommission.nic.in/news/pre_pov2307.pdf

Ramakrishnan PS (2008) Ecology and sustainable development: working with knowledge systems, 2nd rev ed. NBT, New Delhi

Safa MS (2005) Poverty-environment interaction through participatory forest management in Bangladesh: an evidences of poverty reduction

Tmang KM (1994) Integrated resource development of the Sundarban reserve forest, Bangladesh. Food and Agriculture Organisation of United Nations, Khulna

UNDP (1990) HDR 1990 - Concept and Measurement of Human Development. Human Development Report (1990 to present), Human Development Report Office (HDRO), United Nations Development Programme (UNDP), number hdr1990, December

UNDP (1992) Human Development Report. http://hdr.undp.org/sites/default/files/reports/221/hdr_1992_en_complete_nostats.pdf. Accessed 9 Nov 2018

World Bank (2014) Building resilience for sustainable development of the Sundarban: strategy report. World Bank Group, Washington DC. http://documents.worldbank.org/curated/en/879351468259748207/Building-resilience-for-sustainable-development-of-the-sundarbans-strategy-report

Chapter 21
Lake Sustainability and Role of Houseboats: Impact of Solid Waste and Sewage of Houseboats on the Ecology of Dal Lake

Tariq Ahmad Ganaie and Haseena Hashia

Abstract The Dal Lake is the second major freshwater Lake of Kashmir Valley which is known for its crystal waters, scenic beauty and beautiful houseboats. Houseboats are very important for the tourism of Dal Lake. Houseboats and the Dal Lake are widely associated together which have been known as "floating palaces" or "floating hotels". These represent cultural heritage of Kashmir Valley and serve as tourist accommodation and add to the natural grandeur of Dal Lake. Solid waste and waste water together have severely polluted the Lake. The generation of solid waste and sewage from the houseboats and boatmen community are contributing to the pollution of the Lake which results in deterioration of overall ecology of Dal Lake and decreasing occupancy of tourists. This is the outcome of poor management and negligence of authorities (LAWDA) who never fulfill their responsibilities in the conservation of the Lake, solid waste and sewage management. The study has been carried out for estimation of solid waste generation and sewage disposal by houseboats and their impact on the sustainability of the Lake. Consequently results in sedimentation and growth of weeds on a large scale throughout the Lake and pose a serious threat to its sustainability. The data was gathered through structured questionnaires by simple random sampling technique. The impact of houseboats was observed through their waste generation and management. It was witnessed that there is great variation in generation from one category of houseboats to another and to households.

Keywords Houseboats · Sustainability · Solid waste · Sewage · Deterioration · Management

---

T. A. Ganaie (✉) · H. Hashia
Department of Geography, Jamia Millia Islamia, New Delhi, India
e-mail: tariqahmadg7@gmail.com

H. Hashia
e-mail: hhashia@jmi.ac.in

S. Sahdev et al. (eds.), *Geoecology of Landscape Dynamics*,
Advances in Geographical and Environmental Sciences,
https://doi.org/10.1007/978-981-15-2097-6_21

21.1 Introduction

Dal Lake is a Himalayan urban fresh water Lake mainly known for its scenic beauty from tourism point of view all over the world. Floating gardens used for growing vegetables is another typical feature of Dal Lake. Fisheries and lotus stems are the secondary functions associated with it. Dal Lake comprises of three main basins (*Gagribal*, *Lokut Dal* and *Bod Dal*) and myriad of interconnecting channels. Dal Lake is unique in having hundreds of houseboats which are there from the times of Mughals and are the focus of attention of most of the tourists during their visit to the valley. These boats are served by the shikaras which resemble gondolas of Venice but smaller in size but decorated for tourist attraction. Houseboat is a magnificent feature of Dal Lake and is of utmost importance from tourism point of view in Kashmir. For the first-time houseboats were used for accommodating tourists during the Mughal reign. These are distributed unevenly over the Lake. Since, 1990s it is being observed that Dal Lake is not getting attention from its stakeholders. Houseboats have reduced in number because of political instability from 1989 to 2005. But after 2005, again increasing trend of tourists visiting the Lake has put a huge pressure on houseboats and water body especially in the peak summer season from April to August. The concentration of Houseboats in the Lake is one of the major causes responsible for the decreasing depth and degrading quality of water of Lake.

Houseboats serve as a part of heritage of Kashmir and important tourist accommodations. As per some studies and observations almost 60% of the tourists coming to Dal Lake stay in houseboats. Out of total 880 registered houseboats, 638 houseboats are in Dal Lake and Nigeen Lake only. These 638 houseboats are registered with J & K Tourism Department and Houseboat Owners Association.

Houseboats are classified on the basis of the services they provide. There are various types of houseboats with some modifications over one another. The classification has been done by the 'Department of Tourism (J&K)' which is not justifiable on the ground. Houseboats are classified in various categories like: Deluxe, A, B, C and D. The income of deluxe houseboats is thrice the income of D category houseboats. It is because of the comfortable rooms, comfortable bedding, drinking water availability, both hot and cold water availability and other modern facilities like WiFi, telephone, etc. present in the houseboat. Deluxe houseboats are more in number than the D category houseboats. Deluxe houseboats have the locational advantage because they are facing on the front of the Dal Lake along the boulevard road which givers a picturesque view of *Zabarwan* hills, *Shankar Acharya* temple and *Pari Mahal* palace (both located in *Zabarwan* hills). The other reason behind the variations in the rent and income is because of the bureaucracy in the Houseboat Owners Association. The members of this association are also owners of the Deluxe and 'A' category houseboats who have some monopoly over the business. By other associated business, the elite class has better accessibility to better life and other luxury gadgets. They have employed staff to attend the tourists. They do not attend the guests themselves and have their houses built near Srinagar city. Some of these people have settled in

Europe and Australia and have hard grip in the business. On the other hand, the ‘C’ and ‘D’ category of houseboats hardly make their two meals a day.

21.2 Study Area

Dal Lake is an urban freshwater Lake located in the eastern part of Srinagar city (Kashmir Valley, Fig. 21.1) sharing its outline with beautiful *Zabarwan* hills of Greater Himalayas on the east and north-eastern side. It has both inflow and out flow streams and is believed to be fed by a number of underground springs also (Badar and Romshoo 2008). The major portion of the water comes via *Telbal Nallah* which is a snow fed stream flowing from *Marsar* Lake and collects sewage from the adjoining areas during its journey towards the Lake. It accounts for 85% of the total inflow of the Lake while as remaining 15% is supplied by the sewage, springs and other small inlets. It is also connected to *Anchar* wetland and *Nigeen* Lake. The water circulates to other parts of the city and there is confluence of waters between these three water bodies. Geographically it is located between 34° 5′ and 34° 6′ North latitudes and 74° 8′ and 79° 9′ East longitudes. It is fresh water Lake with diverse aquatic biota and phytoplankton. It is located at an altitude of 1583 m with alpine Mediterranean

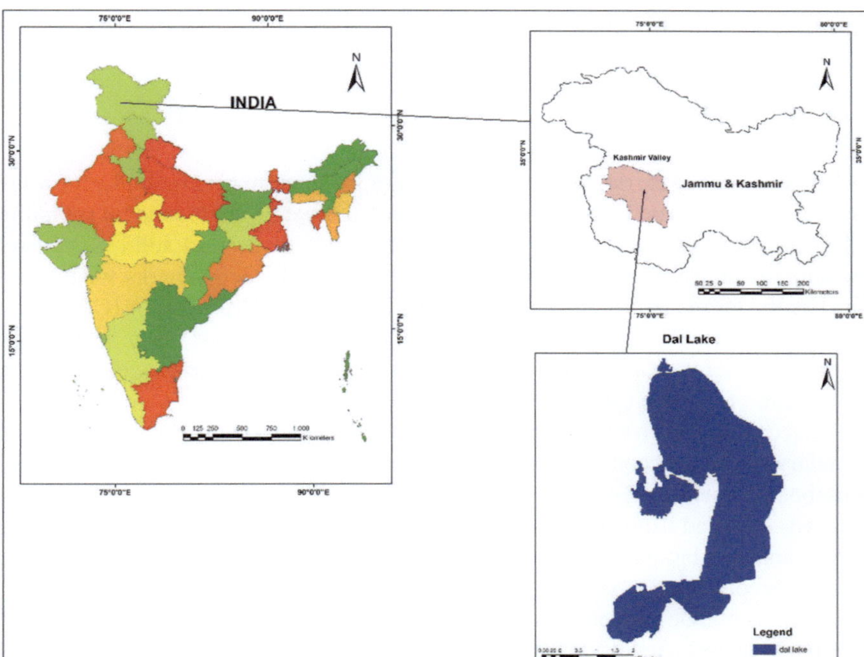

Fig. 21.1 Location map of study area

climatic conditions. The total surface area of the Lake is 10.5 km^2 with a shoreline line of 15.5 kms. The estimated volume of the water is $9.8^3 \times 10^6$ m^3. The volume fluctuates with season because in the summer the Lake receives huge silt which accumulates at the bottom and reduces its capacity to hold water. The length of Dal Lake is 3.87 kms and breadth is 2.15 kms. It includes the marshy and swampy areas adjacent to water body on the south and south-western side. Houseboats are mostly in the southernmost part of Dal Lake near Lock Gate which opens to the river Jhelum. More than 80% of the houseboats in Kashmir are placed in Dal Lake and rest in Nigeen Lake and river Jhelum. These houseboats are spread irregularly in Dal Lake. Almost 5% of the surface water of Lake is covered under houseboats.

21.3 Objectives

1. To analyze the impact of sewage and solid waste on the ecology of Dal Lake.
2. To analyze the sewage and solid waste management infrastructure of houseboats.
3. Assessing the quantity and magnitude of solid waste generation from different categories of houseboats.
4. To study the impact of tourist accommodation on houseboats during peak season of tourism.

21.4 Methodology

This study has been carried out for estimation of solid waste generation and sewage disposal by houseboats in the Dal Lake. The methods were simple random sampling and some statistical techniques. Generation of waste water and solid waste was calculated from every single unit and category wise. The benchmarks were taken from Central Public Health and Environmental Engineering Organisation (CPHEEO) for calculation (CPHEEO, 1974). The impact of houseboats was observed through their waste generation and management. It was witnessed that there is great variation in generation from one category to other and to households.

The necessary data for the current study was collected through primary survey and secondary sources. Secondary data was collected from J & K Tourism Department, Houseboat Owners Association, and SMC, etc. History and development of houseboats was obtained from Houseboat Owners Association. Classification criteria, fee charges and regulations were obtained from J & K Tourism Department. Primary survey was conducted to see the occupancy of houseboats and variation in the occupancy. A systematic random sampling was carried out with 13% sample size. Out of total 640 houseboats inside Dal Lake 65 houseboats were selected. 13 houseboats were surveyed from each category (Fig. 21.2). Per capita solid waste generation, per unit solid waste generation, total generation of solid waste and seasonal variations

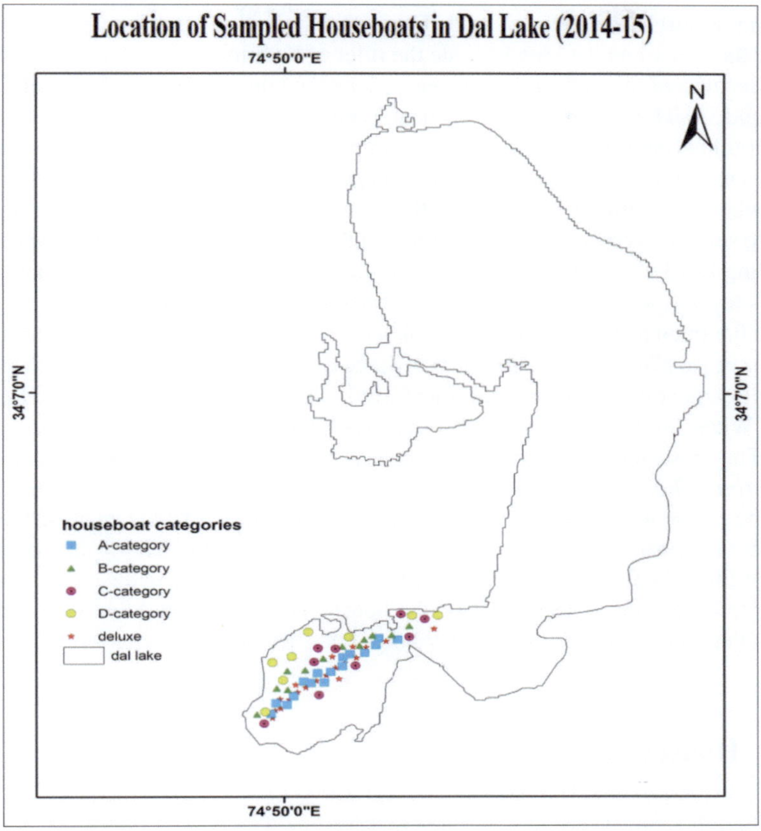

Fig. 21.2 Location of sampled houseboats in Dal Lake

in it were calculated by statistical methods from the structured questionnaire. The units were taken from CPHEEO guidelines that are 400 g/person and 250 g/tourist. Sewage disposal variation in different categories and per unit disposal was also calculated by using CPHEEO guidelines. The per capita sewage disposal taken is 80% of the water supplied.

21.5 History and Development of Houseboats in Kashmir

History reveals that British sold Kashmir to *Maharaja Gulab Singh* 1846–47 as per Treaty of Amritsar. Kashmir was sold only to 75 Lakhs only (Madhok, 1992). *Maharaja Gulab Singh* was general of army of British and purchased for mere sum of money with rivers, lakes, resources and people. Maharaja Singh made it a point that British will try to come back and reside as a state subject or construct any concrete building or purchase a piece of land. Thus, Britishers were kept away from Kashmir during the treaty and the king became the ruler of Kashmir and took its ownership as well. Because of this British people were given the tented accommodation at *Chinar Bagh* and *Sheikh Bagh* at present day near Tyndale Biscoe School, Srinagar. The king did not allow them to make their own camps but later they did with their military might. During this period the only and cheapest mode of transportation was boat till 1923 when car came to Kashmir for the first time. These boats were running from north to south in river Jhelum. All rivers and lakes acted as a medium of transportation including Dal Lake and Nigeen Lake. Dal Lake became the hub of boats and small dongas to transport vegetables, grocery, fruits and people in Srinagar city.

The boatmen formed donga boats which they would move to and fro from *Khanbal (Anantnag)* to *Khadinayar (Baramulla)*. Since English people came from Rawalpindi via bull carts and horses etc. and transferred via carno type of boats called *Donga* boats. Where in the half of the family can reside and English people will spend the time. They also resided in tents in *Baniyari, Ningli, Shadipur, Sumbal, Manasbal, Ganderbal, Shalteng* and rest of the city, which continued to the *Anantnag*. Then with the passage of time the English men with the help of boat men of the time modified the boats. They increased the length, breadth and height of the boat. They also modified the furnishing and architecture of the boats. Then in the later stage after the Victoria fashion of 1880s–1940s the boats were further modified with carvings and modern amenities till present day houseboats came into existence.

21.6 Houseboat and Tourism

Houseboat is a novel idea and heritage of Kashmir. The type of houseboat found in Kashmir is nowhere found in the world (Shah and Shabana 2013). It is famous for its architecture, furnishings, amenities, services, etc. There are five-star- and seven-star houseboats in the world but the stay in Kashmiri houseboats is incomparable as per the tourists. The houseboat is made of Deodar timber. Its floor and roof are also wooden. The roof is decorated with *Kashmiri* art of *Khatambandh* with different decorations presenting *Kashmiri* culture and history. Generally, houseboats have dining halls, lobby or sitting rooms, bedrooms, kitchen and washroom. The cost of houseboats ranges from 50 lakh to 2 crores depending upon the age, type and amenities etc. present. They are well decorated with wooden furniture, paper mashie, etc. and are provided with 24 h water supply and electricity. The owner of the houseboat gives a personal touch and family takes care of tourists by way of guiding them and exploring the far off places, cooking their food and washing their clothes and create friendly atmosphere by remaining duty bound to them. There were 1107 houseboats in 1946 but with the passage of time, taming disaster in tourism, lack of infrastructure resulted in almost lost of 200 houseboats till date. There is an apprehension that more house boats will vanish in coming years because of non-availability of deodar timber, neglectable financial support by the government, ban on reconstruction and repair of houseboats. As per the estimates and reports it is believed that at least 400 houseboats are expected to be lost in the next 20–30 years.

21.6.1 Classification of Houseboats

Houseboats are classified based on the basis of services they provide. There are various types of houseboats with some modifications over one another. The classification has been done by the 'Department of Tourism (J&K)' which is not justifiable on the ground. Houseboats are classified as follows:

- Category-Deluxe
- Category-A
- Category-B
- Category-C
- Category-D.

Different categories of houseboats are randomly placed in the Lake (Fig. 21.3). Different rents and service charges are charged depending upon the category. The registration is being done by the Department of Tourism (J&K) on yearly basis. They are supposed to abide and follow the norms formulated by the said department. The category or classification is transferable on the grounds on regular inspection of the department of tourism.

21.6.2 Occupancy of Houseboats

The occupancy is calculated from the total availability or capacity and the number of tourists received. The maximum numbers of tourists are received in the months of April, May and June. The moderate number is received in July, August, September and October while as least number of tourists are received in December, January and February. Summer season is technically called peak season and winter season is called lean season from the tourists accommodated in houseboats point of view. This is because almost 90% of the tourists visiting Kashmir Valley are received in this season but there is hardly any tourist staying at houseboats during the winter season (Table 21.2).

Due to instability from the last three decades the tourism in Kashmir is in shambles so is the houseboat industry. There are some houseboats that can hardly make their two meals. On the other hand, there is discriminating earning from tourists against the norms. The households earn in the peak season to overcome the burden of lean season. They charge more than the normal fixed charges during the peak season and in lean season charge almost less than the half of it. The variation in occupancy is also because there is a huge flow of tourists to Kashmir Valley for annual *yatra* to *Amarnath* Cave in summer season and the demand for the accommodation is very

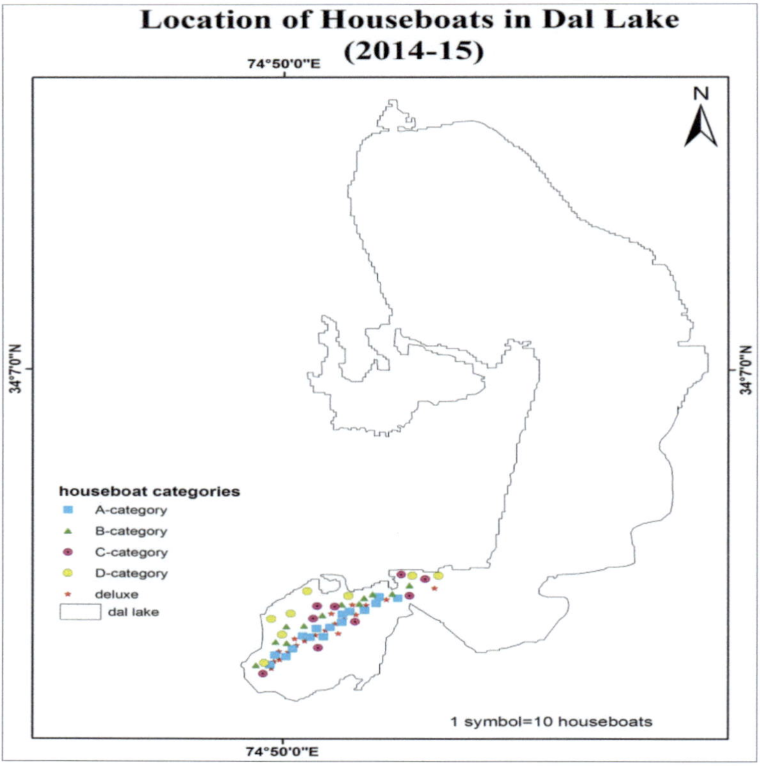

Fig. 21.3 Location of different types of houseboats in Dal Lake

high. It is very difficult to get an accommodation in peak season while as in winter season the link between Kashmir and rest of the world through highway remains closed for most of the times. There are least winter sports in Kashmir to attract tourists despite huge potential. The tourists in winter go directly to the snow laden mountains for winter games and prefer the hotels for accommodation instead of houseboats. The houseboat tariffs are very high compelling medium-income- and low-income tourist to stay either in hotel of guest houses and consequently, there is tough competition among the houseboats, guest houses and hotels. The charges are also less in guest houses than the houseboats and the chances of pollution are minimized.

The above Table 21.1 shows that there is more occupancy in Deluxe and A category houseboats than rest of categories. These two categories have capacity of more than 66% of total capacity in all categories because they have more number of rooms and number of beds. At the same time, this capacity is the reaction of their number as well. There are 350 Deluxe and A category houseboats out of total number of 640 (Table 21.1). The furnishing and facilities available in these houseboats are also more than others.

Table 21.1 Average occupancy of houseboats, 2014–15

Houseboat class	Number of rooms	Average no. of beds	Average occupancy (% age)	Total no. of houseboats	Total tourist capacity
Deluxe	4	12	45	190	2280
A	4	12	45	160	1920
B	3	9	40	120	1080
C	3	6	40	90	540
D	2	4	35	80	320
Total	16	43	41	640	6140

Source Primary survey by the authors, 2014–15

21.7 Solid Waste Generation

The most important issues related to houseboats are the solid waste related problems. It is a big threat not only to houseboats but to the environment of Dal Lake as well. Dal Lake is one of the most polluted Lakes in the world which receives tons of solid waste and sewage from the adjoining areas. It is assumed that houseboats and tourists add significantly to the pollution load of the Lake. Dal Lake is surrounded by residential areas inhabiting more than 60000 people in 24 settlements (Ahmad and Bhat 2008). The waste found in the Lake are polythene, clothes, paper, nylon, vegetables, fruits, etc. from last two decades the depth has drastically reduced due to various reasons but solid waste dumping is the largest contributor to the to the pollution load. Houseboats accounts for population of more than 10000. It is observed that these houseboats collect their solid waste in the dustbins and containers and then these dustbins are carried by the SMC and transported to outskirts of the city (Table 21.2).

Out of total, 662.5 kg of solid waste generated per day (Table 21.3), Deluxe and A category houseboats generate more than 60% of the waste. D category has the least contribution in the waste generation, so far as the tourists are concerned. Table 21.3 given above clearly shows that more than the 66% of the waste generation is from the Deluxe and A category houseboats. At the same time, B, C and D categories are located at the backside of the lake front and have containers available to collect the wastes. The capacity of the containers is 10 kg which have been provided by SMC and there is door to door collection in the houseboats at the backside of the lake front.

In these tourist accommodations 99.14 tons of solid waste is generated, calculated by using the guidelines formulated by CPHEEO (Table 21.4). The per capita generation of solid waste is 0.25 kgs. These are low because tourists do not stay 24 h in the houseboats and meals are also taken outside the houseboats. There is huge contrast in the generation of solid waste in different types of houseboats due to difference in the number of rooms, beds, occupancy rate and locational advantage (Wani and Shah 2013). The waste generated in houseboats is collected in container till the next day the SMC employee comes and collects it. The collection is different in case of houseboats which are in the middle of the Lake, there are some special boats of

Table 21.2 Average occupancy of houseboats, 2014–15

S. No.	Months	Average occupancy of houseboats
1	January	00
2	February	00
3	March	40
4	April	70
5	May	70
6	June	70
7	July	60
8	August	40
9	September	60
10	October	60
11	November	20
12	December	00

Source Primary survey by the authors, 2014–15

Table 21.3 Solid waste generation by different categories of Houseboats (Tourists/Day. kgs), 2014–15

Houseboat class	Total no. of houseboats	Tourists received	Waste generation tourists/day. Kgs
Deluxe	190	1026	256.5
A	160	864	216
B	120	432	108
C	90	216	54
D	80	112	28
Total	640	2650	662.5

Source Primary survey by the authors, 2014–15

Table 21.4 Solid waste generation by different categories of houseboats (Tons/Year), 2014–15

Houseboat class	Waste generation tourists/day/Kg	Quantity of Solid Waste tons/year
Deluxe	615.6	38.385225
A	518.4	32.3244
B	259.2	16.1622
C	129.6	8.0811
D	67.2	4.1902
Total	6769.92	99.143125

Source Primary survey by the authors, 2014–15

Table 21.5 Per unit solid waste generation by different categories of houseboats, 2014–15

Houseboat Class	Quantity of solid waste generation (Tons)	Per unit generation/category
Deluxe	38.385225	0.2020275
A	32.3244	0.2020275
B	16.1622	0.134685
C	8.0811	0.08979
D	4.1902	0.046557778
Total	99.143125	0.154911133

Source Primary survey by the authors, 2014–15

LAWDA department which are used to go to every houseboat for the collection of the waste. Some employees from the LAWDA come in the boats of the department and clean the front side of the Lake so that the Lake will appear clean to the tourists and passers. Inside the houseboat locality, huge heaps of solid waste is found which remain unnoticed by the department for most of the time. At the same time, the passage through the houseboats to the interior of the water body is also choked so it is very difficult to go there and collect the waste from the boats and water body.

The table given above (Table 21.5) clearly shows the average per capita generation of solid waste for all houseboats which is below 0.15 tons/year. But Deluxe and A category generate more than 0.2 tons/year that is more than the average while as B, C and D categories generates less than the average like 0.13, 0.08 and 0.04 tons/year respectively.

21.7.1 Household/Houseboat (Donga) Solid Waste Generation

The most of the solid waste generated in the houseboat community is from the households who live in dongas and have all the business of life in these houses. More than 5000 boatmen live in the Dal Lake. The grand total of solid waste generated by both houseboats and households in the Dal Lake is near about 848 tons/year (Table 21.6 and 21.7). This quantity is very big and cannot be managed by boatmen community by their own will. Again, the solid waste generated by each unit of houseboat categories shows a lot of variation. This is because Deluxe and A category houseboats have also employees and few members. The total average generation of each household is 1.168 but Deluxe houseboats have 1.201 tons per unit of household. Category A also has more than the average per unit generation as 1.347 tons/unit.

The Table 21.7 show that C and D categories have least solid waste generation and A categories have highest per unit solid waste generation. This is because A category and Deluxe houseboats have more occupancy and the employees with family members stay in the houseboats. Other categories have less family members and do

Table 21.6 Solid waste generation scenario by different categories of houseboats, 2014–15

Houseboat class	Total no. of houseboats	Waste generation by residents/day in Kgs	Waste generated/year
Deluxe	190	615.6	228321.5385
A	160	518.4	215630.7692
B	120	259.2	132073.8462
C	90	129.6	90969.23077
D	80	67.2	80861.53846
Total	640	6769.92	747856.9231

Source Primary survey by the authors, 2014–15

Table 21.7 Solid waste generation scenario of households, 2014–15

Houseboat class	Total no. of houseboats	Waste generated/year	Solid waste in households (Tons)	Avg. per capita generation/category
Deluxe	190	228321.5385	228.3215358	1.201692308
A	160	215630.7692	215.6307692	1.347692308
B	120	132073.8462	132.0738462	1.100615385
C	90	90969.23077	90.96923077	1.010769231
D	80	80861.53846	80.86153846	1.010769231
Total	640	747856.9231	747.8569231	1.168526442

Source Primary survey by the authors, 2014–15

not have any employees. At the same time these households have also started to shift from their usual business and left the houses for other jobs. Some of the houseboats are occupied by the rented people and some do not have any residents inside. The disposal site by the SMC for C and D categories is near the banks of Lake. The other sites are inside and solid waste is collected at alternate days by LAWDA.

It is observed that solid waste is majorly contributed by households (Table 21.7). In spite of, the containers provided by SMC to the households in the back side of Lake Front the solid waste heaps are witnessed. It can be assumed that houseboats solid waste is of low quality and can be managed by mechanism by houseboats owner but the waste generated by households remain a challenge because some parts are very inaccessible because they are inside the Lake and there are very narrow streets to approach them. To dispose the water in the water also serves the interest of the people to encroach into Lake and a shareholding LAWDA.

21.8 Sewage Disposal and Management

This is the most common problem of houseboats which is directly linked to the pollution of Dal Lake. The sewage generated is directly disposed-off into the Lake (Fazal and Amin 2013). All the washrooms, toilets, kitchen wastes, domestic garbage, wash basins, etc. goes are directly flushed out into the open water of Lake. It is observed that near the houseboats the water is very polluted and opaque. The color of the water has turned to dark blue in color. The depth near the houseboats has gone down to 1 m and during winters it is even less. The bathroom commodes are directly flushed out into the open water. Some of the houseboats have been provided with septic tanks and pipes but the pipes got choked in few months because it was against the gradient. So far, there has been no mechanism to manage the sewage. The ecology and environment has been destroyed because of the poisonous direct disposal of sewage (Khan et al. 2013). The donga boats which are very closely attached to the houseboats where boatmen live are immensely disposing its sewage in the Lake. These are constructed on the wooden logs which are embedded in the water. Very next to them are the bathrooms and latrines that have a direct vent into the water body. The second harmful substance disposed is the waste water of detergents and soaps which has destroyed the aquatic life of the Dal (Wani 2013).

21.9 Comparative Analysis of Sewage Generation by Houseboats and Households

It is observed that sewage disposed by houseboats categories varies drastically. Deluxe and A categories have more than the 60% of the disposal in case of households and houseboats. Again, this is because there are more family members and employees in these categories. Collectively more than 3,28,000 gallons of sewage is disposed by houseboats and households annually. This sewage contains ammonia, urea, fecal matter, paper, soaps, detergents etc. As a matter of fact, these two types are in the middle of the Lake near the Dal gate which opens into the Jhelum River. So, it is obvious that it also contributes into the pollution of Jhelum River. While comparing the contribution of sewage between households and houseboats households contribute immensely.

Table 21.8 clearly shows that most of the sewage disposed into the water comes from the households of boatmen community. There are 280539 gallons of sewage disposed by households and 48097.5 gallons disposed by houseboats or tourists per year. Only 15% of total sewage disposed comes from the houseboats while 95% comes from the households.

The contribution of sewage per unit category shows that there is less difference in different categories 45% of per unit of total sewage is contributed by Deluxe and A category. This shows that there is more consumption of water because of more number and members of family. Coming to the houseboats category wise 59% is contributed

Table 21.8 Solid waste generation scenario of households, 2014–15

Houseboat class	Total population	Tourist received	Sewage in gallons by houseboats/year	Sewage in gallons by households/year
Deluxe	1564	1026	18621.9	85629
A	1477	864	15681.6	80865.75
B	905	432	7840.8	49548.75
C	624	216	3920.4	34164
D	554	112	2032.8	30331.5
Total	5123	2650	48097.5	280539

Source Primary survey by the authors, 2014–15

by the Deluxe and A category and only 41% by the rest of the houseboats. This indicates that they enjoy much services of water and others and pollute the water on the cost of others.

The average per unit disposal of sewage for households is 438.25 gallons/year and 75.15 gallons/year for houseboats. B, C and D categories have less per unit sewage disposal while as Deluxe and A category have more per unit disposal. The effect of Deluxe and A category is more severe because these categories are in the middle of the Lake but others are mostly near the banks. This sewage is suspended in the water and is carried by the waves to other parts of the Dal Lake. But the immediate effect of this waste is that it increases the BOD level near the houseboats which results into the disappearance of aquatic animals and increases the growth of weeds. It is observed that the area near the houseboats in now more marshy land and some parts have dried up. It is also evident from the site that a concrete jungle of aquatic plants is growing near houseboats which makes backside invisible to the people. The most worried issue is that the people of these houseboats are now claiming of the land in the Dal Lake and assume that it is their own land. This land is dried up and turned into a park for tourists. It has also made a divide in the Lake which separate front side from the backside of the Lake by continues wall of aquatic trees and dry stretch of land.

21.10 Management of Solid Waste and Sewage

The total solid waste generated is 250 tons per year from the households and houseboats but 88% is from the households only. It indicates the major management issue is related to households. Tourists are not big problem for solid waste management because annually they generate 100 tons only which is also in the peak tourist season but households of the boats are permanent residents and continuously generate the waste. Waste water from this community accounts for 328000 gallons per year, out of which 85% is contributed from the dongas which clearly indicates that they are matter of serious concern. There is not any infrastructure provision in the houseboats for sewage and solid waste. The bins have provided by Srinagar Municipal Corporation to some houseboats but majority do not have any containers. The pipelines for the sewage disposal have been provided to 10 houseboats that have become dysfunctional because poor planning and lack of institutional management. The problem lies in the planning bodies and administration that are unable to provide infrastructure for discharge of waste water and solid waste. Dal Lake is low lying as compared to adjoining areas of East and North East which is under *Zabarwan* hills is bordering the Lake on these two sides, in the south and south East is high land and in between is the canal joining the Jhelum River. Because of gradient the water from the houseboats and Dal Lake cannot go against the gradient, which leads to chocking of channels and pipelines of houseboats. Due to interior location, the infrastructure is not available to meet the basic demand. The people are not aware of the benefits of management of wastes. As the solid waste continuously accumulates in the Lake it leads to encroachment by the boatmen community. The Lake has lost 20 km^2 area in the past 20 years due to encroachment and pollution load.

This slowly settles down at the bottom and reduces the depth again. The accumulation of wastes into the water body results into the reduction of the depth after which it becomes completely a dry land with enough vegetation. This paves way for the boatmen community to encroach into the water body and then claim it as their own holding. There is also a conflicting issue that the space between two consecutive houseboats is occupied by the owner and there is no way to go inside the water body through these passages. They are filling these gaps with wastes and then allow trees to grow for their benefits and the authority has banned the local villagers to uproot the weeds which were earlier used as manure for the agriculture.

This is the cyclic process of changing water body into a dry land for encroachment and dumping site for wastes. Both solid waste and sewage is disposed into the water body by houseboats. Most of the waste comes from the local residential areas whether it is solid waste or sewage. The channels for outward movement have been blocked by people for encroachment and due to accumulation of wastes. Then the grave situation of the Lake is observed in the above picture which seems to be uncontrolled and intentional.

21.11 Conclusion

Dal Lake has lost its spatial extent from 36 km^2 due to encroachment and pollution load. The color of the water is deep blue to bluish black which symbolizes the presence of huge organic waste in it. It produces a foul odor due to constant release of methane and carbon monoxide gases. Houseboats are mostly concentrated in the south-eastern tip of Dal Lake. The depth near houseboats is less than 1 m and more than 2 m as we go inside the Lake and other sides of the Lake. It is because of continues disposal of solid waste into the Lake by houseboats and households of these boats. More than 80% of the waste is produced by household but half of the waste is collected in dust bins and taken away by SMC. Houseboats produce less waste but of it mostly disposed into the Lake.

Sewage is the biggest threat to the Lake because it is directly disposed into the water body (Wani 2013) from washrooms and kitchen by the households and households. The sewage disposed by households is 85% out of total sewage disposed by the boatmen community. There is no mechanism in place to check this severe disastrous problem. The associated departments are moving away from their responsibilities and no one claim to be the sole organization of management of houseboats. Though STPs were placed in 10 houseboats few years back but due to negligence and lack of proper planning they became dysfunctional. Now, due to the increasing level of pollution and turmoil there is continuous drop in the tourists that stay in houseboats. The average occupancy of these tourist accommodations is about 40% which is not sufficient to sustain this novel industry. There is an apprehension that in few decades most of the houseboats will be lost especially Deluxe and A category because of lack of business. Some houseboats hardly manage the salary of employees and maintenance cost of houseboats. The historical and cultural asset of Kashmir is losing its credibility and essence because of pollution, competition from guest houses, and lack of institutional framework.

References

Shah AM, Shabana (2013) Houseboat-a component of tourism industry in kashmir. In 3rd International Conference on Business, Economics, Management and Behavioral Sciences (ICBEMBS'2013), pp 90–94. 26–27 Jan 2013

Badar B, Romshoo SA (2008) Assessing the pollution load of Dal lake using geospatial tools. In Proceedings of Taal 2007: The 12th World Lake Conference, pp 668–679

Ahmad P, Bhat GA (2008) Indiscriminate disposal of solid waste choking world famous Dal lake in Kashmir Valley. In The 12th World Lake Conference (2008), pp 1458–1462

Wani D (2013) Microbial assessment and effect of seasonal change on the removal efficiency of FAB based sewage treatment plant. Environmental Engineering & Ecology science, Centre of Research for Development-2013, University of Kashmir, Srinagar-190006

Fazal S, Amin A (2013) Effects of land transformation on water quality of dal lake of srinagar city. J Indian Soc Remote Sens

Khan NA, Khan MY, Raja IA, Bhat AA (2013) Water Chemistry of Famous Dal Lake Kashmir. Nat Sci 2013:22–23

Madhok BR (1992) Kashmir: the storm center of the world. A. Ghosh 5720 W. Little York, 216 Houston, Texas 77091

Wani MA, Shah SA (2013) Magnitude and seasonal variations of solid waste generation in tourist accommodation of Dal lake. Int J Environ Sci 2(1):57–62

Weblinks

Central Public Health and Environmental Engineering Organisation (CPHEEO), (Ministry of Health in 1954), Ministry of Housing and Urban Affairs (1973–74), Goverment of India. http://cpheeo.gov.in. Accessed 12 December 2014

Official website of Jammu & Kashmir Tourism (2014) Department of Tourism, Government of Jammu and Kashmir India. http://jktourism.org. Accessed 11 March 2014

Chapter 22
Ecological Challenges and Vulnerability Assessment for Exploring the Adaptation-Development Nexus for Sustainability in Alaknanda River Basin, Uttarakhand, India

Bindhy Wasini Pandey, Subhash Anand, V. S. Negi, Usha Pathak and Abhay Shankar Prasad

Abstract Climate change is increasingly recognized as critical challenges to hazards, ecological challenges, human well-being, food security, agriculture and sustainable development in the Alaknanda river basin. Climate change has a greater impact on rural livelihoods and its sustainability. Anthropogenic activities are continuously disturbing the natural system of the Garhwal Himalaya and its impact on extreme hydrological events. Human interference, unscientific developmental activities, agriculture extension, tourism activity and road construction are creating hydrological hazards. Soil erosion and landslide have been recognized as major hazards in the high altitude region of Himalaya. The climate variability and its vulnerability studies can help to identify the vulnerability of a region towards climate change which further helps in the formulation of mitigation and adaptation strategies. This research paper proposes a GIS-based analysis for identifying hazards and ecological challenges for the purpose of sustainable development planning. Present research includes a combination of qualitative and quantitative research approaches for livelihood vulnerability assessment and its adaptation for sustainable development determined mainly by a

---

B. W. Pandey (✉) · S. Anand · A. S. Prasad
Department of Geography, Delhi School of Economics, University of Delhi, Delhi, India
e-mail: bwpdsegeo@gmail.com

S. Anand
e-mail: sanandpv@yahoo.co.in

A. S. Prasad
e-mail: abhaydse@hotmail.com

V. S. Negi
Department of Geography, Shaheed Bhagat Singh (Eve) College, University of Delhi, Delhi, India
e-mail: negivirens@yahoo.com

U. Pathak
Department of Geography, Kalindi College, University of Delhi, Delhi, India
e-mail: ushakpathak@gmail.com

S. Sahdev et al. (eds.), *Geoecology of Landscape Dynamics*,
Advances in Geographical and Environmental Sciences,
https://doi.org/10.1007/978-981-15-2097-6_22

weighted matrix index. For this purpose, Hazards Zonation Mapping using Geo-Spatial techniques was carried out so as to classify the land surface into zones of varying degree of hazards. For this analysis, a qualitative research method including participatory research approach (PRA) at village level has been used for ecological challenges and risk assessment. This parameter is important in building resilience capacity and ensuring sustainable development pathways and provides the various mitigation processes from extreme climatic events. This paper has suggested the policy to improve the transfer of scientific knowledge into policy and to increase mutual understanding, partnership, and cooperation for better policy outputs in sustainability and mitigate the ecological challenges and improved the livelihood security. These approaches will be useful in building collaborative arrangements across political and administrative barriers and boundaries to govern at the scale of the sustainability challenge to way towards the Sustainable Development Goal.

Keywords Geo-hydrological hazards · Ecological challenges · Sustainable development goals (SDGs) · Livelihood securities · Disaster management and mitigation

22.1 Introduction

Climate induced disasters such as cloudburst triggered Glacial Lake Outburst Flood (GLOF), flash floods, debris-flow, mass movements and landslides are a very common phenomenon in the Garhwal Himalaya, Alaknanda river Basin, Uttarakhand. There are many instances in the recent past when these catastrophic events caused heavy losses to lives, tremendous damage to property and livelihood. The Alaknanda River Basin located in Uttarakhand witness many of these climatic/natural hazards quite frequently. Especially during the monsoon period of the year, these devastating activities become intense. Moreover, while a fragile environment becomes more vulnerable to disasters in Garhwal Himalaya, natural disasters also degrade the environment and ecological system and create ecological challenges (Singh and Robotham 1995).Every year several incidences of natural hazards are reported from all parts of the Himalayan region causing loss of hundreds of lives (human and animal) along with their assets, houses and livelihood options. Natural causes and anthropogenic activities are continuously disturbing the natural system of the Himalayan environment, the effects of which can be seen in the hydrological behaviour of the river (Negi and Joshi 2004). Preliminary studies conducted on this subject indicate that human interference, unscientific developmental, anthropogenic activities, agriculture extension and unplanned way of road construction are some of the activities which are creating the hydrological imbalances within the Garhwal Himalaya. The mountainous area is more prone to natural and manmade disasters like flash flood, earthquake, landslides and avalanche, etc. (Pandey 2010). High relief, snow-capped summit, deeply dissected topography and geological structure, antecedent drainage, complex, and rich temperate floras in the sub-tropical latitudes give a distinct character of the Garhwal Himalaya (Pandey 2002). The Garhwal Himalaya is considered to be the youngest mountain and is tectonically active. Out of these phenomena, cloudbursts followed by landslides are more prominent. A number of landslides and landslips take place during monsoon season (Pandey and Prasad 2016). It is generally more prone in those areas where the hill is being cut for road construction and proper drainage facilities do not exist. Due to this, soil flows along with the rivers and causes siltation/sedimentation in the lower reaches which cause a flash flood. Soil erosion and landslide are directly proportional to the extent and magnitude of the flood. Whenever soil erosion has increased, the average annual flooded area has been also increased. Because of the developmental works, the Alaknanda basin is characterised by steep slopes and a high rate of soil erosion. In addition to the geological conditions, intense seasonal precipitation, particularly during the summer monsoon, triggers various types of natural hazards (Sati 2013 and Sati Sati 2014a, b). Natural hazards are risks encountered by people in the physical environment which represent a wide deviation from the normal behaviour of a system and have abrupt modifying capabilities and can create imbalance resulting in catastrophic losses of life, livelihood options and properties. Various factors in the form of changing man–land ratio, limited arable land, increasing intensity of land use/land cover and careless application of technology leading to further land loss through soil erosion make the subject of proper land use a complex issue. Frequent occurrences of natural hazards such as landslides, cloudburst, Glacial Lake Outburst Flood (GLOF), heavy snow, avalanche and flash floods are becoming a common feature in mountainous regions (Gardner et al. 1992 and Singh 1992).

The Sustainable Livelihoods Approach (SLA) is a way to enhance the understanding of the livelihoods of rural communities and suggests the adaptation techniques for sustainable development. It can be utilized in regional planning, and in assessing the contribution that existing activities have made for sustaining livelihoods and development. Closest to the humans on the centre of the framework is the resources, assets and livelihood options which include natural resources, technologies, knowledge, skills and capacity, health, access to education, sources of credit and networks of social support (Scoones 2009 and Chamber and Conway 1992). The extent of their access to these assets is strongly influenced by their vulnerability context, which takes account of trends (for example, economic, political and technological), shocks (for example, epidemics, natural disasters, civil strife) and seasonality (for example, prices, production and employment opportunities). Therefore, improved understanding of development policies, land capability, sustainability, land use/land cover planning, the environmental interaction may improve resource management and lead to better policies towards long-term environmental sustainability and livelihood security in the region from the natural extreme climatic events (DFID 1999).This research has discussed and observed out the framework for know-how local knowledge associated with disaster preparedness. It is based on the understanding and transformation the techniques of local knowledge and skill, key dimensions of local knowledge on disaster preparedness and mitigation and the link among local knowledge, disaster preparedness, and livelihoods and poverty reduction for disaster risk reduction.

22.2 Study Area

The Alaknanda river basin extends between 30^{0} 0' N to 31^{0} 0' N and 78^{0} 45' E to 80^{0} 0' E covering an area about 10882 Km^{2}, represents the eastern part of the Garhwal Himalaya, Alaknanda basin in Uttarakhand. It covers eighteen development blocks under six districtsBageshwar, Chamoli, Rudraprayag, Tehri and Pauri Garhwal and Pithoragarh (Fig. 22.1). Alaknanda River meets Bhagirathi River at Devprayag to form a common stream of Ganga River. The Alaknanda Basin is characterized by hilly terrain, deep gorges and river valleys (Bandooni 2004 and Pal 1986). The region is broadly divided into four major divisions (i) The Great Himalayan Ranges (snow-covered regions), (ii) Alpine and pasture land (covered by snow during the four months of winter season) (iii) Middle Himalaya (characterized by high concentration of population) and (iv) River valleys (characterized by mushrooming service centres and institutions).

Among the major rivers of India, the Alaknanda river and its tributaries (Dhauli Ganga, Vishnu Ganga, Nandakini, Pindar, Mandakini and other numerous perennial streams) originate and flow here. The highest mountain peak of the Himalayan ranges such as Nanda Devi, Kamet, Trisul and Chaukhamba are also located in Alaknanda basin.

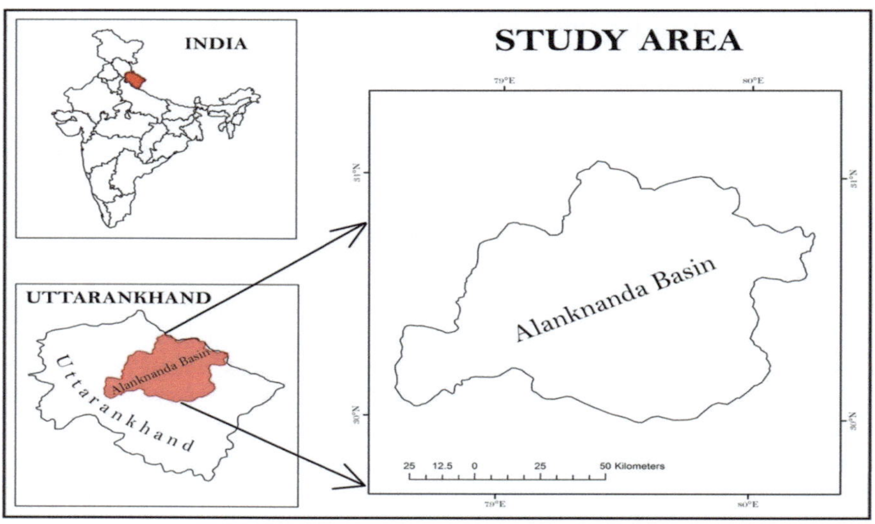

Fig. 22.1 The study area, Uttarakhand, India *Source* Landsat TM, 2017

22.3 Research Methodology and Data Base

The primary data has been collected through qualitative and quantitative methods. The qualitative methods involve interviews and group discussions with local elder people and patrons, group of local communities, local authorities and semi-governmental officers, NGO/development agencies staffs and participant observation. The perceptions of the local community regarding different aspects of sustainability and hazard occurrences and its management have been gathered through a pre-structured questionnaire, interviews and discussions. Informal interviews and discussions with local communities and officials provided information about the locations of prior events and did surveys of historical documents. The analysis is based on maps from Survey of India, and satellite imageries. Various thematic maps pertaining to drainage, geological structure and land use/land cover map were generated with the help of Geo-spatial techniques.

Quantitatively, formal interviews with potential tour operators and travel agents and contingent valuation methods (survey-based interviews with domestic and international tourists and tourist agents). The questionnaire includes questions related to local inhabitants and tourists perception about Natural hazards and livelihood options in the Alaknanda river basin. The interview questions have been grouped into three sets such as (1) sustainability of existing agriculture system (2) responses towards the adoption of diverse cropping pattern (3) reliability of diverse livelihood options (DFID 1999, Agarwal and Joshi 2006). Collected data has been interpreted by using various statistical and cartographical methods. The primary and secondary data were utilized for analysis of livelihood security, vulnerability and adaptation issues for the disaster risk reduction and sustainability. Perceptions of local community regarding different aspects of sustainability, hazard occurrences and its management were recorded through reconnaissance survey and a semi-structured questionnaire. The simple random sampling (SRS) technique was used for the primary survey.

22.4 Results and Discussions

Alaknanda river basin is hilly and mountainous regions with very narrow intermittent valleys. The hills are very steep sloping with occasionally gentle to moderately sloping flatter tops and strongly foot slope (Fig. 22.2). A few intermittent valleys are characterised by very gentle to gently sloping river terraces. On the basis of physical features, Alaknanda basin is divided into three zones viz. Lesser Himalaya, Greater Himalaya and Zanskar Himalaya.

The Garhwal Himalaya has a well-developed topography with gentle slopes and deeply dissected valleys. This area has evidence of recent rejuvenation. The rejuvenation process produced many distinctive features viz. - dissected topography with valleys showing vertical walls in the lower parts and gently sloping and discernible concave hilltop in the upper area, variation in elevation of river beds at the foot of

Fig. 22.2 Deep transverse Gorges, Mandakini River (Alaknanda River Basin) nearby Hanuman Chatti, Uttarakhand *Source* Primary Survey, 2015

Great Himalaya and at the points where the rivers debouch into plains and the altitudinal variations in the surrounding region as projected by the peaks (Pant 2000).The Greater Himalaya is located in the northern part of the Alaknanda river basin. Precipitous scraps and vertical-wailed gorgeous valleys and tumbling and foaming rivers characterise the northern belt of Great Himalaya with its peaks having 6500–7800 m high. Most of the rivers, flowing in the basin, originate from this zone. This has young topography and is tectonically still active and paradoxically is made up of the oldest rock of the Himalaya such as pre-Cambrian metamorphic and granitic gneisses. The peaks are often wedged like in form, well shaped and with incipient ridges and buttresses. There are sheer precipices dropping 2000–5000 m below into the torrents and gorges (Bandooni 2004).The altitude variation of glacial lakes had considered the total mapped lakes and it was found that the 3500–4000 m elevation zone is maximum sensitive for lake change. A number of lakes are represented in Fig. 22.3.

Decade-wise analysis has found that the highest number of lakes was found in the zone of 3500–4000 m in Alaknanda river basin. From 1976 to 2011 there has been an increase of 69 lakes having an area of more than 0.01 km^2 (Fig. 22.3). 18 new lakes grown in 1990 whereas this number increased to 29 in 1999 and 31 new lakes were found in LISS 4 imagery of 2011 (Fig. 22.3). Very less disappearance of lakes was observed in the last 40 years of datasets. That means 3500–4000 m and above elevated area is more vulnerable to Glacial Lake Outburst Flood (GLOF) and flash flood (Fig. 22.4) in Mandakini river basin, a tributary of Alaknanda river basin in Tehri Garhwal and Rudraprayag districts.

The Indian Summer Monsoon is the predominant source of precipitation (rainfall) in Alaknanda river basin with partial contribution from western disturbances

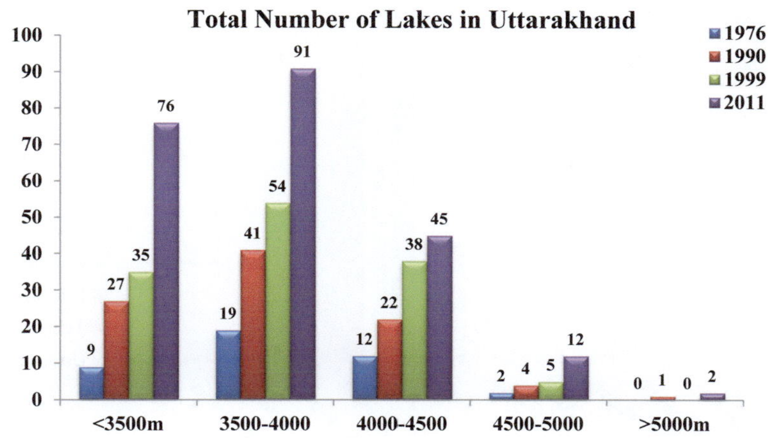

Fig. 22.3 Total number of lakes in Alaknanda River basin, Uttarakhand *Sources* 1976_Landsat MSS, 1990_ Landsat TM, 1999_Landsat ETM, 2011_LISS IV

Fig. 22.4 More Vulnerable to Glacial Lake Outburst Flood (GLOF) and Flash Flood, Nearby Kedarnath Temple, Uttarakhand *Source* Primary Survey, 2016

during the winter season. Winter precipitation generally occurs between December and March when the western disturbances are dominant in the area as they move eastward over northern India. Total summer (JJAS) rainfall for each observation periods between 2007 and 2012 was 1685 mm, 1513 mm, 734 mm, 1662 mm, 1348 mm and 1115 mm for respective years. Based on rainfall data from our observatory at Chorabari glacier, the area received maximum precipitation during the rainy season, i.e. July and August (Fig. 22.5).

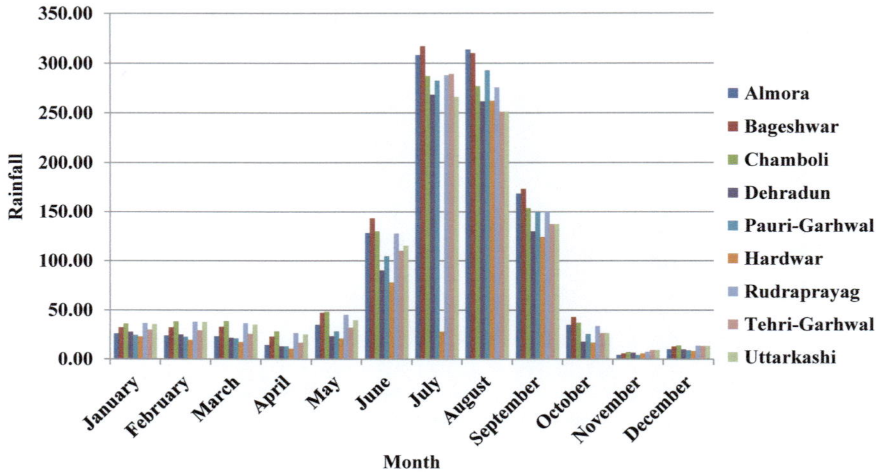

Fig. 22.5 Mean monthly rainfall (cm), Alaknanda River Basin (Mean for 1962–2013) *Source* Disaster Mitigation and Management Centre, Government of Uttarakhand, 2013 and Ministry of Water Resources, 2013

Rainfall variability in the basin is due to the leeward and windward direction of the slope. Joshimath, Karanprayag, Okhimath, and Srinagar are the places located in leeward direction. Consequently, those places receive less rainfall in comparison to the places located in the windward direction as Mandal, Gopeshwar, Gwaldom, Diwalikhal, and water dividing facing east slope. Mandal region receives the highest rainfall (400 cm) and known as Cherrapunji of Garhwal. Average annual rainfall is 125.7 cm. November and December months receive minimal rainfall i.e. 6.78 cm and 21.92 cm respectively. During the monsoon period, mainly two months (July and August), highest rainfall (above 400 cm) occurs. In a few regions, winter and summer seasons are also wet. Heavy rainfall occurs at some places of during July and August when the main rivers and their numerous tributaries flowing above danger marks. Cloudburst, debris flow, soil erosion, landslide, mass movement, and consequently flash flood are very common. The entire region is worst affected due to this catastrophe during monsoon resulting in heavy losses of life and property every year. There are many instances when major catastrophes took place from time to time (Sati 2013).

Distribution of precipitation in the basin is strongly controlled by local orographic effects. Heaviest precipitation occurs along the southern spurs of the main Himalayan range and in the vicinity of more prominent central ranges such as Doodhatoli. The summer monsoon approaches the southern part around the middle of June and extends into the interior until most parts of the basin have it by the third week of the same month. Snowfall in the basin is associated with winter depressions, which start affecting the northernmost part of the basin by November, the remaining part of the valley by December and have their southerly course in the months of February-

March. During exceptionally severe winter snow may occur, as low as 1100 m. But in localities below 2500 m, it rapidly melts away. In the high hills, snow continues to accumulate till April when that sets in. In winter, a large part of the northern area remains buried under a thick blanket of snow.

22.4.1 Ecological Challenger: Extreme Climatic Events and Future Vulnerability

Extreme climatic events and disasters are natural like a cloudburst, flash floods, soil erosion, landslides and GLOF (Glacial Lake Outburst Flood). The long-term man-made disasters are caused by the severe impact that various human activities have on natural processes. Therefore long-term global environmental change has both natural and manmade concerns (Mukherjee 2013). During disaster and post-disaster events, I have identified the high vulnerable issues such as damage to house/properties during a flash flood, drinking water, electricity, emotional well-being, and health, the overall source of income/livelihood options, daily work, livestock loss, and agricultural land. I have categorized them into four categories for vulnerable issues such as high vulnerable (75–100), agree vulnerable (75–100), medium percentage agree (25–50) and low vulnerable (0–25); on the basis of secondary data of disaster management and mitigation centre, Uttarakhand (Fig. 22.6).

The loss of trees has an impact on the income of almost all of the affected households. This loss of trees is not only an economic loss to the household but also changes the visual impact of the village and ecological systems. The flash flood affected the overall affected household income of 97% of households (86% strongly agreed and 11% agreed). This has long-term results such as low yields from flooded fields/land, damage to agricultural land, loss of trees and damage to houses. Furthermore, 84% of participants reported that their daily work was very severely affected by the flash flood and the other 09% were less severely affected. This consequently impacted their household income (Fig. 22.6). 88% of households have strongly agreed and 12% households agreed to damage of households due to natural hazards. The unaffected household was in a way affected. Infrastructure damage impacted on households' quality of life very strongly. The village's drinking water is mainly supplied via pipes from the upper part of the mainstream to different clusters of houses. 76% household strongly agreed and 15% agreed that drinking water is a vulnerable issue during disaster and post-disaster. The flash flood damaged the pipes connections and contaminated the water, and 76% households have been reported to be affected very severely and 15% of households have agreed that they faced problems with drinking water. They had to drink dirty/polluted water which affected their health.

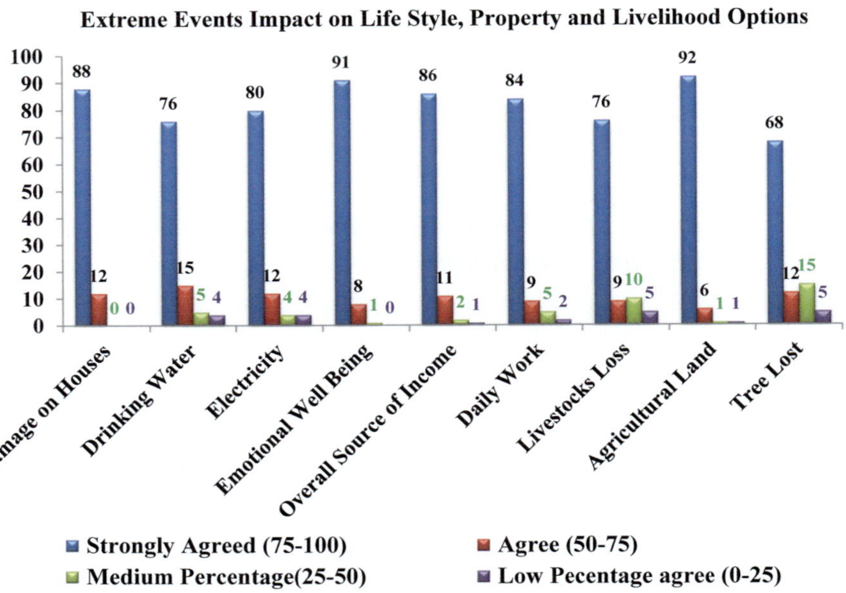

Fig. 22.6 Extreme events impact on life style, property and livelihood options *Source* Disaster Management and Mitigation Centre, Dehradun, 1996–2016

22.4.2 Hazards Zonation Susceptibility in Alaknanda River Basin, Uttarakhand

The distribution of natural hazards and landslide in different absolute relief classes and slope has been shown in Table 22.1 and Fig. 22.7. A maximum number of

Table 22.1 Natural hazards assessment on the physiographic features

	Slope		Absolute relief		Drainage density		Land use	
	Classes	Events	Classes	Events	Classes	Events	Classes	Events
1.	**<15**	0	<1500	10	<1	15	Mixed open forest	95
2.	**15–25**	10	1500–3000	40	1–2	48	Dense forest	34
3.	**25–35**	150	3000–4500	70	2–4	115	Agriculture land	53
4.	**35–45**	59	4500–6000	110	4–6	95	Barren land	61
5.	**>45**	35	>6000	42	>6	51	Settlement	110
Total		254		272		324		353

Source Field Survey and Remote Sensing Data, 2013–2016

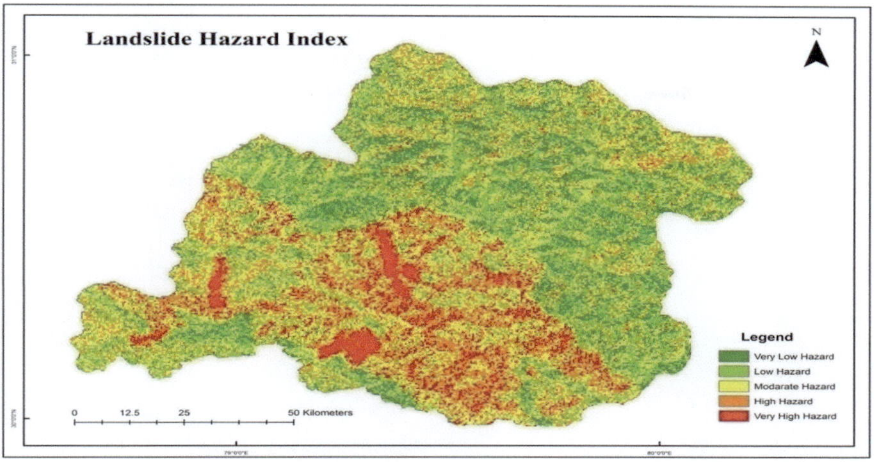

Fig. 22.7 Landslide Hazards Zonation *Source* LANDSAT TM, 2016

landslides to the extent of 110 forming 40.70% of the total landslides is restricted to (4500–6000 m) while only 10 landslides fall in Class-1 (<1500 m) which forms 3.70% of the total number of natural hazards (Fig. 22.7). Lower and lower-middle valley of Alaknanda basin are found between 444 and 1500 m altitude mainly in the southern part of the basin and it occupied about 12% of an area covering parts of Devprayag, Kirtinagar, Kot, Pauri, Khirsu, Jakholi, Augustmuni, Karanprayag, and Narayan blocks. Most of the blocks are in the south-west parts (Pauri, Devprayag, Nandaprayag to Joshimath in the middle Alaknanda basin (Dasoli block).

Upper middle valley and mountains are found in Ukhimath, Dasoli, Ghat, Joshimath, and Kapkot blocks and Mandakini, Kaligangariver in the north-west, along Pindar and Kaphni rivers in the south-east and in the lower part of Girthi Ganga of Dhauli Ganga sub-basin are vulnerable under High vulnerable landslide hazards in Alaknanda river basin.

Upper middle valley and mountains are found in Ukhimath, Dasoli, Ghat, Joshimath, and Kapkot blocks and they are about 16% area of the Alaknanda basin. The altitude of the Alaknanda basin is 3000–4500 m. The main valley is along Mandakini, Kaliganga river in the north-west, along Pindar and Kaphni rivers in the south-east and in the lower part of Girthi Ganga of Dhauli Ganga sub-basin. This Garhwal Himalaya of Alaknanda basin is highlyvulnerable to natural hazards (Bandooni 2004).

The high vulnerable areas of landslide hazards zone are from the central part to south-eastern part (Dasoli, Ghat, Deval and Kapkot blocks). Other two narrow belts are located in the southern block border (Khirsu, Gairsain, and Tharali blocks) and in the south-western (Kirtinagar, and Augustmuni block) part of the river basin. One large continuous belts of this zone extend from north-western part (Ukhimath block) to the central (Dasloi and Ghat block) part of the river basin. Other main areas under this sloping region are Upper Alaknanda basin and Dhauli Ganga sub-basin in the

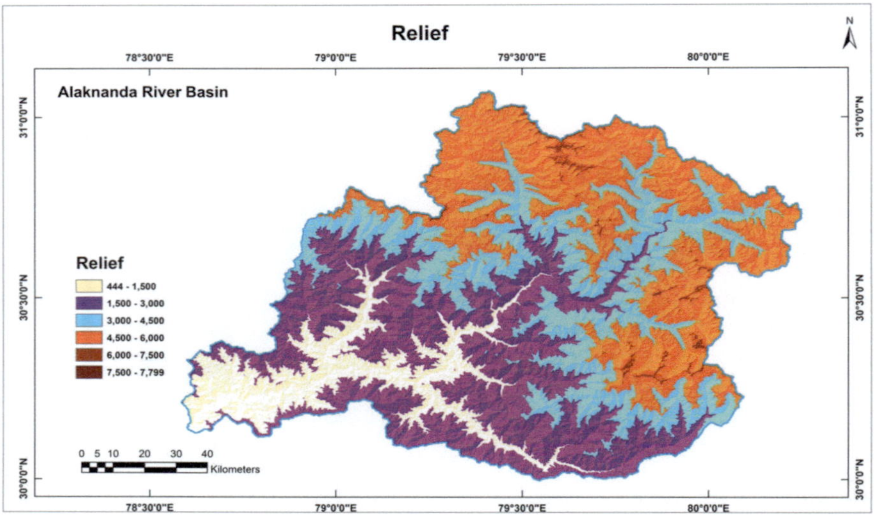

Fig. 22.8 Relief variations in Alaknanda Basin, Uttarakhand *Source* LANDSAT TM, 2016

north (Joshimath block), South-east (Kapkot and Deval blocks) and south central (Karanprayag block) part of the basin (Fig. 22.7).

North and east part of Alaknanda river basin is the belt of the Greater and Zanskar Himalaya where the altitude of the mountain is more than 4500 m from the mean sea level. It occupied more than 50% area of the Alaknanda basin, covering part of the Ujhimath, Joshimath, Munisari, Deval Ghat and Kapkot blocks. The main characteristics of this zone are more vulnerable to natural hazards, permafrost cover of snow, and lack of population and vegetable cover (Fig. 22.8).

The distribution of landslide, soil erosion and natural hazards under various slope classes showed that the maximum number of landslides to the extent of 150 belongs to the slope Class-3 (20–30°) forming almost 59% of the total number of natural hazards (Flash Flood, landslide, soil erosion, cloudburst and Glacial Lake Outburst Flood (GLOF), etc.

The area having 15–25^0 slopes is the zone of gently-moderately to steep slopes. It occupied nearly 12% area of the river basin and is found the maximum in the south-west (Devprayag, Kirtinagar, Pauri and Khirsu blocks) and North-west (Jakholi block), south-east (Deval and Kapkot block) and northern part (Dasoli and Joshimath blocks) of the river basin. The slope ranging between 25 and 45% is known as steep to very steep sloping zone and it occupied 40% area of the Alaknanda basin. The main areas of this slope stretch from the central part to south-eastern part (Dasoli, Ghat, Deval and Kapkot blocks). Other two narrow belts are located in the southern block border Khirsu, Gairsain, and Tharali blocks) and in the south-western (Kirtinagar, and Augustmuni block) part of the river basin.

One large continuous belts of this zone extend from north-western part (Ukhimath block) to the central (Dasloi and Ghat block) part of the river basin. Other main areas under this sloping region are Upper Alaknanda river basin and Dhauli Ganga sub-basin in the north (Joshimath block), South-east (Kapkot and Deval blocks) and south central (Karanprayag block) part of the river basin. The main features of the area are steep slopes, very shallow depth of the soils, rocky and barren lands, less area under forestry and very-very less area under cultivations (Table 22.2). This area is highly vulnerable to natural hazards.

Extreme climatic events, climate and weather change, population growth, loss of biodiversity, environmental degradation, migration and urban sprawl, livelihood insecurity are fundamental and major problems in Alaknanda river basin. Urbanisation and migration processes have significant impacts on mountain environments and communities. Natural disasters in mountains not only damage houses but also have great impacts on the downstream areas, affecting millions of people and resources (Jodha 1992). Garhwal Himalaya of high mountainous area is greatly affected by adverse and destructive natural processes and creates environmental challenges. Over

Table 22.2 People perception regarding techniques of control/magnitudes of Hazards (In Likert Scale)

Techniques	Landslides	Floods	Avalanche	Cloudburst	Rock fall	Soil Erosion
1. Afforestation	12	5	5	5	5	5
2. Control over deforestation	5	5	5	5	5	5
3. Construction of dam and Reservoirs	3	5	0	0	2	2
4. Embankments	3	5	0	1	0	5
5. Control on overgrazing	4	2	1	0	3	3
6. Better drainage techniques on Slope (along the road)	5	2	0	0	2	5
7. Agro-forestry	1	1	1	1	2	3
8. Control on construction of houses in the vulnerable area	3	2	1	0	3	3
9. Better agricultural practices	4	2	0	1	3	3
10. Slope-based construction	5	3	0	0	5	4
11. Check on Urban sprawl/tourism	3	1	1	3	4	5

Source Based on Primary Survey, 2013–2016

the generations, mountain communities have learned how to live or living with the threat of natural hazards and have developed well-adapted and risk-resilient systems (Dhanai et al. 2016, Negi and Joshi 2014). Poor and haphazard construction of mountain roads and houses increases the likelihood of soil erosion and is one of the most important human-induced triggers of soil erosion and landslides (Sati 2008a, b and Sati 2014a, b). Inappropriate agriculture practices and afforestation aggravate the pressure on fragile mountain ecosystems and accelerated the natural hazards (Noor and Rai 2014).

22.4.3 Adaptation Development Nexus for Livelihood Securities and Its Sustainability

On the basis of various types of data structures such as household demographics, livelihood options and patterns, agriculture diversification, health issues, infrastructure, education, water and sanitation, housing and property, coping strategies, covered for the adaptation development nexus. Livelihood securities can contribute to the more effective method for disaster risk reduction. In addition, productive alternative livelihood options can support sustainable income-generating through natural resources. This is important assets for people and communities after a disaster (Prasad et al. 2016).

Uttarakhand Tourism Development Master Plan (UTDMP) for 2007-22 has been developed the eco-tourism and its sustainability, which is coordinated through the Government of India, Government of Uttarakhand, United Nations Development Programme (UNDP) and the World Tourism Organization. The main goals of this plan is to increase remarkable sustainable tourism infrastructure, facilities and products within the prime tourism zones of Uttarakhand state as a development nexus for sustainability and generate sustainable livelihood options and maintain the eco-tourism/ natural tourism concept. **Appropriate Skill Development Models** just like the "Veer Chand Garhwali Programme" should be developed for the regional development; this is the model of the "Village Inherited Tourism Mode", wherein each family has a separate room for travellers, but the authorities will need to provide funds and investment. This model is likewise excellent as a self-employment scheme under the sustainability of livelihood in Rudraprayag, Chamoli and Pauri-Tehri Garhwal districts in Alaknanda basin. To make sure that the advantage of local tourism actually reaches the villagers, it is important to involve local women in restaurant activities, youths as drivers of taxis and guides, and local uneducated men as porters. There is a need to develop human resources, skill and hold capacity building training programmes.

To reduce the sensitivity of the families and communities in the village, this, in turn, may help them to be more efficient in overcoming the shock caused by the catastrophe, disaster and re-establishing their livelihood options and formulate the disaster risk reduction policies. To promote the policy through local community for the capacity building, Self Help Groups (SHGs), Joint Forest Management (JFM) and Community participation in resource management for the sustainability of resource (Gardner and Scally 1992, Singh 2016). Development Strategy for Hill Districts Infrastructure development is a commonplace development agenda to facilitate development in agriculture diversity, natural and organic farming and hybrid seed farming strategies are encouraged for all as a way to create the sustainability.

In Bageshwar district of Alaknanda river basin, following development plans had been applied together with diversifying agricultural products to include off-season vegetables, horticulture, floriculture and fruits (peas, cabbage, beans, tomato and potato). To promote the policy of sustainability through local communities and local resources such as: plantations of chillies, turmeric and other herbal and medicinal plants under the program of agriculture diversification and Promote the eco-tourism for sustainable development.

In Chamoli district of Alaknanda basin, following sustainable plans were implemented consisting of religious/natural tourism to the Valley of the Flowers, Hemkund Sahib, Badrinath, and Kedarnath, Nanda Devi National Park, river rafting, and rock climbing. Develop forest resources as a common property resource for the sustainability such as: Jatropha plantation for biofuel, primal trees for the cosmetics enterprises, and forest-based handicrafts for the development of local or regional communities at the local level through community's participation.

In Pauri Garhwal, following sustainable development plans have been implemented including poultry and wool-based development; Juice Processing Centre under the program of Joint Forest Management (JFM), Sheep, Diversify agricultural products to include herbal and medicinal plants, pulses, bee-keeping and mushroom cultivation start a medicinal plant-based totally pharmaceutical industry for the sustainable development.

In, Pithoragarh, following development plans have been implemented such as poultry and wool-based development, Goat-rearing, Diversify agricultural products to consist of litchi, herbal and medicinal plants, garlic and spices.

In Rudraprayag, following sustainable development plans had been implemented together with religious and natural tourism to Kedarnath; adventure tourism like river rafting, rock climbing. To promote the policy of agriculture diversification such as include herbal and medicinal plants, *haldi* and *coriander.* Promote the forest and Agro-based industries: Bio-fuel, bamboo plantation, and traditionally-grown manual for bakery products Forest-based industry through Self Helf Groups (SHGs).

In Tehri Garhwal of Alaknanda basin, following development plans have been applied such as agricultural diversification towards fruits vegetables, medicinal plants, spices, pulses, herbal and aromatic plants using a cluster approach combined with proper market development can be very successful. Forest and agro-based totally industries such as food-processing industry, forest-based industry, fruit and Juice-processing industry have been promoted for sustainable development through community participation.

In Uttarkashi of Alaknanda basin, following development plans had been carried out inclusive religious and natural tourism to Gangotri, Yamuntori, and many others. To promote the horticulture and floriculture, diversify agricultural products to include fruits and vegetables, development of sheep and goat-rearing and wool-based enterprises has been promoted for the promotion of alternative livelihood options and its sustainability.

In this communication, the Likert scale was employed for analysing the peoples' perception regarding the adaptation strategies to control/mitigate the hazards in Alaknanda basin. This was additionally subjected for analysing and identifying the issues of economic development practices within the village communities. It was also designed to determine and identify the opinion of the challenges. In Likert scale were assigned into five responses. Usually, the most negative responses are enumerated into a given numerical values of 1, whilst the most positive response has a numerical value of 5 such as strong agree (5); agree (4); uncertain (3); disagree (2);strongly disagree (1); (Likert 1932).

The local communities utilize the indigenous ability, skill or adaptation knowledge for the purpose of hazard control and in mitigation techniques to reduce the disaster. The technical expertise, economic status and awareness level are the important tools for disaster risk reduction through which one can assess the status of a community. There are certain numbers of hazards, that cannot be checked absolutely but certainly, their danger intensity may be reduced significantly by using superior technical expertise and efficiency in preparedness the mitigation method. Table 22.2, is an attempt to understand and analyze the level of awareness and technical expertise of the local communities of the Alaknanda basin to mitigate the different kinds of disaster. Every individual disaster is analyzed in relation to its most effective controlling technique.

The measures applied by the local communities have been arranged according to their preference and importance. The control over deforestation and launching afforestation programmes on a large scale were the two most preferred remedies or adaptation techniques, which have an impact on almost every category of disaster. There are certain other adaptation techniques, which have their applicability limited only to a particular disaster. Construction of dam and reservoirs and slope-based construction of roads are other effective techniques used for disaster prevention and mitigation by a large number of respondents. Better drainage technique, control on overgrazing and over urban sprawl and tourism, embankments on the riversides are other highly favoured disaster control techniques replied by a large section of the respondents. Apart from control over deforestation and afforestation techniques, the construction of dams and reservoirs and embankments to riversides are the most important techniques responded by 183 (In Linkert scale 5) and 173 (In Linkert scale 5) respondents respectively. To control the avalanche hazard, the construction of iron poles, check dams on rivulets and the construction of check-dam on slopes are important techniques other than the afforestation programme replied by about 188, 161 and 165 respondents respectively (Table 22.2). As mentioned earlier, in controlling the cloudburst hazard, afforestation is a highly responded technique, in fact, by more than 196 respondents. Respondents have expressed their concern regarding the sprawl of urbanization and tourism and feel that it should be checked immediately; by 47%. While reacting to the prevention techniques, a large number of respondents stated that the cloudburst hazard is a natural phenomenon and it cannot be checked by applying any prevention technique.

22.5 Conclusions

Anthropogenic activities are continuously disturbing the natural system of Garhwal Himalaya of Alaknanda river basin. The important factor causing the flood and also accelerate several hydrological hazards in the basin are heavy rainfall, cloudburst, Glacial Lake Outburst Flood (GLOF), landslides, Slope failure, deforestation, drainage congestion due to urbanisation. These hydrological hazards are mainly responsible for several socioeconomic consequences such as degradation of cultivated land, infrastructure loss, human casualties, loss of road, transport systems, and so on. For any developmental activities/construction in the study area, it is recommended that one adopts the following guidelines to avoid instability and landslides. As a general principle, construction on hills, particularly dams, tunnels, multi-storied buildings and roads should be done on the basis of the landslide hazard Zonation and aspect map. Construction on steep hill slopes should be avoided. Suitable drainage measures are essential since water seepage can cause stability problems. Afforestation of the surroundings using site-specific species can help in the stability of slopes. Construction of buildings adjoining active landslides, on the valley-fill materials of streams and near active tectonic lineaments should be avoided. The natural flow of the streams is obstructed due to the construction of man-made structures that results in diversion of the flow from its natural course. Vulnerability and adaptation assessment such as natural disaster, education, health, agriculture, housing, land reform, social and farm forestry, drinking water, poverty, alleviation programme, social welfare, etc. which play very important and strategic role in the field of disaster management. The application of community participation approach (CPA) is very important in preparing a disaster management plan for the sustainability. This will certainly increase the quality of the plan and its practicability. It is may be helpful to formulate a task force of the able-bodied person of the villages, who can be trained for rescue and relief during disaster management. One more advantageous thing with the local people, local Governance and Panchayats working for the disaster mitigation is that it can provide services without delay, and wait outside help as the immediate initial time after a disaster is very precious for rescue and saving time.

References

Agarwal SK, Joshi AP (2006). Technology on the move in the Himalayan region: from empowerment to sustainable livelihoods. In: Joshi AP, Agarwaland SK, Kumar R (eds) Mountain technology agenda, status, gaps and possibilities. Bishen Singh Mahendra Pal Singh, Dehradun, pp 140–160

Bandooni SK (2004) Land resources management and development in Hill Areas. Research Press India, New Delhi

Chambers R, Conway G (1992) Sustainable rural livelihoods: practical concepts for the 21st century, IDS Discussion Paper, 296, Brighton

DFID (1999). Sustainable livelihoods guidance sheets. Department for International Development. http://www.glopp.ch/B7/en/multimedia/B7_1_pdf2.pdf

Dhanai R, Negi R, Singh S (2016) Sustainability of rural livelihoods through technology adoption: a case from Rudraprayag District, Uttarakhand, India. Int Res J Eng Technol (IRJET), 3(8). https://www.researchgate.net/publication/314154303

Gardner JS, Rowbotham D, DeScally FA (1992) Identification and monitoring of multiple geomorphic hazards in high mountainous area. In: Singh RB (ed) Dynamics of mountain geosystem. Ashish Publication, New Delhi, pp 451–477

Gardner JS, De Scally A (1992) Geomorphology applied to hazard site identification in mountain areas: examples from Kaghan Valley, Pakistan. In: Singh RB (ed) Dynamics of mountain geosystem. Ashish Publisher, New Delhi, pp 214–236

Jodha NS (1992) Mountain perspective and sustainability: a framework for development strategies. In: NS Jodha, M Baskota, T Pratap (eds) Sustainable mountain agriculture, 182. International Center for Integrated Mountain Development (ICIMOD), pp 237–261

Likert R (1932) A technique for the measurement of attitudes. Arch Psychol 140:1–55

Negi GCS, Joshi V (2004) Rainfall and spring discharge patterns in two small drainage catchments in the Western Himalayan, Indian. Environment systems and decisions formerly the environmentalist, vol 24. Springer Publication, pp 19–28. ISSN: 2194-5403 (Print) 2194-5411 (Online)

Mukherjee S (2013). Geological hazards in the Himalayan ecosystem, in the conference climate resilient livelihoods in the Himalaya River Basin, New Delhi

Negi RS, Joshi A (2014) Uttarakhand disaster, one view of livestock management during and after tragedy, Himalayan Vulnerability Uttarakhand, 2013, Sharma et al (eds). Paper 72, IDS, University of Sussex

Noor M, Rai SC (2014) Agricultural diversity and food security in the mountain ecosystem: a case study of Uttarakhand. Concept Publishing Company Pvt. Ltd, New Delhi-110059

Pandey BW, Prasad AS (2016) Land use pattern and hazards risk assessment in Kullu Valley along the Beas River in Western Himalaya. J Land Use Water Manag 16(1 and 2):01–18. (Special Issues-IGC 2016, Beijing, China)

Pandey BW (2010) Hazard ecology: approaches and techniques. Mittal Publication, New Delhi

Pandey BW (2002) Geoenvironmental Hazards in Himalaya: assessment and mapping. Mittal Publication, New Delhi

Pal SK (1986) Geomorphology of river terraces along Alaknanda Valley, Garhwal Himalaya, B. R. Publishing Corporation, Delhi

Pant BR (2000) Geographical personality of Uttarakhand. In: Sati MC, Sati MC (eds) Uttaranchal state: dimensions of development. Indus Publication, Delhi

Prasad AS, Pandey BW, Leimgruber W, Kunwar RM (2016) Mountain hazard susceptibility and livelihood security in the upper catchment area of the River Beas, Kullu Valley, Himachal Pradesh, India. Geo-Environ Disasters 3:3 https://doi.org/10.1186/s40677-016-0037-x

Sati VP (2008a) Natural resource management and food security in the Alaknanda Basin of Garhwal Himalaya, India. In: Peer-reviewed proceedings of 'Tropentag 2008' an international conference, Germany. Stuttgart, Germany

Sati VP (2008b) Natural resources management and food security in the Alaknanda River Basin of Garhwal Himalaya. ENVIS Bull Himal Ecol 16(2)

Sati VP (2013) Tourism practices and approaches for its development in Uttarakhand, Himalaya, India. J Tour Chall Trends, Innov Tour IV(1)
Sati VP (2014a) Landscape vulnerability and rehabilitation issues: a study of hydropower projects in the Garhwal region, Himalaya, Natl Hazards 75(3):2265–2278; https://doi.org/10.1007/s11069-014)-1430-y. (A Springer Publication)
Sati VP (2014b) Towards sustainable livelihoods and ecosystems in mountain regions, environmental science. Springer International Publishers. https://doi.org/10.1007/978-3-319-03533-8
Scoones I (2009) Livelihoods perspectives and rural development. J Peasant Stud 36(1):171–196. (72, Institute for Development Studies, Brighton, UK)
Singh RB (1992). Environment geography of India: challenges and opportunities. Natl Geograph J India 38:87–96
Singh RB (2016) Extreme weather events and the Himalayas: learning from the recent flood disasters, world focus, disaster management and mitigation, Delhi, pp 1–11
Singh RB, Robotham D (1995) The importance of natural hazards research to the Himalayas region, Himachal Pradesh a case in point in R. In: Singh B (ed) Global environmental change. Oxford and IBH publication, New Delhi, pp 185–199

Correction to: Land Use Land Cover Dynamics Using Remote Sensing and GIS Techniques in Western Doon Valley, Uttarakhand, India

Ajay Kumar Taloor, Vaibhav Kumar, Vivek Kumar Singh, Anil Kumar Singh, Ravindra V. Kale, Rahul Sharma, Varun Khajuria, Girish Raina, Beena Kouser, and Naveed Hassan Chowdhary

Correction to:
Chapter 4 in: S. Sahdev et al. (eds.), *Geoecology of Landscape Dynamics*, Advances in Geographical and Environmental Sciences, https://doi.org/10.1007/978-981-15-2097-6_4

In the original version of the book, the following belated correction has been incorporated: The incorrect affiliation of authors "A. K. Singh and R. V. Kale" has now been replaced with the correct affiliations as in below:

A. K. Singh Agromet Advisory Services Division, Indian Meteorological Department, New Delhi 110003, India e-mail: singhanil854@gmail.com

R. V. Kale Western Himalayan Regional Center, National Institute of Hydrology, (Ministry of Water Resources, River Development and Ganga Rejuvenation, Govt. of India) Irrigation & Flood Control Complex, Opposite Military Hospital, Satwari, Jammu Cantt., Jammu 180003, India e-mail: ravikale2610@gmail.com

The erratum chapter and the book have been updated with the change.

The updated version of this chapter can be found at
https://doi.org/10.1007/978-981-15-2097-6_4

S. Sahdev et al. (eds.), *Geoecology of Landscape Dynamics*,
Advances in Geographical and Environmental Sciences,
https://doi.org/10.1007/978-981-15-2097-6_23

Summary

The chapter summarizes the detailed discussions about the geoecological approach for landscape dynamics presented in various chapters in the book. The book is an epitomization of case studies providing new insights as to how landscape patterns and processes impact small creatures and how small creatures in turn influence landscape structure and compositions. The research paper suggests optimization of land use pattern, optimal management of landscape and optimal landscape design and planning. It also suggests the importance of applying landscape ecological principles in biodiversity conservation and sustainable development.

Landscape dynamics refers to every change that occurs in the physical–biological and cognitive assets of a landscape which are driven by extreme climatic events and human intrusions. The change of landscape conditions under the influence of economic activities and catastrophic natural and geographical process create landscape problem as a result to negative ecological, social, economic and other consequences which leads to the damage of structure and functioning of landscape. In this context, landscape dynamics are very important for land management and conservation which involve properties of the stability, persistence resistance, resilience and recovery that operate along with a broad range of temporal and spatial scales. Thus considering the complexity of the processes that occur in the landscape and different scales at which individuals, populations and communities respond to environmental constraints, the investigation of landscape dynamics must be conducted with robust conceptualization and accurate modelling. Development in landscape ecology illustrates the important relationship between spatial patterns and ecological processes. This development requires quantitative methods and massive use of technology to shape integrated land scope management which has been very influential for sustainable development planning. Advanced technology such as remote sensing, geographical information systems, GPS, spatial analysis and modelling approaches are accomplished for ecological understanding of species and ecosystems in landscape ecology.

S. Sahdev et al. (eds.), *Geoecology of Landscape Dynamics*,
Advances in Geographical and Environmental Sciences,
https://doi.org/10.1007/978-981-15-2097-6

The book is an attempt to incorporate the method that links spatial patterns and ecological processes at various spatial and temporal scales. This linkage of time space and environmental change can assist managers in applying plans to solve environmental problems. The book provides methodological tools for studying the effect of spatial arrangement of biophysical and socio-economic component on the sustainability of place. There are research papers more focussed on the use of technology like remote sensing and GIS to build data for assessing sustainable land management and new options for the management of environmental threats.

Climate change is a major component in structuring current research in landscape ecology. A geoecological study was carried out, using historical geohydrometeorological records for dying and dwindling of non-glacial fed rivers under climate change in upper Koshi watershed in the central Himalaya. Other landscape studies maintain that human impact is likely the main determinant of landscape pattern. The human capital is one of the important components for growing industrialisation, innovation, scientific research, education and formation of the higher skilled workers. One study used statistical technique such as cluster analysis for classifying the countries with the same HDI value at different incomes or similar income with different HDI values. Due to innovation, scientific research, education and growing industrialisation, the countries with high HDI and Gross National Income (GNI) per capital have a clusterization in different groups.

Techniques of mapping and GIS can be a great help for planners to identify the potential areas for new planned land use. Using GIS technology land degradation can be identified by mapping the waterlogging and soil erosion. Thus for sustainable management of degraded land quantifying the soil quality indicators are essential. The micro-level soil database on spatial extent may prove a better input in decision-making as well as in proper management of degraded land. Analysis of land use/land cover change detection is one of the important issues highlighted in the book. Land used and land cover change analysis assists decision-makers to ensure sustainable development and to understand the dynamic of our changing environment. In one paper, Landsat satellite imageries of two different time periods, i.e. Landsat thematic mapper of 2001 and Enhanced thematic mapper plus (ETM+) of 2010, were acquired and quantified the land use/cover dynamics of the Western Doon Valley, Uttarakhand from 2001 to 2010 over a period of one decade. The shuttle radar topographic mission (SRTM) digital elevation model (DEM) data have been used for slope analysis.

Urban sprawl results in engulfing of surrounding villages into peri-urban areas, peri-urban areas to town and town into cities. This uncontrolled development can lead to loss of agricultural land open spices and ecological sensitive habitat in and around the urban areas. Understanding the sprawl processes, its dynamics and modelling are useful for effective resource utilization and infrastructure planning. In Shimla city, geoecological study was conducted to highlight the extent of urban sprawl using four temporal satellite images of Landsat thematic mapper. The study highlights the nature rate and location of change and the importance of digital change detection techniques for proper land use planning for sustainable growth of the hilly urban areas. Automatic information extraction of urban features like buildings by using object-based approach and fuzzy logic on high-resolution quick bird imagery has proved very useful and robust method for urban planning.

The book has embraced a wide spatial extent covering Himalaya region, river basin and Coastal land, and provides insight on various contemporary issues ranging from land inventory mapping, hydrological modelling, monitory and concept of modelling of urban sprawl, sustainable economic growth, green building, natural resource management to geospatial techniques and ecological stability. This book forms an essential reference for graduate students, academics, professionals and practitioners in ecology, environmental science, natural resource management and landscape planning and design.

MIX
Papier aus verantwortungsvollen Quellen
Paper from responsible sources
FSC® C105338

If you have any concerns about our products,
you can contact us on
ProductSafety@springernature.com

In case Publisher is established outside the EU,
the EU authorized representative is:
Springer Nature Customer Service Center GmbH
Europaplatz 3, 69115 Heidelberg, Germany

Printed by Libri Plureos GmbH
in Hamburg, Germany